T0402723

Climate Change, Food Security and Natural Resource Management

Mohamed Behnassi • Olaf Pollmann
Himangana Gupta
Editors

Climate Change, Food Security and Natural Resource Management

Regional Case Studies from Three Continents

Editors
Mohamed Behnassi
Research Laboratory for Territorial Governance, Human Security and Sustainability (LAGOS), Faculty of Law, Economics and Social Sciences
Ibn Zohr University of Agadir
Agadir, Morocco

Center for Research on Environment, Human Security and Governance (CERES)
Agadir, Morocco

Himangana Gupta
National Communication Cell, Ministry of Environment, Forest and Climate Change, Government of India
New Delhi, Delhi, India

Olaf Pollmann
SCENSO – Scientific Environmental Solutions
Sankt Augustin, Germany

ISBN 978-3-319-97090-5 ISBN 978-3-319-97091-2 (eBook)
https://doi.org/10.1007/978-3-319-97091-2

Library of Congress Control Number: 2018957615

This Springer imprint is published by the registered company Springer Nature Switzerland AG
The registered company address is: Gewerbestrasse 11, 6330 Cham, Switzerland

About the Publishing Institution

The Center for Research on Environment, Human Security and Governance (CERES)

The CERES, previously the North-South Center for Social Sciences (NRCS), 2008–2015, is an independent and not-for-profit research institute founded by a group of researchers and experts from Morocco and other countries. The CERES aims to develop research and expertise relevant to environment and human security and their governance from a multidimensional and interdisciplinary perspective. As a think tank, the CERES aspires to serve as a reference point, both locally and globally through rigorous research and active engagement with policy-making processes. Through its research programme, the CERES aims to investigate the links between environmental/climate change, their implications for human security and the needed shifts to be undertaken in both research and policy. The CERES, led by Dr. Mohamed Behnassi and mobilizing a large international network of researchers and experts, aims to undertake original research, provide expertise and contribute to effective science and policy interactions through its publications, seminars and capacity building.

Preface

The world is facing increasing challenges due to climate change, resource shortage, food insecurity and resulting land use conflicts. The impact of climate change is multi-sectoral, and it could significantly increase resource conflicts and debilitate the condition in many developing countries across Africa and Asia. Many researchers assume that Africa will be among the regions hardest hit by climate change – although this continent causes the least damage to the climate. Certainly, climate change is taking place – and even progressing faster than expected. We have to acknowledge that climate change is influencing all aspects of human life – the environment, biodiversity and even the security of life.

Twenty-five years after the trend-setting "United Nations Conference on Environment and Development" (UNCED) also known as Rio de Janeiro Earth Summit or Rio-*Conference*, climate research has acquired an extremely political connotation. No doubt, climate research in general has been a crucial factor for the adoption of the *Paris Agreement* (*Accord de Paris*, December 2015). More than 55 countries accounting for more than 55% of global emissions have ratified the Paris Agreement. The Agreement entered into force on 4 November 2016 just before the commencement of COP-22 in Marrakech. This timeline is particularly remarkable considering that the ratification of the *Kyoto Protocol* took more than 7 years.

It is worthwhile to note that many important countries have ratified the agreement, not only China and the USA but also emerging economies such as India and Brazil. The USA, however, has decided against the world to step out of the globally significant agreement to curb climatic change. There is hope that US decision will galvanize the world community further, as Paris Agreement is an ambitious effort to limit global warming well below 2 °C.

Warming of the climate system is unequivocal, and the impacts of unchecked climate change will be disastrous. Therefore, purposeful and rapid climate action from all states is necessary and efforts have to be stepwise tightened. For this, States will have to revise their contributions to the agreement every 5 years as a part of the global stocktaking exercise. The stocktake will show the progress towards target fulfilment in terms of climate protection, adaptation, capacity building, technology transfer and finance. Science will provide the indicators that can be used to measure global progress.

With the Paris Agreement in place, it is essential also to discuss the directly linked Sustainable Development Goals (SDGs). Since the adoption of the SDGs, considerable efforts are being made to achieve them. This volume contributes to the on-going debates in climate and SDGs inlcuding "Zero hunger" (Goal 2), "Good Health and Well-being" (Goal 3), "Clean Water and Sanitation" (Goal 6), "Responsible Consumption and Production" (Goal 12), "Climate Action" (Goal 13), "Life on Land" (Goal 15) and "Partnerships for the Goals" (Goal 17). It explores the linkages between environmental change and food security as well as the relevance and need to consider the management of natural resources, especially water, soil and forest.

The research on climate change and food security nexus has gained prominent importance in the recent years, but there is still much gap in assessing the local and regional conditions that result in failure of certain policies. Rather, it leads to a reduction in the adaptive capacity of indigenous peoples. Compared to other relevant publications on the subject, this book brings to us the solutions for the gravest problems in the most vulnerable region. It brings together case studies from different regions, particularly Asia and Africa, and discusses the local level implementation of celebrated policies. The document follows an interdisciplinary approach linking science to policy making. Forty-one authors have contributed their research to this volume giving useful insights into their region explaining present and possible problems and their innovative solutions.

The volume is divided into 3 parts with 17 chapters. To develop an understanding of the basic problems and the nexus between each sector, soil, water and forest, an interconnectedness between the three parts has been maintained. While the *first* part of the volume talks about food security versus environmental and socio-economic dynamics, the *second* part follows with the linkages between the three sectors. The text in these two parts flows from examining the implications of climate-, water- and food-Nexus for policy making, research and business, including examples and case studies from Yemen, Casablanca, Saudi Arabia and India. It finds solutions for realizing food security through sustainable agriculture or ensuring food security through increasing water productivity. A variety of case studies in part II provide examples of how climate change impacts water supply systems and water requirements and how can water security be assured, also mentioning about impacts of soil degradation on agricultural production. The *third* part focuses on the forestry aspects, the maintenance of which is central to climate adaptation, soil health, agricultural diversity and water health. The detailed structure and approach adopted in this contributed volume is as follows:

Part I *"Food Security Versus Environmental and Socio-economic Dynamics"*

In Chap. 1, Mohamed Behnassi analyses the links between energy, water and food within a climate perspective and their implications for policy-making and governance. As these three interdependent resources are under growing pressure across the world (rising demand, overexploitation and pollutions, climate change, etc.), with many risks of insecurity especially for vulnerable communities in the Global South, it is increasingly imperative to address these resource systems simultaneously. The author focuses on the concept of the water-energy-food (WEF) nexus which has become widely used to help understand the complexities of these interdependent systems and how they can be managed sustainably and equitably to meet growing demand. In addition to addressing conflicts or trade-offs among the water, energy and food sectors, this approach to planning and resource management emphasizes the need to improve efficiencies in resource use to reduce environmental degradation and maximize the social and economic benefits of increasingly scarce natural resources.

In Chap. 2, Baig et al. focus on sustainable agriculture as a means to realize food security in the Republic of Yemen. According to the authors, agricultural extension can play a critical role in addressing the issues faced by the agricultural sector and help in enhancing food supplies. The chapter provides in-depth analyses of the issues and challenges that render the National Agricultural Extension Services (NAES) ineffective and inefficient in the country. The authors say that if appropriate measures are taken in letter and spirit to improve the National Extension Service, it can help the country realize sustainable agriculture.

Mouchrif et al. in Chap. 3 consider increasing water productivity and cereal yields forecasting as major tools for ensuring food security in Casablanca. The chapter is essentially oriented towards the study of water productivity (WP) of rainfed wheat as an indicator of agricultural development related to water management, and winter soft wheat yields forecasting in the rural commune of Ouled Saleh. The results indicate that water productivity is low and has to be improved.

Chapter 4 focuses on the wastage of food as one of the primary causes of food insecurity in the Kingdom of Saudi Arabia. Al-Zahrani et al. mention that about 78% of food purchased in Kingdom of Saudi Arabia (KSA) and United Arab Emirates (UAE) is thrown in the garbage and food leftovers make the single-largest component of the landfills. The chapter examines the factors responsible for food wastage in the Kingdom, besides determining the implications for extension education to reduce such wastage.

In Chap. 5, Baig et al. focus on innovation extension as a means to realize food security in Egypt. The chapter examines the present functioning of extension system and identifies the constraints faced by the agriculture sector and the shortcomings of the agricultural extension department. The authors suggest improvement measures and viable development strategies for improving the working of extension system. Such a case has also been presented for Yemen (Chap. 2), and the results are comparable.

Chapter 6 presents an interesting case in which the food production is increasing but the per capita availability of food is declining. Pal et al. analyse the state of declining foodgrain availability in India despite record agricultural production and rising food subsidy. The authors find that the availability of food declined consistently from its peak of 186.19 kg per person in 1991 to the bottom of 146.51 in 2013. The study also found that the correlation coefficient of per capita net availability with subsidy was low at 0.19 but the correlation of subsidy with agricultural production was high at 0.91. The authors focus on the present government policies and resolutions for food management and the reasons as to why the present policies have not been able to address the problem of food availability in the country.

Finance is also an important player in ensuring food security. Podruzsik and Pollmann in Chap. 7 investigate the importance of the European Union (EU) as an important player in global food security. The chapter gives an overview of the processes of designing and implementing initiatives, building cooperation with governments and participation in partnerships to facilitate change in developing countries, as well as measures taken with the help of financial instruments at EU-level. According to the authors, food security should be guaranteed through investment, training, research and balancing local, regional and global markets and trade. Interestingly, the chapter also focuses on the responsibility of recipient countries.

In ***Part II*** *"Climate, Water, Soil and Agriculture: Managing the Linkages"*

In Chap. 8, Karmaoui et al. analyses the impact of climate change on water demand in the Middle Draa Valley of south of Morocco. The study is based on the outputs of two software runs under socio-economical and climate change scenarios. The result predicts increase in mean temperatures from 1.4 °C to 3 °C and decrease in precipitation by approximately 3.9% to 15% in a number of scenarios considered in the study. The authors conclude that the study area will suffer from a lack of water supply due to the impact of climate change and the increase of water demand, which will be accelerated by high population growth.

Alomran et al. in Chap. 9 determine the water requirements by a date palm in Saudi Arabia. The authors provide information on the determination of monthly and annual water requirements by date palm in eight different regions of Saudi Arabia. They report a decrease of the crop water requirement (CRW) in all sites of study to around 8000 m^3/ha, which is mainly attributed to percentage of shaded area of date palm tree.

Chapter 10 focuses on the contribution of the Corporate Social Responsibility (CSR) to water protection in the Maghreb region. Abdelhafid Aimar talks about the threat to water security in the region and explains that engineering of a new water approach is crucial to reduce social vulnerabilities, secure safe water supplies and ensure water security in the region. With this background, the author examines whether the involvement of the business sector can abate the threats of water deterioration in the region and how CSR could allocate resources to positively and effectively impact the use of valuable water resources, control water contamination risks and sustain water resource management.

Chapter 11 looks at the impact of soil degradation on agricultural production in Africa. Olaf Pollmann and Szilárd Podruzsik describe the potential of soil quality and soil efficiency to create the linkage between agricultural soil treatment and local economic success on the African market. They conclude that it is important for each African country to establish land use concepts to decouple organic farming of small farmers from bigger commercial communities. This decoupling process could keep the soil fertility for small farming communities while other profit-yielding agriculture will support the regional economy.

In Chap. 12, Hind Mouhanni and Abdelaziz Bendou evaluate the feasibility and effects of the reuse of treated wastewater which contains higher concentration of soluble salts compared to groundwater on plant, soil and leachate. The authors conclude that the use of treated wastewater for irrigation presents a risk of salinization for the groundwater, and especially in the case of heavy textured soils. Therefore, caution in the management of irrigation intakes to prevent the accumulation of salts in the rhizosphere must be taken into consideration.

In Chap. 13, Sergiy et al. investigate the eco-service role of soil saprophages in the formation of sustainable man-made ecosystems under forest plantations. The authors found that earthworm eco-service activity had positive impact on environmental conditions of remediated soil. Environmental quality of remediated soil enriched in earthworm casts was confirmed to be improved.

Part III *"Forest Management from a Climate Change and Sustainability Perspective"*

In Chap. 14, Himangana Gupta talks about REDD+ as a multi-sectoral approach and the role of stakeholders in making climate mitigation and adaptation projects a success. The study also integrates expert views of scientists on implementing REDD+ as an effective adaptation mechanism. The author concludes that effective governance, increased stakeholder participation and synergizing the programme with watershed management initiatives can help yield full potential of REDD+.

Chapter 15 captures the conservation aspect of REDD+. Lokesh Chandra Dube analyses the twin objectives of conserving carbon and biodiversity through results based payments coming from the implementation of REDD+ activities in tropical countries. The author has made an assessment of international carbon markets and biodiversity markets to find out the possible ways to amalgamate these two.

In Chap. 16, Lauren Kathleen Sanchez studies the carbon dynamics at Harvard forest and ecological responses to changes in the growing season. The author hypothesized and modelled the phenological and ecosystem function responses to changes in the growing season. The author concludes that both photosynthetic growing season and cumulative ecological degrees were highly correlated to aboveground woody increment, which is an important driver of carbon uptake. Additionally, changes in the growing season were most strongly correlated to soil temperatures below the surface.

Chapter 17 captures the sustainability aspect through participatory governance. Baig et al. talk about ensuring sustainability in forests through the participation of locals. The authors assess the role of forestry extension programmes in educating stakeholders involved in planning, developing policies and undertaking institutional measures aiming at improving the existing forests. They also provide information on the challenges and constraints currently faced by forests and discuss the possible roles of participatory management approach in improving the situation and moving towards a sustainable future.

In a nutshell, this book sheds light on the scientific and policy aspects of the discourse about climate change, food security and natural resource management, and in particular, the perspectives on water, soil and forest. Since the book is a useful collection of case studies from Africa, Asia and Europe, it will encourage cross-continental knowledge sharing. The scope of the book ranges from impacts to mitigation and from in-field experiments to policy implementation. It contributes to the existing knowledge on climate-food nexus and connects climate change to sectors it could impact directly. All chapters and principles in the three parts of this book emphasize local ownership of strategy processes, effective participation from all levels and high-level commitment. They point out the importance of convergence and coherence between different planning frameworks, integrated analysis and capacity development. Besides being relevant for the academicians and scholars working in the field of climate change, forest and agriculture, it aims to catch interest of the policy makers and practitioners to understand ground realities for appropriate action. It is also bound to make an impact on the non-governmental organizations around the world and in the three different continents that this book covers, considering the indigenous and local issues highlighted in this book.

Agadir, Morocco Mohamed Behnassi
Sankt Augustin, Germany Olaf Pollmann
New Delhi, India Himangana Gupta

Acknowledgements

This contributed volume is partly based on the best papers presented during the International Conference on "Human and Environmental Security in the Era of Global Risks (HES2015)", organized on 25–27 November 2015 in Agadir by the Environment and Human Security Program (EHSP); North-South Center for Social Sciences, Morocco (now the Center for Research on Environment, Human Security and Governance (CERES)); the French Institute of Research for Development (IRD), UMR ESPACE-DEV, France; and the Project 4C/IKI-Morocco of the Deutsche Gesellschaft für Internationale Zusammenarbeit (GIZ) GmbH, Germany, in collaboration with the Interdisciplinary Research Centre for Environmental, Urban Planning and Development Law (CRIDEAU); University of Limoges, France; the Universidade Federal do Pará (UFPA) and the Universidade Federal da Bahia (UFBA), Brazil; and the CLIMED Project (ANR: 2012–2015/CIRAD), France.

The content and approach of this volume are mainly based on the empirical research undertaken by the majority of co-editors and chapter's authors in many parts of the world in relation with the core topics covered by this publication.

I have been honoured to chair the HES2015 and to share the editorship of this volume with my colleagues: Dr. Himangana Gupta (Programme Officer, National Communication Cell, Ministry of Environment, Forest and Climate Change, New Delhi, India) and Dr. Olaf Pollmann (North-West University, School of Environmental Science and Development, South Africa). Their professionalism, expertise and intellectual capacity made the editing process an exciting and instructive experience and definitely contributed to the quality of this publication. The editorial team has managed to ensure the depth, relevance and accuracy of the analysis both theoretically and empirically. Based on this, I wholeheartedly thank them for their trust, perseverance and valuable contribution making the whole publishing process a true success.

The chapters in this volume are also the result of the invaluable contributions made by our peer-reviewers, who generously gave their time to provide insight and expertise to the selection and editing process. On behalf of my co-editors Himangana Gupta and Olaf Pollmann, who actively participated in the peer-review process, I would specifically like to acknowledge, among others, with sincere and deepest thanks the following colleagues: Raj Kumar Gupta (Independent Journalist and

Policy Analyst, New Delhi, India), Dr. Pooja Pal (Panjab University, Chandigarh), Dr. Szilard Podruzsik (Corvinus University of Budapest, Hungary); Dr. Carsten Cuhls (Magdeburg-Stendal University of Applied Sciences, Germany); Dr. Oliver Dilly (DLR Project Management Agency, Germany); Dr. Mark Maboeta (North-West University, South Africa); and Dr. Jörg Helmschrot (SASSCAL, Namibia).

I would also like to seize this opportunity to pay tribute to all institutions that made this book project an achievable objective. In particular, we thank the funding institutions of the HES2015 Conference which, in addition to NRCS (now CERES), include the French Institute of Research for Development (IRD), UMR ESPACE-DEV, France, and the Project 4C/IKI-Morocco of the Deutsche Gesellschaft für Internationale Zusammenarbeit (GIZ) GmbH, Germany. Special thanks are due to all chapter's authors and co-authors without whom this valuable and original publication could not have been produced.

Agadir, Morocco Mohamed Behnassi

Contents

Contributors

Aymn F. Abou Hadid Arid Land Agricultural Studies and Research Institute (ALARI), Faculty of Agriculture, Ain Shams University (ASU), Cairo, Egypt

Abdelhafid Aimar Faculty of Economics, Commerce and Management, University of Jijel, Jijel, Algeria

Abdulrasoul Al-Omran Soil Science Department, King Saud University, Riyadh, Saudi Arabia

Fahad Alshammari Ministry of Environment, Water and Agriculture, Riyadh, Saudi Arabia

Khodran H. Al-Zahrani Department of Agricultural Extension and Rural Society, College of Food and Agriculture Sciences, King Saud University, Riyadh, Saudi Arabia

Fouad Amraoui Laboratory of Geosciences Applied to the Planning Engineering (GAIA), Faculty of Sciences, Hassan II University of Casablanca, Casablanca, Morocco

Abdelaziz Babqiqi Regional Observatory of the Environment and Sustainable Development, Moroccan State Secretariat for Environment, Marrakech, Morocco

Mirza Barjees Baig Department of Agricultural Extension and Rural Society, College of Food and Agriculture Sciences, King Saud University, Riyadh, Kingdom of Saudi Arabia

Mohamed Behnassi Research Laboratory for Territorial Governance, Human Security and Sustainability (LAGOS), Faculty of Law, Economics and Social Sciences, Ibn Zohr University of Agadir, Agadir, Morocco

Center for Research on Environment, Human Security and Governance (CERES), Agadir, Morocco

Abdelaziz Bendou Ibn Zohr University, Agadir, Morocco

Oleg Didur Laboratory of Biological Monitoring, Biology Research Institute, Dnipropetrovsk National University, Dnipropetrovsk, Ukraine

Lokesh Chandra Dube NATCOM Project Management Unit, Ministry of Environment, Forest and Climate Change, New Delhi, India

TERI University, New Delhi, India

Samir Eid Ministry of Environment, Water and Agriculture, Riyadh, Saudi Arabia

Loutfy El-Juhany Prince Sultan Institute for Environmental, Water and Desert Research, King Saud University, Riyadh, Saudi Arabia

Himangana Gupta National Communication Cell, Ministry of Environment, Forest and Climate Change, Government of India, New Delhi, Delhi, India

Raj Kumar Gupta Journalist and Independent Analyst in Environment, Food and Social Policy, New Delhi, India

Asaf Hajiyev Parliamentary Assembly of Black Sea Economic Cooperation (PABSEC), Academician of Azerbaijan National Academy of Sciences, Baku, Azerbaijan

Issam Ifaadassan Department of Environmental Sciences (LHEA-URAC 33), Faculty of Sciences Semlalia, Cadi Ayyad University, Marrakech, Morocco

Ahmed Karmaoui Department of Environmental Sciences (LHEA-URAC 33), Faculty of Sciences Semlalia, Marrakech & Southern Center for Culture and Sciences, Zagora, Morocco

Mohammed Yacoubi Khebiza Department of Environmental Sciences (LHEA-URAC 33), Faculty of Sciences Semlalia, Cadi Ayyad University, Zagora, Morocco

Angelina Kryuchkova Laboratory of Biological Monitoring, Biology Research Institute, Dnipropetrovsk National University, Dnipropetrovsk, Ukraine

Yurii Kul'bachko Laboratory of Biological Monitoring, Biology Research Institute, Dnipropetrovsk National University, Dnipropetrovsk, Ukraine

Iryna Loza Laboratory of Biological Monitoring, Biology Research Institute, Dnipropetrovsk National University, Dnipropetrovsk, Ukraine

Mohammed Messouli Department of Environmental Sciences (LHEA-URAC 33), Faculty of Sciences Semlalia, Cadi Ayyad University, Marrakech, Morocco

Guido Minucci Department of Planning and Urban Studies, Politecnico di Milano, Milan, Italy

Abdalah Mokssit National Meteorological Office, Casablanca, Morocco

Abdelhadi Mouchrif Laboratory of Geosciences Applied to the Planning Engineering (GAIA), Faculty of Sciences, Hassan II University of Casablanca, Casablanca, Morocco

Hind Mouhanni Département de Chimie, Centre des Sciences et Techniques, Campus Universitaire Ait Melloul, Ait Melloul, Morocco

Équipe des Matériaux, mécanique et Génie civil – ENSA- UIZ, Agadir, Morocco

Équipe d' Électrochimie, Catalyse et Environnement- FS-UIZ, Agadir, Morocco

Mahmoud Nadeem Soil Science Department, King Saud University, Riyadh, Saudi Arabia

Sergiy Nazimov Laboratory of Biological Monitoring, Biology Research Institute, Dnipropetrovsk National University, Dnipropetrovsk, Ukraine

Oleksandr Pakhomov Laboratory of Biological Monitoring, Biology Research Institute, Dnipropetrovsk National University, Dnipropetrovsk, Ukraine

Pooja Pal University Business School, Panjab University, Chandigarh, India

Szilárd Podruzsik Department of Agricultural Economics and Rural Development, Corvinus University of Budapest, Budapest, Hungary

Olaf Pollmann SCENSO – Scientific Environmental Solutions, Sankt Augustin, Germany

Juhn Pulhin Department of Social Forestry and Forest Governance, College of Forestry and Natural Resources, University of the Philippines, Los Baños College, Laguna, Philippines

Ajmal Mahmood Qureshi Harvard University – Asia Center, Cambridge, MA, USA

Tilak Raj University Business School, Panjab University, Chandigarh, India

Lauren Kathleen Sanchez Yale University School of Forestry and Environmental Studies, New Haven, CT, USA

Maria Shulman Laboratory of Biological Monitoring, Biology Research Institute, Dnipropetrovsk National University, Dnipropetrovsk, Ukraine

Gary S. Straquadine Utah State University – Eastern, Price, UT, USA

Tatiana Zamesova Laboratory of Biological Monitoring, Biology Research Institute, Dnipropetrovsk National University, Dnipropetrovsk, Ukraine

About the Editors

Mohamed Behnassi, PhD, is specialist in environment and human security law and politics. After the obtention of his PhD in 2003 from the Faculty of Law, Economics and Social Sciences, Hassan II University of Casablanca for a thesis titled: *Multilateral Environmental Negotiations: Towards a Global Governance for Environment*, he joined the Faculty of Law, Economics and Social Sciences, Ibn Zohr University of Agadir, Morocco, as Assistant Professor (2014). In 2011, he obtained the status of Associate Professor and in 2017 the status of Full Professor. He served as the Head of Public Law Department (2014–2015) and the Director of the Research Laboratory for Territorial Governance, Human Security and Sustainability (LAGOS) (2015–present). In addition, Dr. Behnassi is the Founder and Director of the Center for Environment, Human Security and Governance (CERES) (former North-South Center for Social Sciences (NRCS), 2008–2015). Dr. Behnassi is also Associate Researcher at the UMR ESPACE-DEV, Institute of Research for Development (IRD), France. In 2011, he completed a US State Department-sponsored Civic Education and Leadership Fellowship (CELF) at the Maxwell School of Citizenship and Public Affairs, Syracuse University, USA, and in 2014, he obtained a Diploma in Diplomacy and International Environmental Law from the University of Eastern Finland and the United Nations Environment Programme (UNEP), Finland. Dr. Behnassi has pursued several postdoctoral trainings since the completion of his Ph.D.

His core teaching and expertise areas cover environmental change, human security, sustainability, climate change politics and governance, human rights, CSR, etc. He has published numerous books with international publishers: *Environmental Change and Human Security in Africa and the Middle East* (Springer 2017); *Vulnerability of Agriculture, Water and Fisheries to Climate Change* (Springer 2014); *Science, Policy and Politics of Modern Agricultural System* (Springer 2014); *Sustainable Food Security in the Era of Local and Global Environmental Change* (Springer 2013); *Global Food Insecurity* (Springer, 2011); *Sustainable Agricultural Development* (Springer, 2011); *Health, Environment and Development* (European University Editions, 2011); and *Climate Change, Energy Crisis and Food Security* (Ottawa University Press, 2011). He has also published numerous research papers and made presentations on these at international conferences. In addition, Dr. Behnassi has organized many international conferences covering the above research areas in collaboration with national and international organizations and managed many research and expertise projects on behalf of various national and international institutions. Behnassi is regularly requested to contribute to review and evaluation processes and to provide scientific expertise nationally and internationally. Other professional activities include Social Compliance Auditing and consultancy by monitoring human rights at work and the sustainability of the global supply chain.

Olaf Pollmann, PhD, is a civil engineering and natural scientist. He holds a doctorate (Dr.-Ing./PhD) in environmental-informatics from the Technical University of Darmstadt, Germany, and a second doctorate (Dr. rer. nat./PhD) in sustainable resource management from the North-West University, South Africa. He is a visiting scientist and extraordinary senior lecturer at the North-West University and CEO of the company SCENSO – Scientific Environmental Solutions in Germany. He is also deputy head of the section "African Service Centers" in West (WASCAL) and Southern Africa (SASSCAL) on behalf of the Federal Ministry of Education and Research (BMBF).

Himangana Gupta, Ph.D., is an expert in climate change and biodiversity policy and diplomacy. In her present job, she is coordinating with scientists and climate experts to compile the Biennial Update Reports and Third National Communication to the UNFCCC. She is a doctorate in environment science with specialization in climate change and biodiversity policy and was a University Gold medallist in masters. She has written research papers in reputed international and national journals on current state of climate negotiations, forestry, industrial efficiency, rural livelihoods and women in climate change mitigation and adaptation.

Abbreviations and Acronyms

AAY	Antyodaya Anna Yojana
ABHO	Agence du basin hydraulique d'Ouarzazate
ACP	African, Caribbean and Pacific countries
AES	Agricultural Extension Service
AFED	Arab Forum for Environment and Development
APL	Above Poverty Line
AREA	Agricultural Research and Extension Authority
BPL	Below Poverty Line
CACP	Commission for Agricultural Costs and Prices
CBD	Convention on Biological Diversity
CDM	Clean Development Mechanism
CERES	Center for Research on Environment, Human Security and Governance
CFRN	Coalition for Rainforest Nations
CIP	Central Issue Price
COP	Conference of Parties
CRW	Crop water requirement
CSR	Corporate social responsibility
CWD	Coarse woody debris
DCI	Development Cooperation Instrument
EEAS	European External Action Service
ESCAP	Economic and Social Commission for Asia and the Pacific
EU	European Union
FAO	Food and Agriculture Organization
FCI	Food Corporation of India
FEC	Food Entitlements Card
FIRST	Food and Nutrition Security Impact, Resilience, Sustainability and Transformation
FSI	Forest Survey of India
FSTP	Food Security Thematic Programme
FWD	Fine woody debris

GCC	Gulf Cooperation Council
GCMs	Global climate models
GDP	Gross domestic product
GFI	Goodness of fit index
GHG	Greenhouse gas
GHI	Global Hunger Index
GPGC	Global Public Goods and Challenge
GW	Groundwater
HCP	Haut Commissariat au Plan
ICDS	Integrated Child Development Services
ICT	Information and communication technology
IFAD	International Fund for Agricultural Development
IFPRI	International Food Policy Research Institute
INFORMED	Information for Nutrition, Food Security and Resilience for Decision Making
INRA	National Institute for Agronomic Research
ISO	International Organization for Standardization
ITCZ	Intertropical Convergence Zone
JICA	Japan International Cooperation Agency
KAN	Knowledge-Action Network
KSA	Kingdom of Saudi Arabia
LAGOS	Research Laboratory for Territorial Governance, Human Security and Sustainability
LAI	Leaf area index
LTER	Long Term Ecological Research
LULUCF	Land Use, Land Use Change and Forestry
MAW	Ministry of Agriculture
MDGs	Millennium Development Goals
MDV	Middle Draa Valley
MENA	Middle East and North Africa
MIS	Marketing information system
MSP	Minimum Support Prices
NAES	National Agricultural Extension Service
NAPAs	National Adaptation Programmes of Action
NASS	National Agriculture Sector Strategy
NCWCD	National Commission for Wildlife, Conservation and Development
NEE	Net ecosystem exchange
NREGS	National Rural Employment Guarantee Scheme
NTFP	Non-timber forest Products
OCTs	Overseas countries and territories
ONEE	Office national d'Electricité et d'Eau potable
ORMVAO	Office régionale de mise en valeur agricole d'Ouarzazate
PDS	Public Distribution System
PRA	Participatory Rapid Appraisal
RED	Reducing Emission from Deforestation

REDD+	Reducing Emissions from Deforestation and Forest Degradation in Developing Countries
RIWA	Rationalization of the Irrigating Water in Agriculture
RMSE	Root mean square error
RMSEA	Root mean square error of approximation
RWP	Rainwater productivity
SAR	Sodium adsorption ratio
SBSTA	Subsidiary Body for Scientific and Technological Advice
SDGs	Sustainable Development Goals
SDSM	Statistical Downscaling Models
SEM	Structural equation modelling
SUN	Scaling Up Nutrition
TBL	Triple bottom line
TPDS	Targeted Public Distribution System
TWW	Treated wastewater
UAE	United Arab Emirates
UGC	University Grants Commission
UNDP	United Nations Development Programme
UNESCO	United Nations Educational, Scientific and Cultural Organization
UNFCCC	United Nation Framework Convention on Climate Change
USAID	US Agency for International Development
USDA	US Development Agency
USGS	US Geological Survey
VPD	Vapour pressure deficit
WBCSD	World Business Council for Sustainable Development
WEAP	Water Evaluation and Planning system
WEF	Water-Energy-Food
WFP	World Food Programme
WP	Water productivity
WTO	World Trade Organization

List of Boxes

List of Figures

List of Tables

Part I
Food Security Versus Environmental and Socio-economic Dynamics

Chapter 1
The Water-Energy-Food Nexus and Climate Perspective: Relevance and Implications for Policy-making and Governance

Mohamed Behnassi

1.1 Introduction

The natural resources used for water, energy and food are under increasing pressure across the world. Growth in both population and incomes is increasing resource consumption globally, stretching towards the planet's ecological boundaries (UNEP 2012; WEF 2016). Climate change exacerbates the pressure on these resources, and makes millions of people, predominantly in developing countries, more vulnerable to insecurity in their availability. According to Scott (2017), climate change has the potential to intensify the risk of insecurity and the significance of interdependencies. Similarly, the three sectors contribute to greenhouse gas emissions while being in the same time vulnerable to climate risks, which affect medium-term water availability, agricultural potential, and the production and consumption of energy (Scott 2017).

Due to these challenges, the food, water, energy and climate security have moved to the top of the global agenda. Addressing them simultaneously while taking into consideration their interdependence is increasingly considered imperative. Yet, in the absence of nexus thinking in planning and policy-making for water, energy and

M. Behnassi (✉)
Research Laboratory for Territorial Governance, Human Security and Sustainability (LAGOS), Faculty of Law, Economics and Social Sciences, Ibn Zohr University of Agadir, Agadir, Morocco

Center for Research on Environment, Human Security and Governance (CERES), Agadir, Morocco
e-mail: m.behnassi@uiz.ac.ma

M. Behnassi et al. (eds.), *Climate Change, Food Security and Natural Resource Management*, https://doi.org/10.1007/978-3-319-97091-2_1

food resources, interactions between the systems have been overlooked. Such siloed approaches have resulted in incoherent policy-making, contradictory strategies and the inefficient use of natural resources (Howells et al. 2013). Certainly, conventional policy and decision-making with regards to each of these areas in isolation is not necessarily anymore the most optimal course of planning or action.

To reverse this trend, the water-energy-food (WEF) nexus approach has become widely used to help understand the complexities of these interdependent systems, and how they can be managed sustainably and equitably to meet growing and competing demand. Understanding and managing the links among food, water and energy is indeed essential when formulating policies for more resilient and adaptable societies (Rasul 2016). Because of increasing resource scarcity and the threat of climate change, policy-makers and planners have given greater recognition to these links in recent years. Proponents of the WEF nexus as an approach to planning and resource management highlight the need to improve efficiencies in resource use to reduce environmental degradation and to address conflicts among the water, energy and food sectors, while maximizing the social and economic benefits of increasingly scarce natural resources co-benefits (Scott 2017).

A nexus approach, which refers to a multidisciplinary type of analysis of the relationship between energy, water, food, and climate change, can help to reduce trade-offs and to build synergies across these different systems, thus leading to a better and more efficient resource use and management as well as a cross-sectoral policy coherence (Halstead et al. 2014). Such a perspective is also a source of transformation and innovation for the research, business, and advocacy spheres. To subscribe in the perspective, actors in these areas have to adapt their values, practices and investments, thus boosting the policy-making processes related to the resource systems mentioned above.

The present chapter is a contribution to the debate about the WEF nexus approach. The analysis aims to: highlight the context within which the approach is evolving and identify the main challenges dealt with; underline its scope and objectives and areas where the approach is highly relevant; and explore its implications for policy-making, the main barriers to its promotion, and the solutions to foster the policy change from a nexus perspective.

1.2 The WEF Nexus Approach: Context and Challenges

1.2.1 Context

The global human requirement for water, energy and food is expected to increase substantially with the increase in world population that is projected to reach 9.6 billion by the middle of this century (UNDESA 2013). There is consensus that this increase will continue during this century and that the bulk of the increase in demand for resources will come from urban areas where currently more than half of the population already lives, although there is still substantial variability in the levels of urbanization across countries (UNDESA 2014). This means an increasing number of people will continue to depend and have an impact on already vulnerable resources to sustain life and economic growth. Many rapidly growing cities, particularly in developing countries, face serious problems related to water, food and energy, with limited or no capacity to respond adequately. These challenges will be further exacerbated by demographic shifts, changing lifestyles, a burgeoning middle class and the growing influence of climate change on the demand and supply chains of these resources (Yumkella and Yillia 2015).

Energy, water and food resource systems, which are highly concerned by these dynamics, are critically interdependent. Energy is needed to produce food and to treat and move water; water is needed to cultivate food crops and to generate many forms of energy; and food is vital for supporting the growing global population that both generates and relies on energy and water services[1]. In addition, land availability is an important element in each of these three resource systems (Halstead et al. 2014).

Additionally, these resource systems are currently in an uncertain shift and transformation with many security implications. Water scarcity and water supply-demand imbalance for instance already affects every continent and it is projected that an increased number of people in the Global South will be living in areas of high water stress with a likely impact on energy and food security. In addition, energy and water are inextricably linked. Despite an energy transition, non-renewable energy sources are still dominating the global energy generation landscape, and these thermal sources of energy generation mostly derived from fossil fuels are at present particularly water-intensive, mainly due to the cooling systems they use that require large amounts of water. A push towards a less carbon-intensive energy sector with a larger share of renewables, stimulated by global mitigation efforts, requires careful consideration of the potential impacts of such energy transition on the other nexus sectors (such as water, land and biodiversity).

Energy and water are also interconnected to food and agricultural production systems which are the largest user of fresh water globally and a key source of both GHG emission and mitigation. An increasing population and shifting dietary trends, especially in developing and emergent countries, mean demand for food and feed crop cultivation is rising. Food production and its associated supply chain account for approximately one-third of the world's total energy consumption. Rising food production has led not only to agricultural land expansion, largely at the expense of forests, but also in many regions to an intensification of agricultural processes on existing land. This expansion and intensification places more stress on agricultural input resources, such as water and energy.

According to current scientific evidence and projections, climate change has the potential to severely impact these resource systems and exacerbate existing resource management challenges (Pardoe et al. 2017). Climate impacts are likely to reduce the agricultural production and to make water stress in many regions worse, threatening the livelihood, food (including access, utilization, and price stability), health and water security of vulnerable communities. The rise in the number of food-

[1]A large of the literature highlights world population and economic growth projections, as well as changing lifestyles and consumption patterns, as the crucial factors leading to an increase in demand for energy, water and food resources in the future (SEI 2011; IEA 2012).

insecure people in the world, coupled with incidences of crop failure due to adverse weather, is making world leaders increasingly aware that future climate scenarios may severely lead to resource scarcity with the potential to limit our ability to feed the growing population during the next decades, therefore contributing to the increase of conflict risks and social instability in many vulnerable regions.

Understanding these drivers of global change in search of solutions is crucial to fully appreciate the value of systems thinking and analysis, which is required to facilitate integrated planning and decision-making. Both water and energy are linked with food production systems, climate change, biodiversity and even the technological solutions and institutional arrangements that are needed to manage and use resources efficiently. How these dimensions and their links are perceived and managed has far-reaching consequences for several global development goals, e.g. poverty alleviation, improvements in health, addressing climate change, energy security, addressing hunger and malnutrition, increasing access to water and sanitation, halting ecosystem degradation, etc. (Yumkella and Yillia 2015).

Even if these interdependencies are currently perceptible and defined as the WEF nexus, these three individual resource systems are unfortunately still organized, managed and researched more or less independently, and in many cases according to differentiated conceptual and policy frameworks. Yet, current pressing challenges, mainly human-induced environmental and climate change, are increasingly generating more risks which have the potential to undermine the viability of these resource systems in interrelated way, thus jeopardizing the human security of many regions, especially in the Global South.

In fact, many nations still pursue isolated sets of policies for water, energy and food, because the relevant institutions often work in isolation from each other. In the same way, a significant portion of the business community and consumers alike are using natural resources in technically inefficient ways without trying to fully appreciate the benefits of the intertwined and interdependent linkages. Alternatively, when those linkages are explored in an inclusive framework, decision makers, investors and civil society are enabled to make the right decisions and find the right solutions that maximize co-benefits and minimize constraints and certain trade-offs. The fact that the so-called bottom billion of the global population who are lacking access to electricity, heating and clean cooking facilities, and those who lack adequate access to food and nutrition, as well as safe drinking-water and improved sanitation, are often more or less the same people attests to the close linkages and the rationale for addressing development goals in a coordinated way.

Additionally, until very recently, governance questions related to the WEF nexus have not had much consideration in the literature, particularly in the context of the institutions and politics governing the WEF sectors (Foran 2015). Indeed, there is very little evidence about how the nexus approach has worked in practice. Leck et al. provide a brief history of WEF nexus thinking (Leck et al. 2015), and Weitz et al. (2016) review the literature on "integrative environmental governance" to fill in some of the gaps.

Based on this, it is becoming increasingly perceptible that every policy option and action adopted with regard to these interrelated systems may meaningfully affect the others, positively or negatively. Thus, it seems growingly imperative and effective to adopt a systems thinking and a 'nexus approach' (Fig. 1.1) for a more coordinated

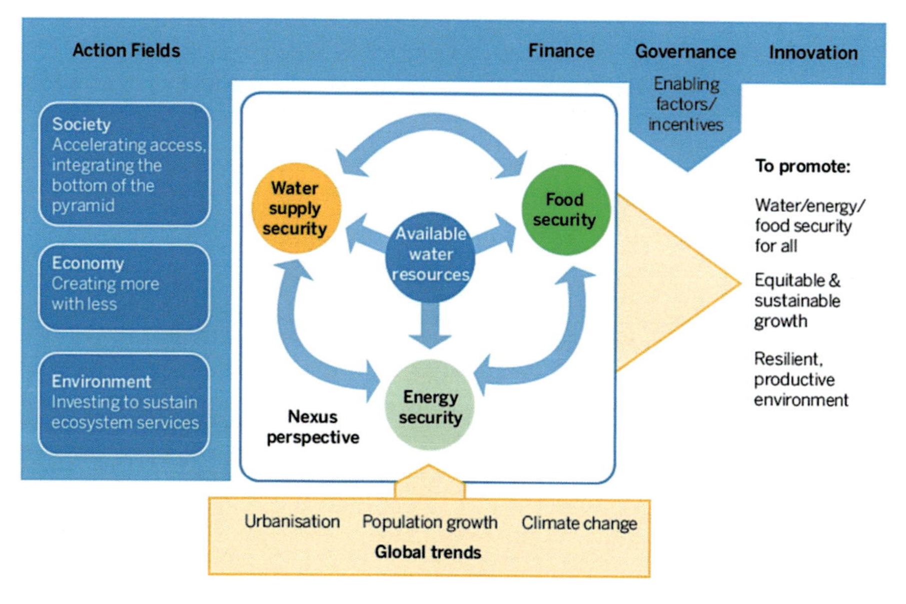

Fig. 1.1 The WEF Nexus Framework. (Source: SEI 2011)

approach to sustainable and equitable resources management (energy-water-food), especially within a climate governance perspective, which in turn requires concerted action in all spheres of influence and at all levels of implementation.

1.2.2 The WEF Nexus Challenges

The WEF nexus is concerned with the challenges surrounding the need to balance growing and competing demands for water, food and energy. In the framework of the Water-Energy-Food Nexus Knowledge-Action Network (Nexus KAN)[2], nexus challenges are defined as current or anticipated threats to equitable and sustainable access to energy, water and food whose causes are embedded in the interactions between these three components. These challenges cover, *inter alia*, the following:

- *Water for food*: consumption of green water and blue water (irrigation) by agriculture; and water use for food processing.
- *Food for energy* (through land): biomass used as domestic sources of energy; and land use for mining, for biofuel, wind and solar energy.
- *Energy for water*: energy is essential for the development and use of water resources ranging from desalination mostly for drinking water, to groundwater pumping for domestic, industrial and irrigation uses.

[2] Future Earth (futureearth.org/sites/default/files/surveywebtext.pdf)

- *Water for energy*: consumptive water use for biofuel production; and non-consumptive water use for hydropower and thermoelectric power.
- *Energy for food*: energy inputs for food production include pumping of irrigation water, mechanization, fertilizer production, food processing, and transportation.

In addition, in an effort to understand the existing interconnections, the Nexus KAN has identified four categories of nexus challenges which tend to reinforce each other (see Table 1.1).

1.3 The WEF Nexus Approach: Rationale and Applications

1.3.1 Scope and Objectives of the WEF Nexus Approach

The core premise of the WEF nexus approach is that the policy objectives in one sector (water, energy or food) can interact with those in other sectors, because they are either preconditions for the realization of another sector's objectives, or one sector (system) imposes conditions or constraints on what can be achieved in the other sectors. In other words, some policy objectives have synergies across sectors, while others require choices to be made between objectives in different sectors. Objectives in different sectors can also reinforce action towards objectives in other sectors (Weitz et al. 2014).

The nexus approach aims at understanding the links between sectors, recognize these in decision-making and promote integrated policy-making (Scott 2017). In the same vein, the WEF nexus promotes a systems perspective, emphasizing holistic and cross-sectoral approaches to decision making and planning (Cairns and Krzywoszynska 2016) at all levels that recognize the value of coordinated approaches (Pardoe et al. 2017). Such an approach is particularly relevant for human security, sustainability and climate change where the connections between water, energy and food are strong.

In addition, the nexus concept embraces socioeconomic and ecological links between the sectors, thus providing analytical frameworks to understand trade-offs and synergies, help increase efficiency, and improve governance in food, water and energy systems (Allouche et al. 2015). The governance context, along with the formal and informal rules that dictate how decisions affecting the allocation of resources are made, determine the outcomes of a nexus approach (Weitz et al. 2016). This context includes civil society and private sector actors, as well as the public sector (Byakika et al. n.d.).

Moreover, the current literature is calling for a better understanding of existing interconnections as a pre-requisite for an efficient governance of these areas (Koop et al. 2017; Weitz et al. 2017). Many researchers assert that the WEF nexus addresses existing governance key gaps by highlighting the need to consider the impacts of policies and actions in one sector on other, inter-related, sectors in order to account

Table 1.1 The WEF nexus challenges as identified by the Nexus KAN

Category 1	**Nexus challenges related to the impacts of food, water and energy production and supply processes on equitable and sustainable access to food, water, and energy:**
	Adverse effects of irrigation through its impact on hydrological cycles and soil fertility.
	Impact of biofuel production through land use change and deforestation.
	Impacts of electricity production on coastal and marine environment.
	Impacts of agriculture (chemical fertilizers, pesticides) and livestock on water quality.
	Impacts of fracking on groundwater.
	Land degradation caused by large-scale dam projects for hydropower and agricultural development.
	Impacts of mining activities on hydrological cycle through water consumption by continual extraction from aquifers and continual discharges into water flow, increasing pollution downstream.
Category 2	**Nexus challenges related to the impacts of resource scarcity on equitable and sustainable access to food, water, and energy:**
	Land degradation due to excessive reliance on biomass caused by insufficient access to electricity.
	Food insecurity caused by insufficient access to water (due to dryness, water salinity or potability, deep water table level, etc.) or energy (infrastructure, distribution, availability of resources for energy production, etc.).
	Food insecurity caused by insufficient access to energy (including energy for fertilizers, mechanization, and irrigation).
Category 3	**Nexus challenges related to the impacts of competing claims over natural resources on equitable and sustainable access to food, water, and energy:**
	Competing claims over water resources for domestic, industrial, and agricultural uses.
	Effects of international land deals (e.g. for biofuel production or mining) on land rights, and in turn on access to resources.
	Competing claims over water and land for energy production, including biofuels, and food production.
	Competing claims over land and water for conservation vs. agriculture.
Category 4	**Nexus challenges related to vulnerability of (equitable and sustainable) access to food, water and energy to extreme events, stresses and shocks (i.e. climate hazards, pollution, sea-level rise, large-scale migration,...):**
	Cascading failures of water and energy infrastructure service in cities following extreme events.
	Impacts of water pollution on urban agriculture.
	Vulnerability of energy production to climate change (for instance, electricity production may be affected by lower summer river flows and higher river water temperatures due to climate change).
	Vulnerability of food production to climate change.

Source: Future Earth (futureearth.org/sites/default/files/surveywebtext.pdf)

for synergies and trade-offs, improving efficiency and reducing the potential for negative side-effects (Leck et al. 2015; Rasul and Sharma 2016).

Nevertheless, despite the relevance of the nexus perspective in tackling key challenges related to water, energy and food in a changing climate, it was criticized by some researchers (such as Weitz et al. 2017; Wichelns 2017) for adopting a narrow focus on the three key sectors while overlooking other influences including wider political and cognitive factors. Other researchers, such as Howarth and Monasterolo (2016), underline the need to understand the governance arrangements through which a nexus approach may be applied.

1.3.2 Areas Where the WEF Nexus Approach is Highly Relevant

Given its relevance to policy making in the WEF nexus areas, the approach is currently advocated by the post-2015 agenda, the Sustainable Development Goals (SDGs), in order to avoid the mistakes made during the formulation and implementation of Millennium Development Goals (MDGs). For instance, the MDGs identified major sector goals, each with its own list of targets, but with little consideration of how efforts to attain a goal in one sector would affect (or be affected by) efforts to achieve goals in other sectors. Consequently, duplication of efforts and limited coordination between sectors, as well as weak partnerships among development partners, led to failure in certain areas. According to Yumkella and Yillia (2015), these mistakes can be avoided in formulating and implementing the SDGs. With the nexus perspective, it is possible in particular to emphasize increased coordination and partnerships between sectors (e.g. energy, agriculture, climate change, water, industry, ecosystems, etc.), as well as between institutional/organizational bodies such as UN entities, governments (both local and central), NGOs, businesses, civil society organizations and so on. By doing so, it is likely to operationalize the concept as a framework for solutions to emerge with multiple gains in several nexus dimensions across various sectors and management entities. By emphasizing the significance of these so-called 'nodes', it is possible to strengthen the interactions and links between key sector actors and in the process reduce the complexity of the nexus perspective.

Indeed, as the SDGs become a new focal point for development planning, it is currently perceived that this will increase recognition of the need for greater collaboration and coordination, especially: to make more informed decisions on goals, targets and indicators; to support the integration of goals across sectors and clarifying how best to allocate resources between competing needs; to make implementation of SDGs more efficient and cost effective by facilitating coordination between international development partners; and to reduce the risk that actions to achieve SDG targets could later undermine one another.

The climate change literature is also embracing the WEF nexus perspective (Azhoni et al. 2017; Conway et al. 2015). For Mohtar and Lawford (2016), in order to ensure that adaptation is effective, strategies need to be coherent and coordinated across water, energy and agriculture sectors. Since climate change is a crosscutting issue, adaptation strategies often require efforts of more than one sector. The integration of climate change into domestic policies, plans and strategies at all levels is an important means through which to encourage action on climate change (both mitigation and adaptation). Studies in policy coherence emphasize the value of policies that are coordinated across sectors so as to avoid climate maladaptation or conflicts between sectors (England et al. 2017).

According to the FAO (2014), the concept of the WEF nexus enables a more coherent or integrated approach to the management of natural resources and can assist in the identification of synergies and conflicts in the pursuit of climate compatible development objectives. Climate compatible development approaches would be enhanced when explicit account is taken of synergies and trade-offs between the WEF sectors. In other words, water, energy and food security are central to the pursuit of climate compatible development (Scott 2017), which in turn reflects synergies and trade-offs among poverty reduction, climate change mitigation and adaptation.

1.4 The WEF Nexus Approach: Implications for Policy-Making, Barriers to its Promotion and Relevant Solutions

Policy is an important tool for integrating issues of water, energy, food, and climate change since it drives overall national and sectoral priorities as well as establishing the frameworks through which cross-sectoral collaborations may be facilitated and 'greater policy coherence' may be promoted (Rasul and Sharma 2016).

Significant barriers to develop a coordinated approach are known to exist, including factors of political economy and incompatibility of institutional structures (Leck et al. 2015), set against the wider context of governance challenges present in many developing countries. Yet, the rate of increasing demand for water, energy and food and the associated pressures emerging through rapidly converging interdependencies emphasize the need for cross-sectoral coordination to avoid significant stress points (Pardoe et al. (2017). In other terms, there is a need for policy approaches to shift from being sectoral in their focus in order to avoid competing and counterproductive actions. In addition to vertical policy coherence, there is a need for horizontal coordination and an effective institutional structure to facilitate this.

For instance, Di Gregorio et al. (2017) suggest that horizontal policy coherence on crosscutting matters such as climate change can be facilitated if climate change is located in a ministry or department with established horizontal connections. Furthermore, this ministry or department should be a powerful and legitimate actor, which often is not the case if climate change is embedded within agencies only responsible for the environment (Cairns and Krzywoszynska 2016)[3]. A coordinated

[3]The National Adaptation Programmes of Action (NAPAs) approach aiming at integrating climate change in domestic policy has encouraged such sectoral mainstreaming but that this often results in plans that are developed by individual sectors, without coordination. Recognizing these shortcomings, the idea of mainstreaming adaptation into national policy is beginning to shift from sec-

approach to climate change across the three highly exposed sectors, often termed the WEF nexus, is crucial given that the impacts of and responses to climate change are generally cross-sectoral (Pardoe et al. (2017). For effective action on adaptation strategies, cross-sectoral coordination needs to be recognized internally as important, and fostered through suitable institutional structures.

Besides situating climate change within a more powerful ministry or department, coordination and collaboration could be fostered through budgets allocated specifically to cross-sectoral projects or simply providing greater annual budget consistency for sectors would also increase confidence to work together on longer-term projects and plans. Data sharing among actors is also a crucial part of the collaborative process. A platform for sharing data among government departments would help foster collaboration and promote efficiency. However, it must be acknowledged that these changes will require political will and trade-offs may need to be addressed on a broader scale (Pardoe et al. (2017).

As mentioned earlier, the nexus perspective emphasizes interlinkages between different sectors and advocates for coordinated approaches that enable feedbacks, trade-offs and synergies across the sectors to be taken into account (Hoff 2011; Kurian 2017). For this to happen, however, Yumkella and Yillia (2015) call for a concerted action in all spheres of influence and at all levels of implementation. The international community could bring actors together and catalyze support for national and subnational governments, and in some cases regional and sub-regional entities. Governments could make and use policy in a much more coordinated way, while the business community could try to adjust production systems for much more efficient resource use. NGOs and civil society organizations could learn to challenge and collaborate with the business sector, science community, and local authorities to help deliver solutions. Financing institutions and development agencies could use their experience in working with governments to promote further replication of nexus-driven initiatives. In addition, individuals and civil society as a whole could try to understand and manage their consumption patterns and the choices they make in an increasingly resource-constrained world.

In order to reinforce these trends and support ongoing initiatives aimed at operationalizing the nexus perspective into concrete actions, Yumkella and Yillia (2015) think that resources need to be mobilized to support planning and decision-making processes across sectors in several countries, especially in those countries where development challenges are huge. In particular, resources are needed to support countries in rethinking existing policies and developing new ones in areas such as water, energy, food, and climate and aligning new policies with those of various nexus dimensions, as well as to ensure the evaluation of trade-offs and synergies as they implement new policies and mobilize financial, institutional, technical and intellectual resources, creating the enabling environment that is required. Countries and partnerships that have established good practices could serve as incubators and disseminators of best-practice cases for integrated policy, technology and innova-

toral (vertically integrated) approaches towards more horizontally integrated approaches which is seen as a positive step towards more effective implementation with risks of maladaptation reduced (Rasul and Sharma 2016).

tion, as well as investor marketplaces for the dissemination of successful innovations across countries and regions. They could develop innovative technical tools and approaches, and policy-oriented material and guidance tools to assess, plan and manage resources in an integrated manner. They could facilitate responsible governance and the broad involvement of stakeholders to optimize desired outcomes, connecting actors working in different sectors. They could develop and promote human and institutional capacities to generate and manage knowledge, as well as providing channels to exchange information, skills and knowledge and build evidence-based platforms for exchange of experiences and expertise, especially knowledge and perspectives on water, energy, food and related development dimensions at the appropriate scale. Finally, the right set of capacity development programmes will increase awareness of the relevance of the nexus perspective to ensure that sector policies are more inclusive and better coordinated to address global development challenges.

Scott (2017) shows that the effectiveness of the horizontal (cross-sectoral) and vertical (between levels of government) coordination, that is essential for a nexus approach, is determined by institutional relationships, which can be influenced by political economy factors. The capacity of governing organizations to understand nexus links and to collaborate with each other is also critical. Scott (2017) suggests that aiming for the ideal of comprehensiveness and integration in a nexus approach may be costly and impractical. Nevertheless, horizontal and vertical coordination are essential. Local-level decision-making will determine how trade-offs and synergies in the WEF nexus are implemented. The capacities of local government organizations and decision-makers need to be strengthened to enhance their ability to adopt nexus approaches and coordinate vertically. For instance, integral to the concept of the WEF nexus is the knowledge of the interdependencies between sectors, which enhances decision-making by identifying potential trade-offs and synergies. However, knowledge of the nexus is still marginally used in policy- and decision-making. Weitz et al. (2016) identify institutional capacity to learn and assimilate this knowledge as a potential barrier to a nexus or integrated approach.

Lele et al. (2013) argue that given the unique regional and sector challenges of food, water and energy security, their nexus must be deconstructed to find effective, contextualized solutions. And governance challenges are at the heart of the nexus in each region. Governance is defined in various ways, but, with a few notable exceptions, the definitions have undergone relatively little analysis. In turn, governance issues are imbedded in policy, institutional, technological and financing options exercised at the global, regional, national and local levels. Furthermore, strong interactions between levels prompt policy responses to specific events and outcomes. The current governance arrangements, where they exist at all, are woefully inadequate to address the challenges. They are imbedded in a lack of strategic clarity, and among stakeholders there is an unequal distribution of power, voice and access to information, resources and the capability to exercise a sound influence which will produce equitable and sustainable outcomes. Often there are huge tradeoffs between the short-term wins of individual stakeholders and long-term holistic solutions.

In line with the assumptions made by Scott (2017:12), the adoption of a WEF nexus approach in policy-making and planning faces some critical barriers.

Institutional relationships, often influenced by political economy factors, can make vertical and horizontal coordination or integration particularly challenging. The capacity of governance organizations to understand nexus links and collaborate with each other is also critical. A WEF nexus approach recognizes that trade-offs between water, energy and food systems – that is, choices between objectives in different sectors – are inherently associated with the competing demands for water, energy and food resources for human and economic development. In contexts of increasing resource scarcity and climate change, these trade-offs become more significant. For the WEF nexus, governance is concerned with how these trade-offs are decided, informed by knowledge of the links between sectors. Win–wins (synergies) may also be possible, but trade-offs are the source of contestation. The notion of a trade-off implies that someone or something (e.g. an ecosystem) loses out. Therefore, decisions about trade-offs are political. Asymmetry of knowledge about links in the nexus could reinforce inequalities of access and political influence. A nexus approach should be used to make decisions about trade-offs transparent, ensuring that the cross-sectoral consequences of a decision are understood.

Despite the critical barriers to the mainstreaming of nexus approach, it is increasingly urgent to manage the interactions between resource systems, otherwise this may lead to new and complex societal risks. For example, cascading failures may occur when failure in a component of one system (e.g., electricity generation) leads to increased risks of failures in a component of another system (e.g., water treatment). Food, energy, and water resources are increasingly linked, for instance through global markets, as illustrated by joint price movements between world food and oil prices. Furthermore, changes in components of one system may alter thresholds and tipping points in the other systems. As these systems feedback on one another, the behavior of the overall system is difficult to predict based on an understanding of the individual components. Finally, the scale and scope of these systems-of-systems means that they are together subject to a wider range of stressors from, for instance, global climate change impacts, such frequency and intensity of droughts and floods, or economic or political shocks. Thus, knowledge of the linkages, synergies, and conflicts in the food-energy-water (FEW) nexus is urgently needed to provide evidence-based decision-making for policies in each sector that are most likely to produce positive effects in the other sectors as well as for developing innovative approaches to nexus governance. To achieve such integration, policy makers need to incorporate information about impacts of WEF interactions and the robustness of policy decisions across a range of future conditions. A nexus perspective thus requires new tools and methods to improve risk management (White et al. 2017).

1.5 Conclusion

Through this chapter, the analysis contributes to highlighting the vital importance of the WEF nexus approach, its implications for policy making and the ways to overcome the critical barriers to its mainstreaming. I believe that understanding the linkages and interdependencies in the food-energy-water nexus is necessary to identify risks and inform strategies for integrated nexus governance that supports multisector resilience, security, and sustainability. The non-consideration of the current understanding about the interdependence between food, energy, and water systems, and the persistence of policy, planning, and management decisions that perceive and manage each sector in isolation, may lead to underestimate the tradeoffs and interactions. The complexity of each individual system and the difficultly to identify the interconnections, make it challenging to assess multiple systems in an integrated manner, especially with current scientific methods and modeling capabilities. Nevertheless, from a precautionary point of view, these challenges should not prevent sector actors from endeavoring to understand and manage these intricacies because the interactions between these resource systems may lead to new and complex societal risks (White et al. 2017).

References

Allouche, J., Middleton, C., & Gyawali, D. (2015). Technical veil, hidden politics: Interrogating the power linkages behind the nexus. *Water Alternatives, 8*(1), 610–626.

Azhoni, A., Holman, I., & Jude, S. (2017). Adapting water management to climate change: Institutional involvement, inter-institutional networks and barriers in India. *Global Environmental Change, 44*, 144–157.

Byakika, S., Schreiner, B., & Sullivan, A. (n.d.). *Incentives for integrated WEF nexus planning in Kenya: Activating knowledge and networks*. Cape Town: Pegasys Institute.

Cairns, R., & Krzywoszynska, A. (2016). Anatomy of a buzzword: The emergence of 'the water-energy-food nexus' in UK natural resource debates. *Environmental Science & Policy, 64*, 164–170.

Conway, D., Archer van Garderen, E., Deryng, D., Dorling, S., Krueger, T., Landman, W., et al. (2015). Climate and Southern Africa's water-energy-food nexus. *Nature Climate Change, 5*, 837–846.

Di Gregorio, M., Nurrochmat, D. R., Paavola, J., Sari, I. M., Fatorelli, L., Pramova, E., et al. (2017). Climate policy integration in the land use sector: Mitigation, adaptation and sustainable development linkages. *Environmental Science & Policy, 67*, 35–43.

FAO. (2014). *The water–energy–food nexus at FAO: Concept note*. Rome: Food and Agriculture Organization of the United Nations. www.gwp.org/globalassets/global/toolbox/references/the-water-energy-food-nexus-at-fao--- concept-note-fao-2014.pdf

Foran, T. (2015). Node and regime: Interdisciplinary analysis of water–energy–food nexus in the Mekong Region. *Water Alternatives, 8*(1), 655–674.

Halstead M., Kober T. & van der Zwaan B. (2014). Understanding the Energy-Water Nexus, ECN-E--14-046. http://www.idaea.csic.es/sites/default/files/Understanding-the-energy-water-nexus.pdf

Howarth, C., & Monasterolo, I. (2016). Understanding barriers to decision making in the UK energy-food-water nexus: The added value of interdisciplinary approaches. *Environmental Science & Policy, 61*, 53–60.

Howells, M., Hermann, S., Welsch, M., Bazilian, M., Segerstrom, R., Alfstad, T., Gielen, D., Rogner, H., Fischer, G., Velthuizen, H., Wiberg, D., Young, C., Roehrl, R., Mueller, A., Steduto, P., & Ramma, I. (2013). Integrated analysis of climate change, land-use, energy and water strategies. *Nature Climate Change, 3*(7), 621–626.

International Energy Agency (IEA). (2012). *World Energy Outlook 2012*.

Koop, S. H. A., Koetsier, L., Doornhof, A., Reinstra, O., Van Leeuwen, C. J., Brouwer, S., et al. (2017). Assessing the governance capacity of cities to address challenges of water, waste, and climate change. *Water Resources Management, 31*, 3427–3443.

Kurian, M. (2017). The water-energy-food nexus trade-offs, thresholds and transdisciplinary approaches to sustainable development. *Environmental Science & Policy, 68*, 97–106.

Leck, H., Conway, D., Bradshaw, M., & Rees, J. (2015). Tracing the water–energy–food nexus: Description, theory and practice. *Geography Compass, 9*(8), 445–460.

Pardoe, J., et al. (2017). Climate change and the water–energy–food nexus: Insights from policy and practice in Tanzania. *Climate Policy*. https://doi.org/10.1080/14693062.2017.1386082.

Rasul, G. (2016). Managing the food, water, and energy nexus for achieving the sustainable development goals in gb. South Asia. *Environmental Development, 18*, 14–25.

Rasul, G., & Sharma, B. (2016). The nexus approach to water-energy-food security: An option for adaptation to climate change. *Climate Policy, 16*(6), 682–702. https://doi.org/10.1080/14693062.2015.1029865.

Scott A. (2017). Making governance work for water–energy–food nexus approaches, Working paper, CDKN.

Stockholm Environmental Institute (SEI). (2011). *Understanding the nexus*. Background paper for the Bonn2011 Nexus conference – The water, energy and food security nexus: Solutions for the green economy.

Lele, U., Klousia-Marquis, M., & Goswami, S. (2013). Good governance for food, water and energy security. *Aquatic Procedia, 1*, 44–63.

UNDESA. (2013). World population prospects. The 2012 revision. Volume I: Comprehensive tables. Retrieved online (May 2015): http://esa.un.org/wpp/Documentation/pdf/WPP2012_Volume-I_Comprehensive-Tables.pdf

UNDESA. (2014) World urbanization prospectus. United Nations Department of Economic and Social Affairs (UNDESA). Retrieved online (April, 2015) http://esa.un.org/unpd/wup/Highlights/WUP2014-Highlights.pdf

UNEP. (2012) *Global environmental outlook 5: Environment for the future we want*. Nairobi: United Nations Environment Programme. www.unep.org/geo/assessments/global-assessments/global-environment-outlook-5

WEF. (2016). *The global risk report 2016* (11th ed.). Geneva: World Economic Forum.

Weitz, N., Strambo, C., Kemp-Benedict, E., & Nilsson, M. (2017). Closing the governance gaps in the water-energy-food nexus: Insights from integrative governance. *Global Environmental Change, 45*, 165–173.

Weitz, N., Huber-Lee, A., Nilsson, M., Davis, M., & Hoff, H. (2014). *Interactions in the SDGs: A nexus perspective*. London: Independent Research Forum.

Weitz, N., Strambo, C., Kemp-Benedict, E. & Nilsson, M. (2016) *Governance in the water–energy–food nexus: Insights from integrative environmental governance*. Paper for the European Consortium for Political Research (ECPR) General Conference, Prague, 7–10 December 2016.

White, D. D., Leah, J. J., Maciejewski, R., Aggarwal, R., & Mascaro, G. (2017). Stakeholder analysis for the food-energy-water nexus in phoenix, Arizona: Implications for nexus governance. *Sustainability, 9*, 2204. https://doi.org/10.3390/su9122204.

Wichelns, D. (2017). The water-energy-food nexus: Is the increasing attention warranted, for either a research or policy perspective? *Environmental Science & Policy, 69*, 113–123. https://doi.org/10.1016/j.envsci.2016.12.018.

Yumkella, K. K., & Yillia, P. T. (2015). Framing the water-energy nexus for the post-2015 development agenda. *Aquatic Procedia*, 5/8–5/512.

Mohamed Behnassi, PhD, is specialist in environment and human security law and politics. After the obtention of his PhD in 2003 from the Faculty of Law, Economics and Social Sciences, Hassan II University of Casablanca for a thesis titled: *Multilateral Environmental Negotiations: Towards a Global Governance for Environment*, he joined the Faculty of Law, Economics and Social Sciences, Ibn Zohr University of Agadir, Morocco, as Assistant Professor (2014). In 2011, he obtained the status of Associate Professor and in 2017 the status of Full Professor. He served as the Head of Public Law Department (2014–2015) and the Director of the Research Laboratory for Territorial Governance, Human Security and Sustainability (LAGOS) (2015–present). In addition, Dr. Behnassi is the Founder and Director of the Center for Environment, Human Security and Governance (CERES) (former North-South Center for Social Sciences (NRCS), 2008–2015). Dr. Behnassi is also Associate Researcher at the UMR ESPACE-DEV, Institute of Research for Development (IRD), France. In 2011, he completed a US State Department-sponsored Civic Education and Leadership Fellowship (CELF) at the Maxwell School of Citizenship and Public Affairs, Syracuse University, USA, and in 2014, he obtained a Diploma in Diplomacy and International Environmental Law from the University of Eastern Finland and the United Nations Environment Programme (UNEP), Finland. Dr. Behnassi has pursued several postdoctoral trainings since the completion of his Ph.D.

Chapter 2
Realizing Food Security Through Sustainable Agriculture in the Republic of Yemen: Implications for Rural Extension

Mirza Barjees Baig, Ajmal Mahmood Qureshi, Gary S. Straquadine, and Asaf Hajiyev

2.1 Introduction

The Republic of Yemen with its coordinates 12° and 17°N Latitude and 43° and 56°E Longitude, located in Western Asia, lies in the southwest Arabian Peninsula. The country has rugged surface features composed of mountains, hills, plateaus, plains, and valleys (FAO 2009). Spreading over an area of 536,000 km^2, comprising many islands in the Red Sea and Arabian Sea. The country is bordered by Saudi Arabia to the north, Oman to the east, the Arabian Sea to the south and the Red Sea to the west. The country has been divided into 20 governorates, which are further divided into districts.

With Yemen's estimated population of 26.2 million (World Bank 2015), about 80% live in the rural areas (IFAD 2015). Agriculture is the prime sector in economic development, contributing about 17.5% towards GDP (NASS 2012) and the main source of employment accommodating almost 54% of the population (FAO 2009). Yemen is a low-income country and GDP amounts to USD 33.76 billion with GDP per capita averaging USD 1361 in nominal terms (World Bank 2011; IFAD 2013). Playing an important role in the economy, agriculture sector generates about 20% of

M. B. Baig (✉)
Department of Agricultural Extension and Rural Society, College of Food and Agriculture Sciences, King Saud University, Riyadh, Kingdom of Saudi Arabia
e-mail: mbbaig@ksu.edu.sa

A. M. Qureshi
Harvard University – Asia Center, Cambridge, MA, USA

G. S. Straquadine
Utah State University – Eastern, Price, UT, USA
e-mail: gary.straquadine@usu.edu

A. Hajiyev
Parliamentary Assembly of Black Sea Economic Cooperation (PABSEC), Academician of Azerbaijan National Academy of Sciences, Baku, Azerbaijan

M. Behnassi et al. (eds.), *Climate Change, Food Security and Natural Resource Management*, https://doi.org/10.1007/978-3-319-97091-2_2

the total internal revenues (FAO 2009). In addition, the sector creates significant employment opportunities in the other sectors like: transport, processing, and trading that may reach to 54% (Irin 2007; NASS 2012). Agriculture also plays an important role in food security, improves the trade balance and results in an integrated rural development. In addition, the agriculture sector helps in stabilizing the population balance by reducing internal migration and mitigating the resultant social and economic problems (NASS 2012).

Farming in Yemen, in spite of harsh and difficult environment, has a long history of contributions to the rural economy. Despite numerous constraints, the agriculture sector has considerable potential to produce enough food to feed its people and make it an engine of growth, if farming systems are re-introduced to the farmers with some improvements.

The purpose of the chapter is to identify the problems and issues faced by the farming systems and discuss the possible role of agricultural extension to elevation crop productions. In the light of the analyses of the identified shortcomings, various viable remedial measures to improve the farming systems and the national extension service have been presented. Also, we anticipate that sustainable agriculture backed by an efficient extension service could help enhance crop yields, improve food situation, and ensure food security in the country.

2.2 An Overview of Agriculture Sector in Yemen

Agriculture is the key sector, which can ensure food security and also be the main source of income for the majority of people. The sector helps to improve the trade balance and promotes integrated rural development. In addition, agriculture sector helps in stabilizing and lowering rural migration to the cities, thus mitigating resultant social and economic problems. About 80% of the population lives in rural areas (IFAD 2015) and some 22 million inhabitants residing in rural areas are involved in agriculture or agricultural related activities (FAO 2009). According to the recent FAOSTAT estimates, about 65% population lives in the rural areas as depicted in Fig. 2.1. Information regarding the evolution of population and labor force size, agricultural growth has been presented in Tables 2.1 and 2.2.

Semi-arid environment and varied topographic features of the country provide favorable conditions for a wide variety of crops to grow. The total agriculture area is estimated at 1668, 858 ha, of which, 1,132,910 ha (68%) is cultivated while the uncultivated area is 535,948 ha (32%). However, sorghum, maize, millets, pulses, wheat, barley and millions of mango trees are sustained by only 3% arable land (FAO 2009). Vegetable crops are raised on the fertile soils in many agro-ecological zones of the country on an estimated area of about 57,000 ha. Potato and tomato crops cover almost 50% of this area. More than 20 species of vegetables are grown mainly under irrigation system. The vegetable cultivation, including several exotic varieties on the irrigated areas are expanding and the ground water depletion is also the main cause of this kind of expansion (FAO 2009). The growth rate in agricultural

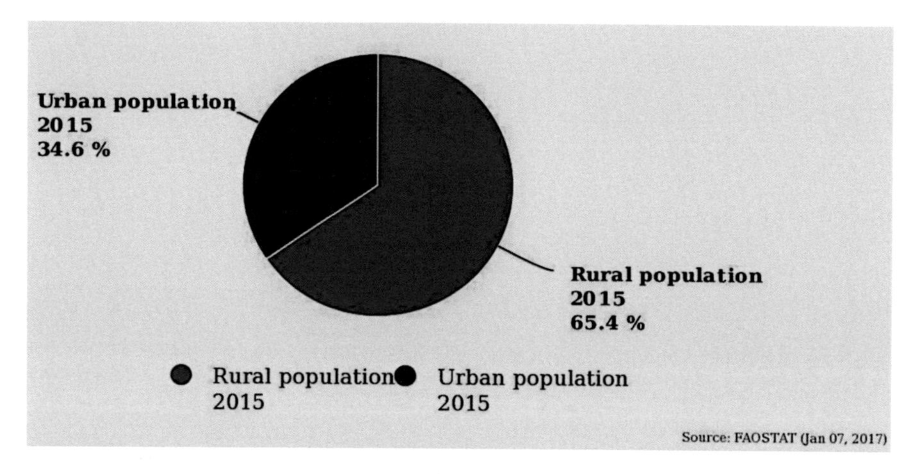

Fig. 2.1 Rural and Urban Population in Yemen 2015. (Source: FAO 2016) Available at: http://www.fao.org/faostat/en/#country/249

Table 2.1 Evolution of population and labor force size

	Size [millions]				Annual growth rate [%]		
	2000	2005	2010	2015	2000–2005	2005–2010	2010–2015
Total population	17.52	20.14	22.76	25.54	2.83	2.48	2.33
Total labor force	3.91	4.78	5.64	6.57	4.1	3.36	3.1
Labor force in agriculture	1.87	2.09	2.19	2.21	2.25	0.94	0.18

Source: FAOSTAT (2015)

Table 2.2 Evolution of population and labor force composition

	Share [%]				Annual growth rate [%]		
	2000	2005	2010	2015	2000–2005	2005–2010	2010–2015
Rural population [% of total population]	73.73	71.06	68.26	65.35	−0.73	−0.8	−0.87
Labor force in agriculture [% of total labor force]	47.85	43.84	38.81	33.69	−1.74	−2.41	−2.79
Females [% of labor force in agriculture]	33.99	38.97	40.21	40.53	2.77	0.63	0.16

Source: FAOSTAT (2015)

sector averaged only 2.4% per year, compared to the population growth rate of 3.7%, one of the highest in the world. The demand for vegetables and fruits are met through indigenous productions but only 40% of the domestic demand for grains (FAO 2009).

The country that once used to be self-sufficient in cereals now imports 75% of its food requirements to fill the gap in local food production within the country. In the past, Yemen was famous for producing good quality coffee being the main cash crop of the past now has been replaced by qat, a mild stimulant regularly chewed by about 70% of Yemeni men.

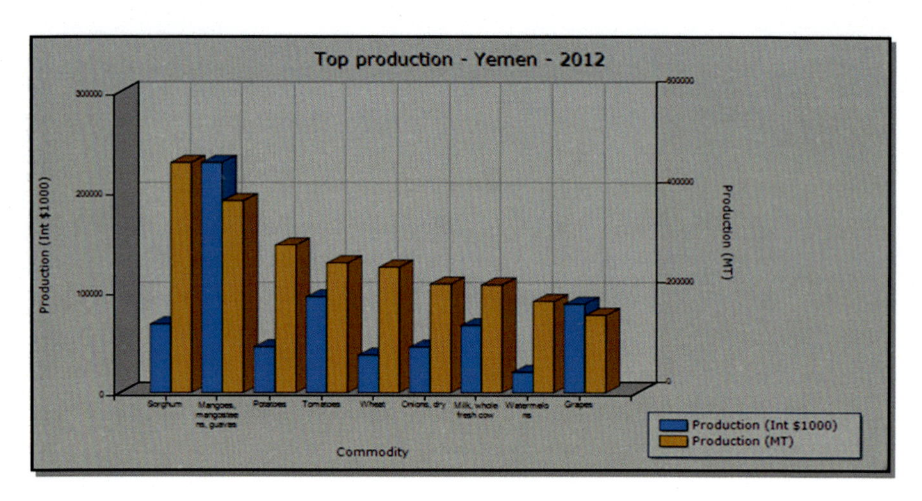

Fig. 2.2 Top production of various agricultural commodities and their respective values. (Source: FAO 2015)
Available at: http://faostat.fao.org/site/339/default.aspx

Agriculture sector is negatively impacted and not in a position to realize its potential due to lack of resources, and external factors such as climate change, social conflicts, and lack of security. On the other hand, most of the farmers are poor and survive at subsistence level and rural women, though play a significant role in farming, have not yet come into mainstream development. The country needs a development plan, and a long-term agriculture policy to achieve food security and reduce rural poverty. According to a comprehensive study by the International Fund for Agricultural Development (IFAD 2009), poor agricultural production could be a result of low and weak technological base, resulting in modest annual growth of less than 4% between 2003 and 2007.

However, the agricultural productivity, particularly those of crops and livestock is 50% lower as compared to the other Middle Eastern countries placed in similar environments. A comprehensive scenario on agriculture and its allied components is highlighted in Figs. 2.2, 2.3, and 2.4 and Tables 2.3 and 2.4.

The agriculture sector consumes up to 90% of available water in Yemen (NASS 2012). A significant volume of water estimated to be 50–65% is being wasted due to inefficient irrigation systems. Low agricultural productivity, water scarcity, climate change, insufficient off-farm economic and employment opportunities, high rural population growths together with high dependency ratios constitute critical negative factors affecting rural areas with increased rural poverty. Limited resources, especially water, and lack of access to basic services are major factors for migration from rural areas especially mountainous villages and settlements are being increasingly abandoned in search for employment opportunities in the urban areas (IFAD 2013).

Most farms are extremely small with typically low household farm incomes (GAFSP 2013). Unlike most of the world, economic dependence on agriculture in Yemen has been growing because of reduced opportunities in the industrial and

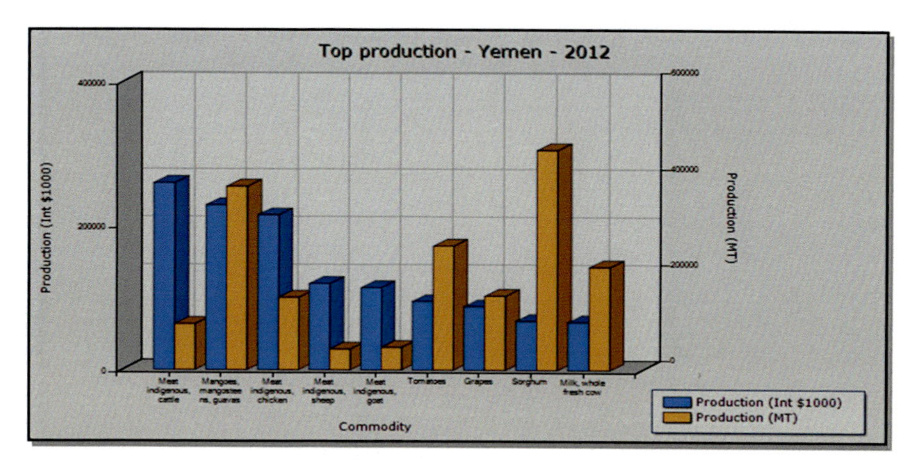

Fig. 2.3 Top production of various agricultural commodities and their respective values. (Source: FAO 2015)
Available at: http://faostat.fao.org/site/339/default.aspx

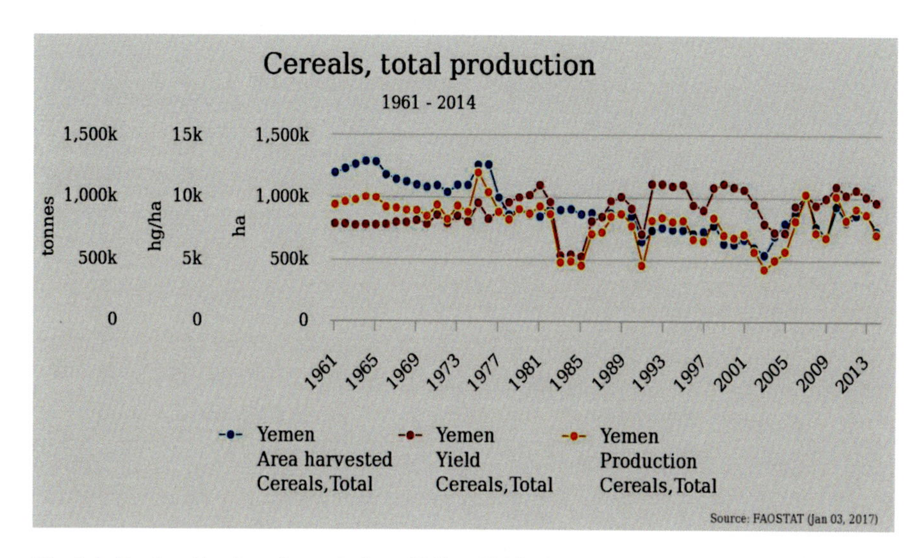

Fig. 2.4 Total production of cereals from 1961 to 2014. (Source: FAOSTAT 2016)
Available at: http://www.fao.org/faostat/en/#country/249

services sectors. The agriculture sector has realized its potential, registering annual growth of over 5% for grains, and impressive increases in coffee and honey production. As depicted in Table 2.4, FAO (2016) reported that the total cereal production was estimated at 653,000 tonnes, including 300,000 tonnes of sorghum and 165,000 tonnes of wheat, however both the cultivated land and agricultural production declined as compared to 2014 (FAO 2016). Despite the negative effects of scarcity of resources, the potential exists to achieve significant gains in agricultural

Table 2.3 Production top 20 food commodities (MT) with their respective values in Int. $1.000

Rank	Commodity	Production (MT)	Production (In Int $1.000)
1	Meat indigenous, cattle	96,302	260,149
2	Mangoes, mangosteens, guavas	383,107	229,545
3	Meat indigenous, chicken	151,402	215,659
4	Meat indigenous, sheep	44,088	120,043
5	Meat indigenous, goat	47,850	114,653
6	Tomatoes	258,654	95,589
7	Grapes	154,869	88,526
8	Sorghum	459,241	68,710
9	Milk, whole fresh cow	215,321	67,193
10	Eggs, hen, in shell	63,420	52,600
11	Potatoes	294,686	45,954
12	Onions, dry	216,739	45,522
13	Wheat	250,264	37,741
14	Tobacco, unmanufactured	23,251	37,033
15	Bananas	127,468	35,899
16	Chick peas	58,560	28,347
17	Dates	55,181	28,181
18	Oranges	122,000	23,577
19	Milk, whole fresh sheep	56,637	22,055
20	Coffee, green	19,828	21,302

Source: FAOSTAT (2015)
Available at: http://faostat.fao.org/site/339/default.aspx

Table 2.4 Production of top three crops 2014–2015

Yemen cereal production				
2010–2014 Average		2014	2015 Estimate	Change 2015/2014
	000 Tonnes			Percent
Sorghum	432	342	300	−12
Wheat	234	192	165	−14
Millet	86	74	80	8
Others	108	93	108	16
Total	860	701	653	−7

Note: percentage change calculated from unrounded data
Source: FAO/GIEWS country cereal balance sheets
Available at: http://www.fao.org/giews/countrybrief/country.jsp?code=YEM

productivity (NASS 2012). The country has a strong history of producing a wide variety of agricultural products in difficult environments to support the agriculture sector and the rural economy. Significant hikes in agricultural productivity are possible, even if resource constraints limit these gains. In this chapter, it is argued that the country needs to identify the constraints resulting in lower yields. The most important issues and challenges are discussed in the following sections.

2.3 Challenges to the Agriculture Sector

Agriculture in Yemen faces numerous constraints namely stressed land, scarce water resources, expansion in cultivation of qat cultivation on fertile and productive land, inadequate marketing systems, human resources with low education and working ability, lack of infrastructure and production technologies, and insufficient availability of inputs. These constraints prevent agriculture sector from making significant contributions to rural incomes, national GDP, and addressing the trade imbalance in food items. According to FAO, agriculture sector in Yemen is beset with numerous constraints (Box 2.1). Yemen is essentially a rural economy and agriculture sustains 80% of its population. Development indicators for the country are presented in Table 2.5 (Annex).

2.3.1 Water Resources

Lack of water is a crucial issue in Yemen. Half of the country does not have access to adequate water and sanitation; as such, water is already a cause of conflict within the country. Agriculture uses more than 90% of the country's scarce water resources. Most of the crops are that of qat, a non-food crop. Though the per capita average share of renewable water resources is one tenth of the average in most Middle Eastern countries and one 50th of the world average yet, even at low usage rates, demand increasingly exceeds fresh water supply (USAID 2016). Compared with an average of 1250 m^3/cap/year for the Middle East and North Africa, availability of 150 m^3/cap/year makes Yemen one of the most water deficit countries. The level is less than one tenth of the water threshold i.e. 1700 m^3/capita/year. In highland governorates, less than 20% of communities have access to safe drinking water from public water supply systems whilst 60% of settlements rely on unprotected springs and wells and 20% on cisterns, streams and tanks. Total water demand which stands at 3400 million m^3 per year exceeds renewable resources of 2500 million m^3 per year, thus leading to a steady decline in groundwater levels, varying between 1 m/year in the Tuban-Abyan area and 6–8 m/year in the Sana'a basin. The public supply of water, where available, is unreliable and inconsistent (IFAD 2013) (Fig. 2.5).

Water shortage for irrigation is the most serious problem preventing agriculture from realizing its potential in the country. Rains and groundwater are prime water sources (FAO 2009). The problem is getting aggravated, as the renewable water resources are finite and limiting on daily basis and inadequate to meet rapidly but ever-increasing demand. In the absence of any perennial river in the country, groundwater being the only reliable resource is supporting the major economic activities, including agricultural production. The current level of depleting groundwater extraction in the country greatly exceeds recharge in most of the aquifers, especially in northern areas and the inter-mountain plains, including Sana'a basin. Over-pumping has caused a marked decline in water levels and has deteriorated the water quality as well as it undermines the sustainability of the resource base (NASS 2012; IFAD 2013).

Box 2.1: Problems and Issues Faced by Agriculture in Yemen

Among the wide range of constraints affecting agricultural production, the most prominent are:

- Renewable water resources are finite and limited, while demands for water are growing;
- Land is a limited resource;
- Excessive land fragmentation of agricultural land-holdings on mass scale;
- Productions from the traditional subsistence agriculture and other prevalent agricultural systems are decreasing, particularly on terraces due to the growing migration of male workers;
- Wide spread rural poverty among the farmers, practicing subsistence farming;
- Lack of coordination among the various agencies engaged in the agricultural development and rural infrastructure development;
- An uncoordinated development of rural infrastructure;
- Lack of infrastructure facilities;
- Lack of production technologies;
- Low productivity, resource constraints;
- Insufficient availability of inputs;
- Lack of credit facilities;
- Lack of funding and limited support from the government to the farmers;
- Previously launched development projects were unable to meet farmers' expectations;
- Climate change and environmental issues;
- According to the Global food security index 2016 (an annual measure of the state of global food security) Yemen ranked 100th among the 113 countries with the score of 34.0. FAO (2016) reported that more than half (51%) of the population (14.4 million Yemenis) are food insecure. Food insecurity remains a major challenge;
- Key role played by rural women in agriculture remains neglected and is largely unacknowledged; and have not yet come into the mainstream development;
- Agriculture sector has not realized its full potential due to lack in export opportunities;
- Absence of quality and safety standards, hindering farmers to get premium prices for their produce and products;
- Inadequate marketing systems, limited marketing opportunities for the small farmers, provided by traditional retail and wholesale markets,
- Lack of farmers' associations and organizations result in challenges for the farmers at all levels in marketing their produce;
- Qat is displacing coffee and other food crops due to its increasing production that is competing for limited groundwater resources (Source: FAO 2009, 2016; NASS 2012; GFSI 2016)

Table 2.5 Development indicators of Yemen World Bank 2015 (Annex)

Indicator	2013	2014	2015
Population, total	25,533,217.0	26,183,676.0	–
Population growth (annual %)	2.6	2.5	
Surface area (sq. km)	527,970.0	527,970.0	
Population density (people per sq. km of land area)	48.4	49.6	
GNI, Atlas method (current US$)	33,318,506,441.7		
GNI per capita, Atlas method (current US$)	1300.0		
GNI, PPP (current international $)	93,320,014,470.7		
GNI per capita, PPP (current international $)	3650.0		
Life expectancy at birth, total (years)	63.6		
Mortality rate, under-5 (per 1000)	46.0	43.8	41.9
Prevalence of underweight, weight for age (% of children under 5)	39.9		
Immunization, measles (% of children ages 12–23 months)	78.0	75.0	
Primary completion rate, both sexes (%)	70.1		
Gross enrolment ratio, primary, both sexes (%)	101.0		
Gross enrolment ratio, secondary, both sexes (%)	49.2		
Gross enrolment ratio, primary and secondary, gender parity index (GPI)	0.8		
Prevalence of HIV, total (% of population ages 15–49)	0.1	0.1	
Forest area (sq. km)			
Annual freshwater withdrawals, total (% of internal resources)	169.8		
Improved water source (% of population with access)			
Improved sanitation facilities (% of population with access)			
Urban population growth (annual %)	4.3	4.2	
GDP at market prices (current US$)	35,954,502,303.5		
GDP growth (annual %)	4.2		
Inflation, GDP deflator (annual %)	7.9		
Time required to start a business (days)	40.0	40.0	40.0
Mobile cellular subscriptions (per 100 people)	69.0	68.5	
Internet users (per 100 people)	20.0	22.6	
High-technology exports (% of manufactured exports)	0.4		
Overall level of statistical capacity (scale 0–100)	52.2	55.6	55.6
Merchandise trade (% of GDP)	60.1		
External debt stocks, total (DOD, current US$)	7,646,774,000.0	7,710,440,000.0	
Net migration			
Personal remittances, received (current US$)	3,342,500,000.0	3,350,500,000.0	
Foreign direct investment, net inflows (BoP, current US$)	−133,570,895.6	−738,028,978.9	
Net official development assistance and official aid received (current US$)	1,003,530,000.0		

Source: World Bank (2015) (Available: http://databank.worldbank.org/data/reports.aspx?source=2&country=YEM&series=&period)

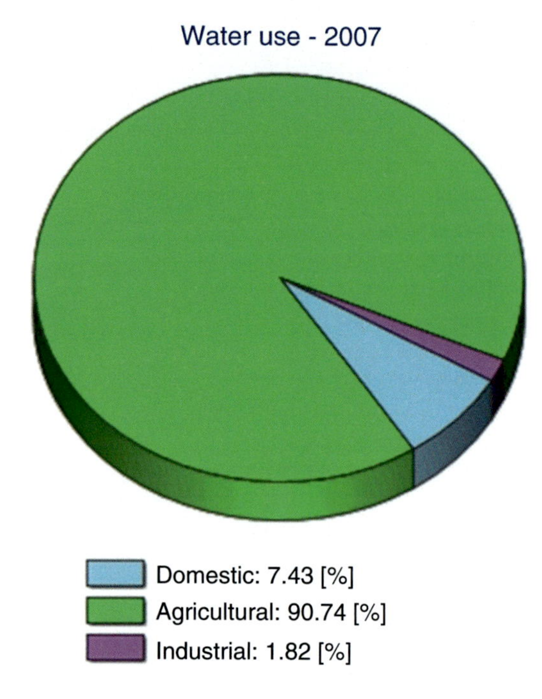

Fig. 2.5 Water use by various sectors in 2007. (Source: FAO 2015) Available at: http://faostat3.fao.org/home/E

Yemen is an arid country and its agriculture sector consumes almost 90% of the potable water, and does not actively promote sustainable water saving techniques. Urban supplies and agricultural production have been under increasing threat from unregulated and widespread use of pumped irrigation supplies by the private sector. The situation calls for immediate regulatory measures to reverse the trend in over-exploitation of water resources. Various strategies have been drafted and are in place, however, the measures taken so far are quite insufficient to make agriculture sustainable and ensure sustainability of water resources of the country. Almass and Scholzb (2006) maintain that Yemen's agriculture sector and its water resources are in real stress. Almost half of the population, about 25 million masses, do not have access to safe water and sanitation (USAID 2016) (Fig. 2.6).

2.3.2 Rainfed Agriculture

Rainfall and underground water resources like wells and springs are major sources to sustain agriculture on roughly 77% of the total cultivated area. Just about half of the total rainfed area receives rainfall less than 350 mm, which could be considered below the minimal amount needed for rainfed agriculture (FAO 2009). While the development of rain-fed agriculture has not yet received as much attention as it deserves, the pressure on groundwater resources has been ever increasing. The re-charge of water is far less than the water being pumped out. The groundwater con-sumption for agricultural purposes is as high as 70%, and according to an estimate

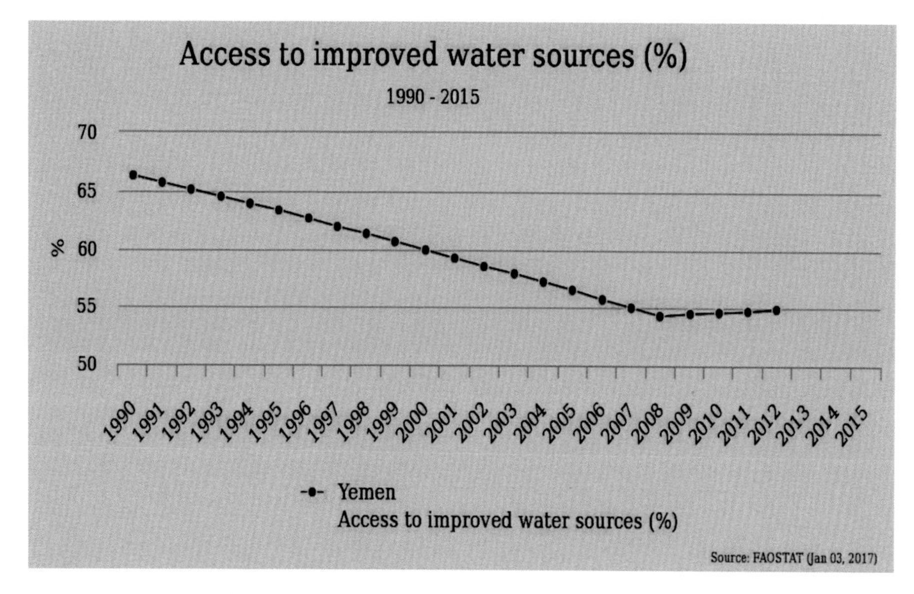

Fig. 2.6 Access to improved water sources (%) 1990–2015. (Source: FAO 2016) Available at: http://www.fao.org/faostat/en/#country/249

over 30% of that consumption goes into the cultivation of qat alone (New Agriculturalist 2010). Adaptive research on the rainfed areas can help sustaining agricultural crops by identifying new varieties and agronomic packages for producing grain and fodder crops that could utilize less water, but are more drought resistant and greater salt-tolerant resulting in higher yields. However, to bring the ideas to the farmers, an aggressive extension program to disseminate information on high yielding varieties and appropriate agronomic techniques to boost production of rainfed grains and forages would be required.

Rainfed farming remains intrinsically risky enterprise due to the rising temperatures and declining, erratic and variable rainfall. In this situation, soil moisture conservation becomes even more important strategy to adapt, because if rains fail (and fail more frequently), crops will also fail, especially if within season spells of drought are prolonged. Supplemental and full irrigation can mitigate longer within season and inter-annual drought, but under climate change conditions, irrigation water supply will also be less secure (Hugh et al. 2011).

2.3.3 *Poor Land Resources and Small Landholdings*

Agriculture is dominantly practiced on small landholdings throughout Yemen, though tenants or sharecroppers cultivate many holdings. Almost 60% of all rural households have some land, although 44% have less than 1.0 ha; yet an average landholding size in coffee growing areas is as low as only 0.3 ha (IFAD 2009). A survey conducted by IFAD (2013) revealed that some 44% of rural households own

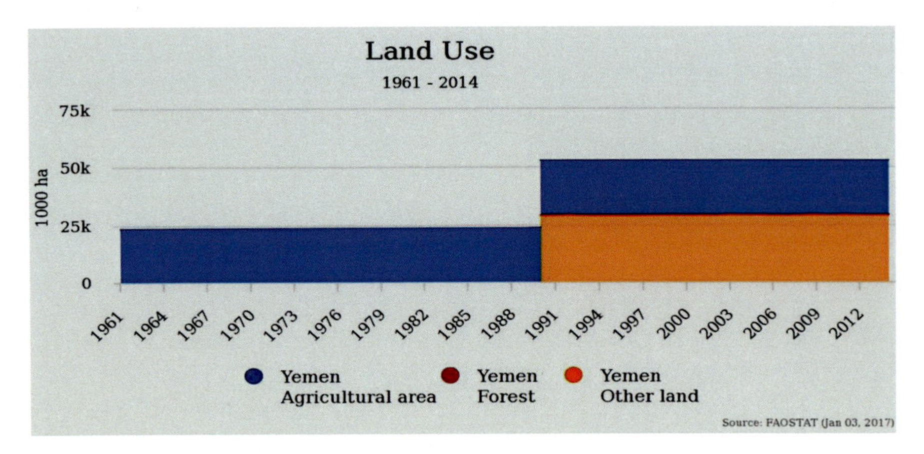

Fig. 2.7 Agricultural area in Yemen (1961–2014). (Source: FAO 2016) Available at: http://www.fao.org/faostat/en/#country/249

farms with area less than 1.0 ha and about 40% of rural households are landless. Twenty-five percent of the landless farmers rear livestock, whereas almost 10% are fishermen and only 5% hold government jobs or are involved in off-farm income generating activities (Fig. 2.7).

In mountainous areas, agricultural fields consist of small terraces, and seemingly difficult slopes are farmed. In the coastal plains (the Tihama, which covers a vast majority of Hodeidah governorate), the land tends to be owned by large landholders; small farmers are interspersed among large farms and a majority of the rural poor work as casual daily laborers or as tenants and sharecroppers. Many of the landless poor find work on large commercial farms. Small-scale animal rearing is also practiced by the landless (IFAD 2013). However, limited access to land and smallholdings even with vivid increases in agricultural productivity cannot generate sufficient incomes for the farmers. Even though there are annual rainfall fluctuations, arable land in Yemen is spread over an area of 1.45 million hectares. Rainfed agriculture is being practiced on about 51% of cultivated land, some 30% is irrigated using groundwater pumped from wells, roughly 10% land is brought under cultivation by employing spate irrigation, about 6% land is irrigated from dams, and almost 3% lands receive irrigation through other sources (IFAD 2013) (Fig. 2.8).

An area of about 1.31 million hectares (94% of arable land) was cultivated in the year 2009–2010 out of which 52% were cereals, 13.8% were fruits and vegetables, 12.5% were fodder crops, and qat was grown on 11.7% of the land. Other cash crops (coffee, cotton, sesame, and tobacco) occupied 6.7% of the land while the legumes were only 3.3%. About 20 million hectares was kept as a grazing land. Four hundred and twenty thousand hectares was irrigated with groundwater resources, almost 11 times more than in the 1980s i.e. 37,000 ha. Notably rainfed cultivated area reduced to about 695,388 ha, while in the 1980s, it was about 1.06 million hectares. The remainder was flood/spate irrigated (IFAD 2013). The recent FAOstat, 2016 estimates on lands and their utilization indicate that permanent crops, forests. Arable

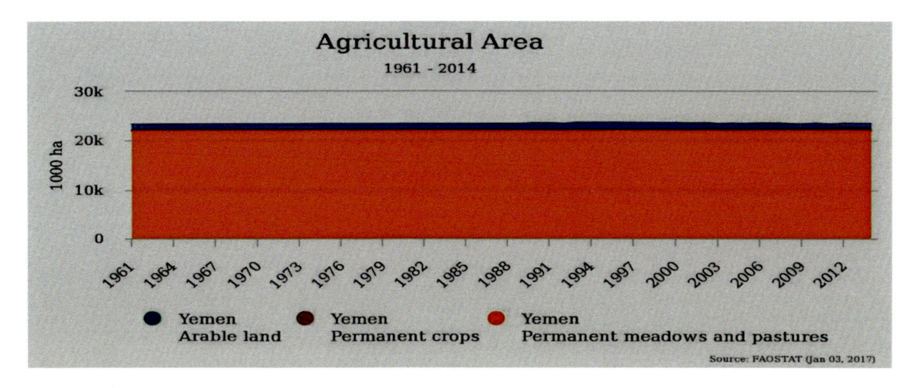

Fig. 2.8 Agricultural area in Yemen (1961–2014). (Source: FAO 2016) Available at: http://www.fao.org/faostat/en/#country/249

Table 2.6 Land use in Yemen in 2014

	(Area in 1000 ha)
Country area	52,797
Land area	52,797
Agricultural area	23,546
Arable land and Permanent crops	1546
Arable land	1248
Permanent crops	298
Permanent meadows and pastures	22,000
Forest	549
Primary forest	0
Other naturally regenerated forest	549
Planted forest	0
Other land	28,702
Area of arable land and permanent crops under protective cover	0.29
Total area equipped for irrigation	680

Source: FAO (2016)
Available at: http://www.fao.org/faostat/en/#country/249

land, permanent meadows and pastures, and other lands cover 0.6%, 1.0%, 2.2%, 41.7% and 54.5% respectively (Table 2.6).

2.3.4 Desertification and Agriculture

Desertification affects around 30 million hectares of the land accounting for over 50% of total land. The hazards of mounting desertification are quite high. Desertification is caused by water erosion, overgrazing, depletion of tree cover,

Table 2.7 Types of land degradation

Type of land	Area (Ha)	Area (sq km)
Desert land	4,856,897	48,569
Chemical degradation of land (Saline agriculture land)	37,090	371
Sand dunes	5,815,937	58,159
Land degraded by heavy winds erosion	475,246	4752
Land degraded by light winds erosion	102,943	1029
Physically degraded lands	12,717	127
Rocky land	28,196,804	281,968
Mountainous terraces	661,504	6615
Naturally stable land (forests and trees)	272,154	2722
Land affected by water light erosion	643,960	6440
Land affected by water medium erosion	1,846,813	18,468
Land degraded by water heavy erosion	2,579,835	25,798
Wetlands (Sabkha)	48,346	483

Source: Millennium Development Goals Needs Assessment, Yemen Country Report; Yemen AREA –ACSAD (2002)

abandonment of terraces, changes in socioeconomic factors and farming practices that determine unsustainable practices for land cultivation (ICARDA 2012). It is estimated that 95% of Yemen's agricultural land remains at risk of deterioration, threatening the government's goal of attaining food self-sufficiency. At present, Yemen imports about 75% of its food, according to government statistics (Irin 2007).

Internal migration from rural areas to cities adversely affects agricultural production, which in future, may result in food insecurity. According to a study, about 85% of Yemen's agricultural land is deteriorating due to water shortages, partly caused by the widespread cultivation of Qat, which requires a lot of water, adding to the growing threat of desertification (Irin 2007). Good quality fertile lands, which account for about 13.6% of the total arable land, are shrinking (Irin 2007).

Models suggest that temperatures in Yemen will steadily rise, and that there is likely to be an increase in variability and intensity of rainfall. Yemen is already experiencing higher incidence of floods. Climate change and variability will thus add to other natural resource challenges to create a need for a wide range of adaptive measures (Table 2.7).

2.3.5 Food insecurity

Yemen is ranked as the 11th most food-insecure country in the world, with one in three Yemenis suffering from acute hunger (IFPRI 2011). Over half the rural population (51%) is food-insecure as compared to 27% of the population in urban areas (CFSS 2012). According to the National Agriculture Sector Strategy (NASS) (2012), some 24% of the households are food insecure. In 2012, about 46% of

Yemenis – around 10.5 million people – did not have adequate food and almost half of all households (45%) are now buying food on credit. Some 6.4 million food insecure masses (37% of the total rural population) live in rural areas. According to a UN report (2015), about 13 million people in the country are food insecure.

In another estimate by IFAD (2015), almost 12.9 million people across the country are food insecure, out of which about 6.1 million are in need of food on Emergency basis, while 6.8 million were placed in crisis situation. The level of food insecurity increased by 21% compared to 2014. The factors like disturbance of markets, diminishing employment opportunities and declining rural livelihoods, can further deteriorate the food security situation if external conflicts and internal insecurity continue to persist.

Overall, 13% of children under 5 are acutely malnourished. Due to chronic malnutrition among children, nearly 60% of children are suffering from stunted growth and severe (life threatening) stunting affects one third of all children in the country. Stunting growth is predominantly affecting two children out of three in rural areas. The poor mountain agriculture areas of the highlands are unsuitable for agriculture and two thirds of Yemenis living on the fringes of mountains are food insecure (IFAD 2013).

2.3.6 *Food Imports*

Farmers on small landholdings are unable to feed the ever-increasing populations of the country. Therefore, the country is very much reliant on food imports, which account for nine-tenths of its total food requirements (UNDP 2015). Yemen is largely dependent on imports to satisfy its domestic consumption requirement for wheat, the main staple (FAO 2015).

However, the National food security depends on the financial capacity of government to import food commodities to compensate for production shortfalls (IFAD 2013). Almost 80% of cereals consumed in the country are being imported to meet food requirements (Government of Yemen 2013). Yemen imports 70% of all cereals, 90% of wheat and 100% of rice to feed its requirements (IFAD 2013). The wheat import dependency is about 95% and in the last 5 years, an average of 2.8 million tonnes per annum of wheat was imported annually out of a total domestic wheat utilization of about 3 million tonnes.

According to recent FAO statistics, the import requirement for cereals in the 2014 was estimated at about 4.5 million tonnes, including 3 million tonnes of wheat, 0.7 million tonnes of maize and 0.4 million tonnes of rice. This compares with 4.4 million tonnes of cereals imported in 2013 (GIEWS 2015).

However, due to the ongoing conflict and the resultant restrictions, only 15% of the pre-crisis volume of imports is reaching Yemen. The situation has largely affected business activity and the reaching of goods into the country. Consequently, some 75% businesses are struggling to maintain their regular provisions and supplies (UNDP 2015) (Fig. 2.9 and Table 2.8).

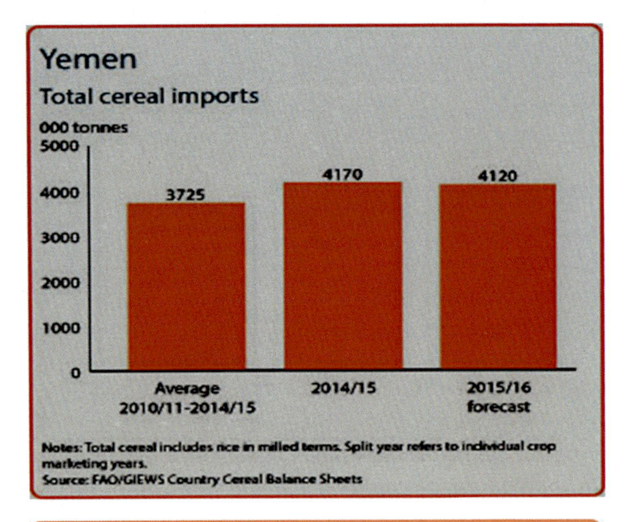

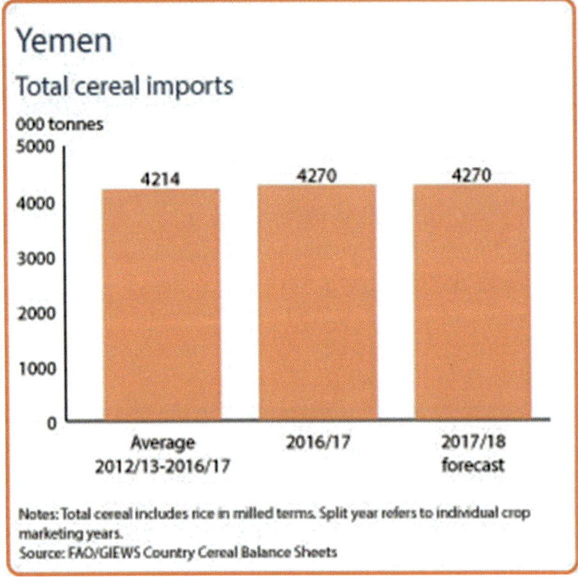

Fig. 2.9 Imports of top three crops 2014–2015 (Source: FAO 2016) Available at: http://www.fao.org/giews/countrybrief/country.jsp?code=YEM

2.3.7 Gender Issue and Women Participation

Yemen faces an issue of high gender inequality. Strict gender segregation is a way of life in the country (IFAD 2013). With a value of 0.128, women in Yemen rank at 75th position among the 75 countries in terms of empowerment (IFAD 2007). Women grossly contribute to farm labor and take care of small livestock herds. In addition, they are in charge of household activities (cooking, fetching water,

Table 2.8 Imports of top ten commodities – their quantity (t) and value [1000 USD] 2011

	Commodity	Quantity [t]	Value [1000 USD]
1	Wheat	2,686,857	961,895
2	Sugar refined	456,013	353,866
3	Maize	438,635	153,262
4	Sugar raw centrifugal	207,857	159,677
5	Palm oil	152,410	155,811
6	Beverage non-Alc	144,323	58,790
7	Cake of soybeans	111,773	51,058
8	Fruit juice nes	85,027	Not available
9	Chicken meat	81,683	166,014
10	Pastry	56,685	Not available

Source: FAO, 2015 (Available at: http://faostat3.fao.org/browse/area/249/E)

collecting fuelwood, washing, child rearing etc.) In general, women are not recognized as farmers (except when left alone to manage the farm) and they do not have access to resources, inputs, and services (such as training, needed to strengthen their skills and increase their efficiency). Even though their share of work in the household and on farm is by far the most significant, women are not decision makers and are not fully involved in development programs, mainly in rural areas. However, the role of women, as an economically productive and effective force is slowly getting recognition. Women in rural areas are unable to fully participate in activities that are aimed at reducing disparity and empowering women. Several women NGOs, associations and advocacy groups have started working on the issue at local, national and regional levels (IFAD 2009, 2013).

2.3.7.1 Neglected Role of Women and Relation to Nutrition

Women have always played a vital role in Yemeni agriculture (particularly livestock) and this role has increased with male migration to the cities for work. However, women have inadequate access to the resources and services, therefore rural women deserve to be empowered to influence authorities and participate in decision-making. In fact, real development of agriculture and improvement in household food security and nutritional status is impossible without women's empowerment. Women empowerment can be enhanced through the inclusion of women households systematically in all committees, and by recruiting female agricultural extension staff.

2.3.8 Poverty in Yemen

Yemen is the poorest country in the Arabic Peninsula (IFAD 2013). While in 2008 about 40% of rural people were living below the national poverty line of USD 2 equivalent per day, the triple crises of recent years (food price crises, fuel price crisis, and global financial crisis) increased the incidence of rural poverty to 48% by 2010. Following the political and social violence of 2011–2012, it is now estimated that rural poverty has increased to over 60%; which implies that as of today around 15 million rural people live in poverty and food insecurity. In addition, a large segment of population living marginally above the poverty line remains highly vulnerable to economic and natural disasters (IFAD 2013).

2.3.9 *Malnutrition*

In 2012, some 10.5 million people (46%) did not have enough food (NASS 2012). Yemen is faced with the issue of malnutrition as over 50% of children are chronically malnourished, and as high as 60% are stunted. Global acute malnutrition levels of over 30% were recorded in 2011. Based on Global Hunger Index calculated by the International Food Policy Research Institute (IFPRI), 2011, Yemen ranks among the ten hungriest countries of the world. Over 80% of food-insecure people are living in rural areas, and at the same time rural-urban disparity is escalating. The dwellers of the highland and mountainous areas are the most vulnerable as compared with the other areas of the country (IFAD 2013). Linkages between production, guidance and nutrition are of paramount importance. Combating malnutrition requires specific agricultural extension and educational initiatives, most appropriate of them include: creation of awareness on nutrition, nutrition-specific issues, school-food program, vegetable home gardens, backyard poultry etc. (NASS 2012; IFAD 2013).

2.3.10 *Climate Change and Agriculture*

Climate change models unanimously project that temperatures will increase across Yemen over the next few decades by levels higher than the projected global average.

While these models agree on temperature, they disagree on the direction of change in precipitation levels. Most of these models project a modest increase in precipitation, while others point to a reduction in precipitation.

The discrepancy is mainly attributed to the location of Yemen in the Inter-Tropical Convergence Zone (ITCZ) which has highly unpredictable climate patterns. It is however projected that climate change will drive up the variability of precipitation leading to more frequent and intense rainfall events and more extended droughts. These projected changes are generally expected to aggravate existing problems facing socio-economic development and present additional development challenges to international development agencies including IFAD.

In the absence of adequate adaptation measures, extreme rainfall events could cause devastating flash floods that erode fertile soils, destroy crops, buildings and infrastructures and take lives of unprepared rural victims. They pose significant risk to the livelihood of rural communities – particularly the poor, women and the marginal communities.

Droughts will accentuate the severe water scarcity in Yemen and will have a detrimental impact on agriculture, particularly rain-fed agriculture which constitutes the economic mainstay for the majority of the rural population, reducing the area of viable rain fed land.

Recently Al-Kharasani (2014) conducted a study on climate at the Taiz meteorological station and analyzed the data that indicated an increase of 1.04 °C in maximum air temperature and 2.5 °C in minimum air temperature during the period from 1983 to 2013. Based on the findings of study, it is anticipated that climate change would affect the agriculture sector due of the change in temperature. Climatic data of both air temperature and rainfall has shown that the air maximum temperature has increased by 1.04 °C during the period from 1983 to 2013. This rise may cause drought in some areas resulting in the loss of some agricultural crops. The authors believe that the rise in temperature could be due to the increasing greenhouse gases in the atmosphere as a result of anthropogenic activities.

2.3.11 Qat: Rival of Food and Cash Crops

Qat (*Catha edulis*) is a recreational drug, a semi-narcotic plant. As a mild stimulant Yemenis chew its leaf for relaxation (IRIN 2012). It is cultivated on an estimated area of 154,000 ha that represents about 22.3% of irrigated area. The crop consumes almost 40% of the potable water. It is mainly cultivated in the highland where water scarcity is critical. Reducing qat production, and promoting water reuse of treated wastewater for irrigation could mitigate the water crisis. However, this is currently socially unacceptable (Almasa and Scholzb 2006).

Qat is a, favorite among the farmers for its ever-increasing demand and guaranteed marketing at lucrative price. The farmers even get indirect government subsidy on diesel used for pumping out the ground water. The government is taking measures to discourage the qat cultivation, because apart from huge water consumption, it has become a social problem, not to mention the increasing threat of food security as the land being used for qat cultivation could be used for growing food crops. So far, the success has been limited in convincing farmers to grow alternate crops.

Being a profitable crop, it is increasingly displacing food crops, and also utilizing limited groundwater. About one third of Yemeni agriculture is devoted to producing qat. Its production is profitable but replaces the production of food crops or export crops, and its consumption can be a social and health problem. However, it is also a mainstay of the rural economy in the highland areas where it is grown, with over 25% of farmers, including very poor ones, growing the crop. Farmers view qat as a powerful agent for bringing cash back from urban to rural areas. No doubt, qat has its role in the rural economy and poverty reduction, however strict implementation of appropriate measures would reduce, and ultimately phase out the crop to get rid of the menace caused because of it. Farmers are to be educated to produce market-driven alternatives and other high value crops such as coffee and almonds instead of wasting scarce water resources on producing qat.

A report suggests that one in every seven working Yemenis is employed in producing and marketing qat, making it the second largest source of employment, exceeding even employment in the public sector (World Bank 2006). Qat cultivation is widely spread in the Sana'a Governorate. While the use of qat is acknowledged as a social plague, absorbing up to 28% of the earnings in low-income groups, its production and trade undoubtedly provides quick returns and employment (IFAD 2009). Irrigated qat production has been increasing at an annual rate of 3.5% between 1970 and 2006 and the associated increase of groundwater withdrawal has reached an estimated 837 million m^3/year (i.e. 84% of the renewable groundwater resources). In addition, uncontrolled pesticides are being used on qat, causing serious health hazards and contamination of water resources.

2.3.12 *Political and Governance Issues*

Following the events of 2011 and the dismissal of the previous regime, there has been a popular consensus on a 2-year transition period (2012–2014) and on a national dialogue conference to frame a new constitution and to prepare elections to a fully democratic government. Till date, this process has not met significant success. Yemen is experiencing governance issues due to political unrest. Although certain donor-funded projects are ongoing, the uncertainty prevails. Development activities cannot be carried out satisfactorily in the certain southern governorates due to security risk caused by militants.

2.4 Coffee Production

Yemen has long been famous for producing the best coffee in the world, known as Arabic coffee – *Coffee Arabica.* Yemeni coffee trade found great success because of its high and unique quality, which distinguished it from all other coffee types in the world. Yemeni coffee enjoys a highly lucrative market niche worldwide for its excellent quality (second only to Jamaica Blue Mountain) and receives higher price as compared to coffee produced in other countries (about twice the price of quality coffee from Costa Rica, Guatemala and Kenya). Some 11 countries import green coffee beans from Yemen on a regular basis due its better taste and premium quality (IFAD 2009). With the value of about USD 20 million at present, Yemen exports its coffee from 4000 to 6000 MT/year. However, in the 1950s the export was about 12,000 MT/year. About 100,000 small farmers are involved in coffee production, representing about 9% of all farming households (IFAD 2009).

Table 2.9 Coffee export prices in the selected countries, USD/MT

Year	Jamaica	Yemen	Kenya	Guatemala	Brazil
2003	10,333	**3328**	1426	1198	951
2004	21,140	**4319**	1804	1573	1240
2005	20,975	**4454**	2554	2299	1861
2006	18,841	**4224**	2772	2278	1985
Average	17,822	**4081**	2139	1837	1509

Source: FAOSTAT (2012)

After enjoying the monopoly and great prosperity for three centuries, the Yemeni coffee has gone through trade setbacks and declines due to some unfortunate negative events. Trade started to decrease gradually, and in the nineteenth century, Yemen lost many global coffee markets. The main reason is that coffee production shifted to other regions of the world that competed with Yemen at the time. Some local factors have negatively impacted export and production of coffee in Yemen. A recent study conducted by Al-Zaidi et al. (2016) established the need for launching of agricultural extension and capacity building programs for the coffee growers to narrow knowledge gap and enhance their skills on coffee production. They further advised to use Television and Radio programs for this purpose as the farmers preferred these information media to have the needed information on this strategic crop (Table 2.9).

2.4.1 *Why Coffee Production Declined?*

There are many factors responsible for decline of coffee production which, besides others, include, irregular water supply, drought to which coffee trees are highly sensitive, insufficient strategic focus, inadequate investment across the value chain, limited research, lack of extension support for the coffee growers, absence of transparent marketing systems, collapse of the standardization system and, lack of branded names. These problems combined have resulted in low productivity and reduced revenues. However, international demand remains high and can absorb significantly increased quantities, provided that quality is improved and consistently maintained.

2.4.2 *Qat Production vs. Coffee*

Today, coffee plants of the past have been replaced with the semi-narcotic qat trees, remarkably dwindling coffee lands. Estimates for 2012 showed that the planted area of coffee in Yemen declined to 34,900 ha, as opposed to 162,500 ha for planting qat. Coffee production amounted to 19,800 tons, as compared to 190,800 tons of qat (Al-Monitor 2014). As a result, Yemen has been deprived of a significant source of income that could have been as important as other commodities such as wheat and cotton.

Despite all the obstacles surrounding coffee production, coffee is next to oil, and is Yemen's main export. Coffee cultivation is spread throughout most regions of the country. Around one million farmers are engaged in coffee production, although they employ the same primitive ancestral methods to ensure same quality and taste despite the severe water scarcity (Al-Monitor 2014).

Yemeni coffee fetches premium price due to its high-quality, unique flavor and refined taste. In spite of climatic and soil factors which are favorable for production, farmers employ old traditional production practices. Therefore, the country needs to chalk out a comprehensive strategy to facilitate its active participation in the international conferences and exhibitions to learn the best practices in order to augment coffee production at all the stages from production to marketing to compete with other coffee producing countries. It is imperative to adopt the best international practices, which are adopted by other countries associated with the coffee business. For regaining the lost position, it is important to create an agency that could primarily focus on the development of coffee sector. Such an organization will encourage farmers to grow coffee, manage water conservation, facilitate distribution of high quality seedlings of to the farmers, and most importantly employment of competent and qualified technical staff.

2.4.3 *Marketing*

Marketing of agricultural products in Yemen faces numerous challenges. Small-scale farmers have little access to extension advisory services that could provide vital information to them on marketing their produce. Unfortunately, an effective Market Information System (MIS) does not exist in the country. Due to the limited marketing opportunities provided by traditional retail and wholesale markets, and the lack of associations and organizations, small farmers are at great disadvantage. Besides the fact that the existing markets for agricultural products are inefficient, numerous products do not meet international specifications. The value chain is not well developed for many products, as the processing, packaging, and storage industries are at a low stage of development. Markets operate freely with little government intervention. However, still additional government support is needed in the areas of extension, rules to strengthen farmer cooperatives and associations, setting and enforcing grades and standards to improve quality and provision of credit to support the growth of the processing, packaging and storage industries (IFAD 2013).

2.5 Strategies for Realizing Sustainable Crop Production

The strategies for efficient use of limited water resources to make agriculture profitable, economically efficient, equitable and sustainable include:

Enhancing Crop Production

- Institutional strengthening to promote efficient use of water;
- Conserving water resources to realize sustainable agriculture;
- Increasing farmer incomes through increased water use efficiency;
- Enhancing resource availability, sustainability and quality through water harvesting and watershed management;
- Productivity of agriculture on the rainfed areas can be improved by restoration of the terraces, managing watersheds and water harvesting;
- Productivity of rainfed agriculture can be improved by employing more efficient agricultural water management practices through the adoption of modern irrigation techniques, agronomic packages and water harvesting practices;
- Adoption of crop production packages to cater to the arid environment and resilient in the face of prospective climate change;
- An increased recognition of the role of rural women in meeting food needs, improving nutrition and protecting the environment;
- Improving productivity in irrigated agriculture through modern irrigation techniques and advisory services;
- A strong focus on improving productivity and sustainability of livestock production, as this sub-sector has growth potential and as livestock are the principal asset and economic activity of the poorest and of the landless; and
- Diversification of cropping patterns into new or revived cash crops (coffee, honey, almonds, oil crops etc.) and into more nutritious foods to help improve and diversify household nutrition, and development of related value chains, and parallel reduction of area planted to Qat.

Addressing Environmental Issues, Managing Watersheds and Controlling Erosion

- There is a need to design programs addressing environmental issues with the participation of local communities and mass media. In order to address environmental issues and biodiversity it is important to build extension programs based on the rich indigenous knowledge and the traditional approaches in conservation and utilization of resources which were practiced for centuries and are fading away with the passage of time need to be adopted.
- Surface water should be managed economically and efficiently.
- Efforts be made towards the protection of renewable natural resources from degradation, pollution and depletion and the conservation of biodiversity at the level of the watershed;
- Clear and comprehensive coordination mechanisms are to be established among the stakeholders and the concerned organizations for developing appropriate interventions and constructing suitable structures capable of combating or at least minimizing the erosion hazard and sedimentation into the downstream areas.

Combating Climate Change

- It is not easy to combat climate change. However, some measures like promotion of climate resilient agricultural approaches and technologies could reduce its harmful effects. Some approaches include: implementation of the disaster-risk community action plans; adoption of sustainable agricultural practices to promote the drought/heat tolerant crops; improvement of water management for both drinking water and irrigation through the construction/rehabilitation of resilient water infrastructure and restoration of natural resources and improving the communities by managing natural assets (IFAD 2013).

Empowering Women

- Promote women's skills in coffee growing, beekeeping, tunnel horticulture and micro enterprises;
- Provide technical, financial, managerial training for women;
- Improve domestic water supply;
- Increase entrepreneurial women's access to financial and business services;
- Develop women's agricultural skills, to empower them economically;
- Ensure women's access to savings and credit from MFIs;
- Encourage women representation on management boards of associations.

2.6 The National Extension Service in Yemen

Agricultural Extension Service (AES) is the responsibility of the government and is administered through the Ministry of Agriculture. The role of extension services is usually restricted to the delivery of information to the farmers on the new technologies and government programs. However, its working needs to be improved by creating awareness on sustainable development of the local areas as well. However, the potential role of extension personnel, as vehicles to convey information from rural communities to scientists, planners, and policy-makers, is too often disregarded and most probably inexistent. Therefore, in order to realize sustainable participation of extension personnel in creating awareness, a two-way information flow needs to be ensured (NAP 2000).

MAI reportedly has more than 1300 extension agents and technical experts located throughout the country. Nevertheless, this system is clearly not effective at providing farmers with information and necessary skills to improve their coffee cultivation and marketing. Agricultural extension workers are personnel who are responsible for meeting the goals of extension system. However, there have been less data on the roles and performance of extension workers in the country, even though there are sporadic studies on criticism that extension was not being able to perform the necessary changes in the rural community (Sallam and Akram 2005).

2.6.1 Evolution in the Extension System

Realizing the weakness of Extension systems, several international organizations and donors, especially the World Bank helped strengthening extension services during the 1980s. However, later, the government's priorities changed and extension has been suffering setbacks since 1990s. Until recently, Agricultural Research and Extension Authority (AREA), an autonomous body, remained responsible for both research and extension in Yemen, but the Authority was criticized for focusing on research and ignoring extension. Now, as a result of re-organization, the function of extension has been given to the General Directorate of Extension and Training located within the Ministry of Agriculture and Irrigation. Extension has been decentralized along with other technical departments in Yemen (Qamar 2012).

2.6.2 Agricultural Blocks and Extension Centers

Agricultural Blocks (district level extension offices) and Extension Centers (village level extension offices) have been established in most of the governorates during 1980s when several donor agencies particularly the World Bank focused on strengthening extension services at the request of the government of Yemen. There were four to six Extension Centers under each Agricultural Block. However, as the projects ended and the government's priority for extension gradually declined, the maintenance of the centers lacked due attention and the infrastructure started deteriorating making extension staff and the farmers disappointed. Qamar (2012) noted that out of the total of 467 Agricultural Blocks and Extension Centers, only 392 were functional while some 75 were non-operational due to various reasons including precarious physical condition, lack of operational budget, going away of frustrated resident staff, lack of equipment and transport facilities, and the buildings of the extension offices were encroached by others.

2.6.3 Status of Agricultural Extension

Though agricultural extension has made use of Public media (Radio, TV channels and newspapers) to create awareness on the environment and on the conservation of natural resources for the future generations yet still seems to be in its initial stages, however, slowly gaining acceptance and momentum. Also, programs to address environmental issues with complete participation of local communities and by using different mass media are the need of the day. The school and university curricula need to be revised and updated to accommodate the present challenges of conserving the fragile environment and sustaining the utilization of limited available resources. Extension activities need to be updated and revived to focus on agricultural and environmental issues in simple and attractive manner, based on the rich indigenous knowledge, traditional approaches and modern scientific practices to realize conservation and utilization of resources. Raising awareness need to be developed and incorporated in the national campaigns and awareness programs at the national, regional and local levels (FAO 2009).

2.6.4 Challenges Faced by Agricultural Extension

Over the last decade, extension service started experiencing some challenges due to socioeconomic changes and agricultural sector reforms taking place in the country. In spite of decentralization in Yemen, only 3.2% of farming households receive any type of extension support. Even in the Internal Plateau Zone, where extension services are most accessible, only 6.6% of households benefit from extension. According to country report on plant genetic resources (FAO 2009), capacity building programs face problems at the individual, institutional and system levels. Reasons behind the unsatisfactory performance of extension are low salaries, weak management, absence of monitoring, evaluation and accountability, negligible budget and transportation facilities, mostly outdated technical information, and technical directorates running independently their own fragmented extension activities. Other issues include:

(a) *Poor linkage between Research and Extension*

Several authors, studies and reports confirm that there are no longer any clear linkages between researchers and extension agents. So, the latter receive few if any refresher courses and on-the-job training (Muharram and Alsharjabi 2001; USAID 2005; NASS 2012).

(b) *Extension workers lack logistic support*

Due to lack of logistical support and with the few vehicles, and nearly no operating funds, extension agents rarely make visits to the farmers, making minimal impacts on the famers and the producers, and have not developed relationships of trust with them (USAID 2005).

(c) *Lack of operational funds*

Extension, like most other technical departments of the Ministry suffers from acute shortage of operational funds. Under the circumstances, donor-funded projects are welcomed as at least temporary source of funds. As indicated earlier, dozens of Agricultural Blocks and Extension Centers, which are located at crucial district and village levels are either non-functional or have been abandoned due to lack of basic residential facilities and their maintenance as there are no funds available.

(d) *Extensionists do not use innovative ways*

Extensionists often adopt academic style for dissemination that fails to make any impact on the farming communities in part because it is dependent on paternalistic lectures rather than visual aids and demonstrations.

(e) *Decentralization has caused negative impact*

Like in many developing countries (Indonesia, Philippines, Tanzania, Uganda, Pakistan, etc.) where decentralization has weakened extension services due to political interference of elected local government officials, Yemen's extension services have also suffered. Proper utilization of budget allocated for extension purposes has become an issue to divert these funds to non-extension activities.

Box 2.2: Problems and Challenges Faced by the Agricultural Extension in the Yemen
Institutional and Organizational Defects:

- Extension department is located at the Ministry of Agriculture and Irrigation (MAI) and the Ministry has massive mandate in agriculture and irrigation;
- Low operational budget allocated for agricultural extension activities accompanied with the absence of transparency leads to misuse of limited budget.
- Limited government financial resources are unable to sustain the inherited inappropriate organizational set up;
- Extension organizations do not make clear job descriptions available to the extension staff and professionals and provide insufficient information on activities to be performed;
- Lack of adequate institutional and technical support for public extension services.
- Inappropriate selection and use of extension methodologies and approaches;
- Prevailing of weak management and supervision at national and regional levels;
- Lacks legislation that could organize various agricultural extension professionals engaged in practicing and services delivery.
- Supervision, monitoring, evaluation and accountability of the extension staff are absent at national level;
- Distribution of available extension and resources at local, national, regional and in different regions, areas, and agri-ecological zones throughout the country are poor and imbalanced.
- Weak coordination among extension services at regional and national levels as well as between extension and other relevant agencies;

(continued)

Box 2.2 (continued)

- Extension services employs on traditional extension approaches. Most recent extension approaches such as participatory extension, rural advisory extension, farming system extension, farmer-to-farmer extension or extension through paraprofessional local agents have not been put into practice;
- Though proved very viable, vibrant and sustainable, yet the republic does not make an extensive use of participatory extension systems.

Extension Staff

- Extension staff are not educated enough to undertake extension activities;
- Low human resources capacity;
- Low salary scale and incentives for agricultural extension personnel;
- Rewards, punishment system and accountability features are absent in the extension service;
- Limited number of female staff in the extension service including public, cooperative and private agencies;
- Absence of suitable plans for the professional development for agricultural extension to make them aware of new policies and newly emerging changes;
- Machines and equipment especially transport vehicles are not available to enable extension staff to perform their duties;
- Low organizational status/position is attached to extension staff as compared to other professionals of other disciplines;
- Extension employees, both technically and managerially are unable to meet the needs of a renewed mandate in agriculture.
- Extension staff prefers to work and/or live in Agricultural Extension Centers (ECs) and Blocks (ABs) as they believe that they would be deprived of basic services away from populated areas as some ECs and ABs are not furnished;
- Extension professionals are assigned multiple job assignments that adversely affect their working in extension;
- Lack of monitoring and evaluation system of the ongoing research-extension farmers
- (R-E-F) coordination linkage mechanism;
- Participation of female and extension agents in specialists research-extension farmers
- (R-E-F) coordination linkage is very low.

Dissemination of Technologies to Extension

- Extension staff are not well equipped with the skills and knowledge on disciplines like: rain-fed farming, livestock, improved management of natural resources, gender and rural women, agricultural cooperatives, post-harvest technologies and marketing of agricultural commodities and socio-economic aspects – market analysis;

(continued)

Box 2.2 (continued)

- Adequate and suitable package of farming technologies and recommendations for various farming systems and ecological zones covered by extension especially for the rain-fed areas are not available;
- Extensionists rely and use mostly outdated technical information;
- Extension service neither gets benefit of current extension techniques, nor any updated technical knowledge regarding current agronomic, business, or management techniques.

Financial and Material Resources:

- Low salaries, weak management;
- Lack of government funding and negligible budget – with only 1.0% of the budget, MAI is expected to serve 54% of the workforce engaged in agriculture;
- Lack of transportation facilities,
- Technical directorates running independently their own fragmented extension activities;
- Public extension system does not function well due to lack of resources (Source: Muharram and Alsharjabi 2001; NASS 2012)

(f) *Training Options for Extension Professionals*

Pre-service education in agricultural extension is offered at agricultural faculties of major universities, i.e. Sana'a University, Aden University and Dhamar University. However, it is equally important that extension agents and officers are exposed to "In-service trainings" on a regular basis. In order to improve the working of extension agents, it is important to equip them with the basic scientific extension methods, and upgrade their scientific and technical abilities to work with research scientists and the farmers through refresher courses and updated on-the-job training to cater to issues of the day (USAID 2005).

(g) *Lack of electricity facilities in the villages*

Not all the villages have been electrified so far. In such cases, Radios instead of television are being used by the farmers. To solve the problem of power supply, extensionists make use of generators to compensate frequent power outages. Cell phones are quite common in rural areas so these should be used.

(h) *Information and Communication Technology (ICT)*

Extension offices, especially those in the governorates and districts, do no employ the modern tools of information technology to supplement their field extension activities. However, a well trained and experienced staff at the Directorate of Information of the General Directorate of Extension and Training do have necessary equipment and studio for preparing television and radio programs besides a large variety of audio-visual equipment, consequently are using radio and television programs to carry out extension functions and deliver extension messages on a regular basis for the farming communities. Certain equipment though is now obsolete and needs replacement with modern ones.

(i) *Relationship between Agricultural development and Extension Education*

The most viable option to increase productivity in the agriculture sector would be to revisit the government's extension and research capacity. Analysis of current service provision has found it lacking to non-existent in most parts of Yemen. In addition, research and extension do not work together as they should, where advances in research are to be passed on to the farmer through the extension system. Public extension service needs restructuring, capacity building, introducing effective extension techniques, strengthening linkages to research, and recruiting a new male and female extension agents having a positive, service oriented attitude to replace the existing, non-productive and extension agents with low performance. It is important to equip the new recruited employees through capacity building of is needed with updated skills on financial management, efficient water usage, water saving agronomic techniques, introduction of new crops and varieties, and association or cooperative development. These measures taken to improve extension services would have positive impact on agricultural production.

2.7 Strategies for Improving the Working of Agricultural Extension

First of all, it is important to identify the problems faced by the farmers and the Extension Service. Restructuring of extension system would help overcoming the problem of organizational differences, addressing discrepancies, modifying the present institutional position and form of extension in various agencies.

2.7.1 Improving the Working of the Extension Organization

- Identification of extension organizational constraints in different agencies and areas by using the diagnosis tools of Participatory Rapid Appraisal (PRA) would improve its working. Such tools help define the problems, set priorities, and devise possible viable solutions with the assistance and involvement of all concerned parties.
- Organization defects can also be removed by conducting a comprehensive evaluation study on the presence of extension staff in different areas representing various extension agencies; and extension approaches adopted by them to assist the farmers under different farming systems. Such measures could help in the redistribution of extension staff, resources and facilities. It will also help in improving extension approaches and methodologies; formulating human resources devel-

opment plan; staff rationalization, and possibly in developing recruitment guidelines for extension especially female staff.

2.7.2 *Elevating the Existing Generally Weak and Transforming Low Organizational Status*

- Elevating the existing weak system and transforming low organizational status of extension into a more realistic position that matches with and reflects its importance;
- Supporting extension research and studies aiming at: testing and evaluating extension approaches and methodologies applied in the field; and revising and formulating local versions of tested approaches and methods to develop the package of technology that could be replicated under farmer's field conditions.
- Further follow on studies must be conducted to investigate the adoption and impacts of commercialization of advocated farming technologies.
- Attitude of the researchers, extension staff, and farmers towards new extension approaches and methods deserve continuous and perpetual evaluation to check distortions and make improvements.
- There is a need also to foster and institutionalize relationships concerning extension with various relevant agencies under which extension is operating considering legal, organizational and financial dimensions affecting such relationships, including gender issues and rural women development.
- Extension departments be given more authorities and control over the allocated budget.

2.7.3 *Redefining the Extension Role and Responsibilities*

Such an initiative would also necessitate redefining extension roles and responsibilities at national and regional levels. A readjustment must consider various internal and external changes and resultant new trends and policy guidelines especially program budgeting, the needs for rules and regulations including job description and criteria for occupying extension posts at all levels and working circles.

2.7.4 *Improving the Working Living Conditions of the Extension Workers*

- Agriculture extension is viewed as a fundamental pillar in agricultural and rural development, food security, improving living standards; farmer and rural economics and shielding the future challenges like free market system including the agreements of the World Trade Organization (WTO).

- Based on its constructive roles, it seems imperative to activate and revitalize the role of agricultural extension and its national institutions on a country level. Its rehabilitation, strengthening and developing its capabilities in various aspects would enable it to function efficiently and effectively meeting the future challenges and resulting agricultural development.

2.7.5 Strengthening the Linkages Between Research and Extension

At present, extension activities are carried out under numerous agencies. The reorientation of all these agencies seems imperative. It is important to involve all levels of management and staff with the aim of enabling them comprehend and deal with the new government policies and directions. Further, revisiting, restructuring and renovation of the existing Extension System would help overcome the organizational differences and discrepancies affecting the present institutional position and form of extension in various agencies (Muharram and Alsharjabi 2001; NASS 2012). Another viable option would be, though presents a tremendous challenge, but also a tremendous opportunity, to start over creating a new extension service that could be public, private, or a combination of the two. Though the country needs complete overhauling yet based on the identified constraints and the obstacles, its working can be improved by:

- Updating extension activities in agriculture and rural areas;
- Allocating adequate annual operational budget and making sufficient funds available by employing easy procedures for the efficient functioning of agricultural extension and carrying out extension activities;
- Improving the salary scale and incentives of extension personnel.

2.7.6 Strengthen the Links and Establish Coordination Among Agricultural Organizations

All the organizations working in research, development sector and extension outreach programs need to establish working relationships and beneficial coordination with each other to make the information flow efficient among them and finally to the farmers. Establishing suitable coordination linkage mechanisms among extension services and the cooperative and the private sectors, with flexible and varied options would help achieving the aims and interests of various agricultural and rural development partner organizations.

Establishment and strengthening of partnerships among the farmers, private sector, government, and donors is necessary and could be realized through:

- Enhancing cooperation and coordination between agricultural extension and all other government and civil agencies relevant to the context of agricultural and rural development;
- Developing linkage mechanisms and partnerships with the private and cooperative sectors to lessen the stress on the government resources and multiply the extension efforts in order to achieve the aims and interests of various participating agricultural and rural development organizations;
- Developing and activating existing institutional bodies (i.e. extension services, technical committees, coordination units, national extension committee, the multi-disciplinary teams etc.) to make extension effective and efficient;
- There is a need to define the roles and responsibilities of all the stakeholders and partners as outlined in National Extension Strategy.

2.7.7 *Moving Towards Innovative Rural Extension*

First step in any extension program is to create awareness among the farmers regarding what to deliver and how to deliver so that farmers could help to help themselves. Once farmers get to know extension, then there is a need to revitalize the extension service, which will be the key to improving the productivity of agriculture, and raising rural incomes.

On the basis of identified constraints and obstacles, the following suggestions are proposed to overcome the problems and make agricultural extension effective and efficient by:

- Developing an active and typical information service that caters to the information needs of extension personnel in all fields relating to their work, in a suitable and timely manner by employing the right channels with proper contents systematically and regularly;
- Adopting and employing best proven extension techniques, to transmit an updated technical knowledge regarding current agronomic practices, business, or management techniques.
- Arranging trainings on "How to do Agri Business and making more profits by smart marketing of agricultural products. Currently Extension service does not provide trainings to the farmers as well the extensionists although it is also the part of the extension service.
- At this point, cash infusions, and a few training courses for underpaid and unmotivated extension staff, will not contribute towards the development of the agriculture sector.
- Develop and launch awareness and extension education and capacity building programs involving all the concerned parties.

2.7.8 Making Use of Communication Media

Numerous means of public media (Radio, TV channels and Official newspapers) are available in Yemen to create awareness on the importance of the environment and the need to conserve our natural resources for future generation. However, such endeavors are in their initial stages, and are gaining momentum slowly.

2.7.9 Adopt Participatory Approach

While working for the development of agriculture, NASS consultations made field visits to meet the stakeholders. The stakeholders emphasized the successful working of participatory approaches during the past decade. All of them believed that a highly participatory approach involving communities is the most effective way to ensure sustainability and improvements in production and food security.

Stakeholders emphasized that by participation they mean involving the selected communities in the full cycle of activities starting with the decision-making process for investments and services, followed by supervision, management and physical implementation of the activities, monitoring physical and financial aspects.

While working in the field, NASS consultants were informed by the stakeholders about the proposed key elements essential for the participation. The mechanisms include: setting up of community and producer associations at the local level; cooperatives need to strengthen partnership at the community level and empowerment of farmers in marketing and purchases; community mobilizers to act as the interface between communities and community planning and the services responding to community demand; community-based extension and animal health workers; and ensuring joint and participatory Monitoring and Evaluation (Government of Yemen 2013).

Community involvement in planning and implementing development projects from the outset is essential for their success, future ownership and sustainability. Appropriate guidance during participatory planning process by trained facilitators is a key factor of success of project activities.

Enthusiasm for community-driven approaches to development has been successfully demonstrated in the ongoing programs. Appropriation by well-trained communities of their Community Action Plans, infrastructure and productive investments is a guarantee for long-term sustainability.

Based on the past experiences as reported by the Government of Yemen (2013), encouraging and promoting the direct involvement of the beneficiaries in planning, implementation and of the projects in all the steps and activities would result in positive outcomes.

2.7.10 Importance of the Indigenous Knowledge

The combination of indigenous knowledge and modern approaches works better than trying them in isolation. Indigenous knowledge is based on the life-long experience of locals whereas modern technical practices are based on tested scientific principles.

Indigenous knowledge regarding conserving and managing natural resources, practicing traditional agriculture and managing grazing lands can add value to the modern scientific technical know-how. While developing such practices, it is important to attach due consideration to the Indigenous knowledge while addressing technical environmental issues.

2.7.11 Up-Date Curricula

FAO (2009) stressed the need for revising the curricula at the school and university levels and update it to accommodate the challenges of conserving the fragile environment and sustaining the utilization of limited natural resources available.

2.7.12 Reliable Monitoring and Evaluation System

- Developing a reliable monitoring and evaluation system with reasonable appreciation and incentives of promotion for extension staff is important. A comprehensive evaluation of extension in different extension agencies, and the approaches applied under different farming systems needs to be conducted. Such initiatives will help the fair redistribution of extension staff, resources and facilities to meet the priorities and satisfy the new goals set by the extension.
- Ensuring the active participation of government, private and cooperative sectors and all other key stakeholders in implementing, monitoring and/or evaluating the plans and making recommendations for realizing sustainable agricultural development.

2.7.13 Logistic Support

Required and needed support like machines, equipment and furniture for extension work in selected areas must be provided. Also, developing an active and efficient information service that caters to the information needs of extension personnel in all fields relating to their work, in a suitable and timely manner through the right channels with proper contents systematically and regularly will help.

2.7.14 *In-Service Trainings*

Well-planned and focused meetings and orientation training-cum-workshops for the extension staff would educate them on the new government policies with regards to extension and the role of extension. In-service trainings for reorientation of all extension staff to enable them to grasp and deal with the new government policies and directions must also be conducted.

2.7.15 *Extension Needs to Make Farmers Market-Oriented*

If the farmers make more profits based on the information and guidance provided by extension, they will have an enhanced trust in extension. However, small scale farmers do have little access to extension advice on markets for their products, and on the other hand, an effective Market Information System (MIS) does not exist in Yemen to benefit and assist the subsistence farmers.

2.8 Conclusions and Recommendations

In Yemen, many factors are responsible for negatively impacting agriculture which includes: increased land degradation and soil erosion, changes in water availability, biodiversity loss, more frequent and more intense pest and disease outbreaks as well as un-expected natural disasters. In addition to agricultural issues, Yemen also suffers from political unrest and faces many other socio-economic challenges like high levels of poverty, rapidly increasing population, severe resource constraints, and food security concerns. However, by devising various strategies, agriculture sector can be improved and to some extent, it is capable of addressing the challenges faced by the country.

- The discussion made in the chapter leads to establish that sustainable agricultural development can enhance domestic agricultural production, alleviate poverty in rural communities, preserve environment and conserve dwindling natural resources if supported and backed by the appropriate policies.
- In order to combat poverty, it is important to focus on developing the value chain for: (i) high-value commodities (e.g. coffee, honey, horticultural products) by involving the private sector; and (ii) fisheries. Both the sectors carry the significant potential to reduce poverty reduction and enhance economic growth. High-value crops could be suitable alternatives for the small farmers presently growing qat.
- In order to make agriculture sector sustainable, the crops should be grown based on the availability of resources and their suitability to their corresponding environment. Efficient and sustainable use of scarce natural resources like water and land can elevate and improve crop yields and make agriculture sector sustainable

and economically viable. Sustainable agriculture can be realized by using the agricultural inputs such as seeds, fertilizers, agricultural mechanization, irrigation systems judiciously and efficiently; making the modern scientific knowledge on agricultural techniques, management of crop and yield, and animal health improvement available to the farming communities through organized agricultural extension; creating marketing opportunities and appropriate environment that could provide means and tools for large and small farmers (NASS 2012).
- Livelihoods of rural communities can be elevated by providing them with improved inputs and developing their knowledge. It is argued that food requirements of increasing population cannot be fully met by developing agriculture sector, as arable land resources are limited and insufficient to expand agriculture base. Certain NGOs and individual experts are involved in extension activities like training of farmers, funding of small projects in livestock, agriculture terraces, pastures management, bee-keeping, etc. Their participation should be encouraged.
- The country is in dire need of multi-sectoral initiatives to address the issues like food security, climate change, and poverty reduction. However, the development of agriculture sector from technical, economic, and policy perspectives, can be easily integrated into multi-sectoral strategies.
- Implementation of the National Water Sector Strategy with true letter and spirit would boost agriculture due to continued accessibility of water.
- The development of coffee production and its marketing on scientific lines would enhance exports and generate revenue in the country that can boost the national economy to some extent. More efforts should be focused to bring behavioral change in the rural communities to grow food commodities instead of growing qat. It is recommended that qat production must be replaced with the coffee crop to earn foreign exchange.
- Agricultural Extension Service serves through the Ministry of Agriculture and is responsible to deliver information on new technologies and government programs to the farmers. However, its working needs to be improved by creating awareness to realize sustainable development. Extension workers are the key players in creating awareness, however, they are unable to do their jobs due to the lack of basic facilities like furniture, logistic support vehicles and operating funds. In addition, they are entrusted with non-extension and multiple tasks not falling under the mandate of the Extension service. Therefore, it is important to give them a clear line of action and their performance must be evaluated through a reliable monitoring and evaluation system. Extension staff deserves reasonable appreciation and incentives of promotion based on their performance. Participatory approaches in Yemen involving communities have proved the most effective way to ensure sustainability, improvements in production and food security. Therefore, it should remain the working extension strategy to realize sustainable agriculture development in the country.

Acknowledgements The authors are extremely grateful and express their gratitude to the Saudi Society of Agricultural Sciences for extending us all the possible help and sincere cooperation toward the completion of this piece of work and research.

References

ACSAD. (2002). Land degradation in republic of Yemen: Summary report of land degradation project in republic of Yemen. The Arab Center for the Studies of Arid Zones and Dry Lands (ACSAD). Ministry of Agriculture and Irrigation, Sanaa (Yemen). Ministry of Tourism and Environment, Sanaa (Yemen). Available at: http://www.acsad.org

Al-Kharasani, M. A. A. (2014). Climate change and agriculture in Yemen. *African-Asian Journal of Rural Development, 47*(2), 61–80.

Al-Monitor. (2014). *Yemen's coffee revival. Al-Monitor: The pulse of the Middle East.* Available at: http://www.al-monitor.com/pulse/culture/2014/02/yemen-revive-historic-coffee-trade.html. Assessed on 2 Apr 2015.

Al-Shehri, S. (2011). *Yemeni coffee crop: Historic fame, production areas and the famous types.* Available at: https://ar-ar.facebook.com/notes/feecoffee//297430906966087. Accessed on 25 Dec 2014.

Al-Zaidi, A. A., Baig, M. B., Shalaby, M. Y., & Hazber, A. W. (2016). Level of knowledge and its application by coffee farmers in the Udeen Area, Governorate of Ibb – Republic of Yemen. *The Journal of Animal & Plant Sciences, 26*(6), 1797–1804.

Almasa, A. A. M., & Scholzb, M. (2006). Agriculture and water resources crisis in Yemen: Need for sustainable agriculture. *Journal of Sustainable Agriculture, 28*(3), 55–75. https://doi.org/10.1300/J064v28n03_06.

CFSS. (2012). The state of food security and nutrition in Yemen. Comprehensive Food Security Survey (CFSS). World Food Program (WFP). Report Available at: https://www.humanitarianresponse.info/sites/www.humanitarianresponse.info/files/assessments/CFSS%2C%20YEMEN%2C%202012.pdf

FAO. (2009). Country second report on the state of world's plant genetic resources for Food and Agriculture in Yemen (1996–2006) (pp. 1–48).

FAOSTAT. (2015, 2016). Country profile–Yemen. Food and agriculture organization of the United Nations. Available at: http://www.fao.org/faostat/en/#country/249

FAO. (2015). Millions of Yemenis face food insecurity amidst escalating conflict. Food and Agriculture Organization of the United Nations. Available at: http://www.fao.org/news/story/en/item/283319/icode/

FAO. (2016). *FAO warns of rapidly deteriorating food security in Yemen.* Available at: http://www.fao.org/news/story/en/item/380653/icode/

GAFSP. (2013). Proposal for Yemen: Smallholder agricultural productivity enhancement program. Phase-1 Roll out of National Agriculture Sector Strategy (NASS) for Yemen's Efforts towards achieving food security under Global Agriculture and Food Security Program (GAFSP) Sana'a. June 4th, 2013.

GFSI. (2016). *Global food security index. An annual measure of the state of global food security* (p. 9). Washington, DC: The Economist Intelligence Unit Limited 2016.

GIEWS. (2015). *Country brief – Yemen.* Available at: http://www.fao.org/giews/countrybrief/index.jsp

Government of Yemen. (2013). Proposal for Yemen: Smallholder agricultural productivity enhancement program phase one roll out of national agriculture sector strategy (NASS) for Yemen's efforts towards achieving food security under global agriculture and food security program (GAFSP) Sana'a June 4th, 2013.

Hugh, T., Burke, J., & Faurès, J. M. (2011). *Climate change, water and food security.* Rome: Food and Agriculture Organization of the United Nations 2011, 174 pp., Softcover, isbn:978-92-5-106795-6.

ICARDA. (2012). *Combating land degradation in Yemen – A national report.* A review of available knowledge on land degradation in Yemen – combating desertification in Yemen – Oasis Country Report 4 (pp. 1–81). Available at: www.icarda.org

IFAD. (2007, December 11–13). *Country strategic opportunities paper (COSOP) 2008–2013.* Document EB 2007/92/R.18. Rome.

IFAD. (2009). Enabling poor rural people to overcome poverty in Yemen. International Fund for Agricultural Development, Rome. Available at: https://maintenance.ifad.org/documents/38714170/39150184/Enabling+poor+rural+to+overcome+poverty+in+Yemen.pdf/0b2360c9-68b6-4fb7-bf3b-8c702fd65569

IFAD. (2013). Rural growth programme (pp. 1–189). Detailed design report Main report and appendices. Document Date: 20 Aug 2013. Project No. 1672 Near East, North Africa and Europe Division Programme Management Department.

IFAD (International Fund for Agricultural Development). (2013). Harvesting water to increase productivity. Seeds of innovation: NENA and CEN Regions. Rome. www.ifad.org/operations/projects/regions/pn/infosheet/IS4_Sudan.pdf.

IFAD. (2015). A gender balanced model for community development model. Seeds of Innovation. Near East North Africa and Europe Office of IFAD (International Fund for Agricultural Development). Fact Sheet. Available at: https://www.ifad.org/documents/38714170/39150184/gender+balanced+eng.pdf/ec76284c-adf3-417b-8d6dc695d0b48316

IFPRI. (2011). Yemen national food security strategy. Overview and Action Plan. International Food Policy Research Institute, Washington, DC 20006–1002 USA. Ministry of Planning & International CooperationFood Security Strategy Program, Sana'a - Republic of Yemen. Available at: http://ebrary.ifpri.org/utils/getfile/collection/p15738coll2/id/124805/filename/124806.pdf

IRIN. (2007). *Yemen: Land degradation threatening farmers, says senior official*. News published on Oct 18, 2007 in IRIN International News. Available at: http://www.irinnews.org/report/74843/yemen-land-degradation-threatening-farmers-says-senior-official.

IRIN. (2012). *YEMEN: Time running out for solution to water crisis*. An article published on August 13, 2012. Available at: http://www.irinnews.org/report/96093/yemen-time-running-out-for-solution-to-water-crisis

Muharram, I. A., & Khalil, M. A. (2001). The situations of Agricultural Extension in Yemen: A quick Overview. The Yemeni Journal of Agricultural Research and Studies, Issue no. 5, Agricultural Research and Extension Authority (AREA), Dhamar -The Republic of Yemen.

NAP. (2000). National Action Plan. Published by the Government of Republic of Yemen, Sana. The Republic of Yemen.

NASS. (2012). A promising sector for diversified economy in Yemen: Strategy jointly prepared in cooperation with the United Nations Development Program, Economic Diversification Support Program (Agriculture Sector) and the Ministry of Agriculture and Irrigation. Ministry of Agriculture and Irrigation. National Agriculture Strategy – 2012–2016. The Republic of Yemen.

New Agriculturist. (2010). *Country profile – Yemen*. Available at: http://www.new-ag.info/en/country/profile.php?a=1371. Assessed on 5 Apr 2015.

Qamar, M. K. (2012). Extension in Yemen.Global Forum for Rural Advisory Services. Available at: http://www.g-fras.org/en/world-wide-extension-study/130-world-wide-extension-study/asia/western-asia/330-yemen.html#history.

Sallam, M., & Akram, B. (2005) *Agriculture extension situation in Dhamar province*. Dhamar Rural Development Project, Ministry of Agriculture and Irrigation, Yemen.

UNDP. (2015). *Yemen conflict paralyzes economic activity, puts women businesses at risk*. http://www.undp.org/content/undp/en/home/presscenter/pressreleases/2015/11/16/yemen-conflict-paralyzes-economic-activity-puts-women-businesses-at-risk.html

USAID. (2005). *Moving Yemen coffee forward assessment of the coffee industry in Yemen to sustainably improve incomes and expand trade*. Publication produced for review by the United States Agency for International Development. Prepared by ARD, Inc.

USAID. (2016). *YEMEN country development cooperation strategy 2014 – 2016*. Available at: https://www.usaid.gov/results-and-data/planning/country-strategies-cdcs

World Bank. (2002). *The Republic of Yemen – economic growth: sources, constraints and potentials*. Washington, DC: The World Bank.

World Bank. (2006). *Country assistance evaluation. independent evaluation group*. Washington, DC: World Bank.
World Bank. (2009). Yemen: reducing serious water stress. Water in Middle East and North Africa. http://go.worldbank.org/SOS3F9MVY0
World Bank, FAO (Food and Agriculture Organization of the United Nations), and IFAD (International Fund for Agricultural Development). (2009). *Improving food security in Arab Countries*. Washington, DC: World Bank.
World Bank. (2011a). World Development Report 2011: Conflict, Security, and Development. Washington, DC.
World Bank. (2011b). World development indicators database. Accessed 10 Sept 2014. http://databank.worldbank.org/.
World Bank; United Nations; European Union; Islamic Development Bank; and Yemen, Ministry of Planning and International Cooperation. (2012). *Joint social and economic assessment for the republic of Yemen*. Washington, DC: World Bank.

Dr. Mirza Barjees Baig is working as a Professor of Agricultural Extension and Rural Development at the King Saud University, Riyadh, Saudi Arabia. He has received his education in both social and natural sciences from USA. He completed his Ph.D. in Extension Education for Natural Resource Management from the University of Idaho, Moscow, Idaho, USA. He earned his MS degree in International Agricultural Extension in 1992 from the Utah State University, Logan, Utah, USA and was placed on the "Roll of Honour". During his doctoral program, he was honored with "1995 Outstanding Graduate Student Award". Dr. Baig has published extensively in the national and international journals. He has also presented extension education and natural resource management extensively at various international conferences. Particularly issues like degradation of natural resources, deteriorating environment and their relationship with society/community are his areas of interest. He has attempted to develop strategies for conserving natural resources, promoting environment and developing sustainable communities through rural development programs. Dr. Baig started his scientific career in 1983 as a researcher at the Pakistan Agricultural Research Council, Islamabad, Pakistan. He has been associated with the University of Guelph, Ontario, Canada as the Special Graduate Faculty in the School of Environmental Design and Rural Planning from 2000–2005. He served as a Foreign Professor at the Allama Iqbal Open University (AIOU) through Higher Education Commission of Pakistan, from 2005–2009. Dr. Baig the member of IUCN – Commission on Environmental, Economic and Social Policy (CEESP). He is also the member of the Assessment Committee of the Intergovernmental Education Organization, United Nations, EDU Administrative Office, Brussels, Belgium. He serves on the editorial Boards of many International Journals Dr. Baig is the member of many national and international professional organizations.

Prof. Ajmal Mahmood Qureshi serves as the Senior Associate, at the Asia Center, Faculty of Arts and Sciences, Harvard University, Cambridge, MA, USA. Previously, he has served at the United Nations as the FAO Resident Representative (Ambassador) for over seven years in China, multiple accreditations to North Korea and Mongolia, regarded as one of the most challenging assignments in the UN System. He has represented the Organization to the host government, diplomatic missions, bilateral and multilateral donors, international organizations, NGOs and other stakeholders. His services have been recognized at the highest administrative and political levels. He was awarded the title of "Senior Advisor and Professor at Chinese Academy of Agricultural Sciences". He successfully implemented a large array of technical assistance projects throughout the country including Tibet. In addition, through a large number of regional projects, Chinese expertise was used in training and capacity building of countries in the Asia Pacific Region. He was the pioneer to launch of Food Security Project in Sichuan Province that was named as an ideal model and was being replicated in over hundred countries. He also assisted in the preparation of China's Agenda 21, holding of the Roundtable on Agenda 21 and was the keynote speaker on the "Environmental

Issues". In the restructuring of the FAO, Prof. Qureshi contributed to the preparation of the Organization's strategic plan in the context of China's priorities.

As Chairman of the Donors Group in China, Prof. Qureshi helped mobilize millions of dollars for the agriculture, forestry and fisheries sectors of China. For over eight years, at the Ash Centre of Democratic Governance and Innovation at Harvard Kennedy School, he has worked on the issues of governance and democracy. Prof. Qureshi has served in North Korea, Mongolia, Uganda and Represented the United Nations to the host governments, diplomatic missions, bilateral and international organizations and all other stakeholders. He was the Member of the High Level Steering Committee advising the President of Uganda on the implementation of the Plan for Modernization of Agriculture. Through his offices, he had close interactions with the multilateral donors; inter alia, USAID, World Bank, UN System, Asian Development Bank, SIDA, AUSAID and NORAD. As Head of International Cooperation, he led the country's delegation to the Governing Boards of Rome based UN Organizations: FAO, WFP and IFAD. Prof. Qureshi also organized and assisted in holding numerous international events like: FAO Ministerial level Regional Conference for Asia and Pacific in Islamabad, attended by 28 Ministers. He has the honor to serve as the Consul General (Head of Mission) of Pakistan in Istanbul, Turkey.

He received his education in Pakistan, France and the USA with major focus on political science, international relations and development economics. At the Harvard Kennedy School, he dealt with issues of governance, democracy and innovation; and under the leadership training programs, inter-acted and collaborated with senior officials from China, Viet Nam and Indonesia. He has published extensively in Harvard journals, bearing on China and DPRK's Food Security. He has also been named as the "Most Distinguished Alumni" at the Boston University. He is also the Member of the Board of Directors of the United Nations Association of Greater Boston (UNAGB).

Prof. Dr. Gary S. Straquadine serves as the Interim Chancellor, Vice Chancellor (Academic Programs) and Vice Provost at the Utah State University – Eastern, Price – UT USA. He did Ph.D. from the Ohio State University, USA. Presently he also leads the applied sciences division of the USU-Eastern campus. He is responsible for faculty development and evaluation, program enhancement, and accreditation. In addition to his heavy administrative assignments, he manages to find time to teach some undergraduate and graduate courses and supervise graduate student research. Being an extension educator, he has passion for the economic development of the community through education and has also successfully developed significant relations with agricultural leadership in the private and public sector. He has also served as the Chair, Agricultural Comm, Educ, and Leadership, at The Ohio State University, USA. Before accepting the present position as the Vice Provost, he served on many positions as the Department Head; Associate Dean; Dean and Executive Director, USU-Tooele Regional Camp and the Vice Provost (Academic). His professional interests include extension education, sustainable agriculture, food security, statistics in education, community development, motivation of youth and outreach educational programs. He has also helped several under developing countries improving their agriculture and educational programs. He has a very strong passion for the healthy ecosystems for the heathy communities; higher education and International Development Programs.He has beeb honored with numerous awards and honour for making significant contributions to the society and science. His research has been published as the book chapters in the prestigious books and scientific articles in the high impact journals.

Dr. Asaf Hajiyev is a Professor at the Baku State University and the Executive Member (Academician) of the Azerbaijan National Academy of Sciences. He holds the Dr. Sci. degree from Bauman Moscow State Technical University, and has done Ph.D. and post-doctoral research at Lomonosov Moscow State University. He has the vast research and teaching experience. He serves as the Chair of the Department of Probability and Statistics, Chair, of the Department of Controlled Queues, Institute of Control Systems at the Azerbaijan National Academy of Sciences; Chair of the Department of Theory of Probability and Mathematical Statistics, Baku State University; and the Senior Scientific Researcher, Department of Probability Statistics, Royal Institute Technology in Stockholm. Being a renowned researcher, he has served at the several universities around the world including China, Germany, Italy, Portugal, Sweden, Turkey, and USA. He serves on the editorial boards of many prestigious national and international academic journals. He has been an organizing member and the Keynote speaker at the numerous international conferences. He has been honored with many prestigious awards like: Azerbaijan Lenin Komsomol Prize Winner on Science and Engineering; Grand Prize at the International Conference "Management Science and Engineering Management" Macao; and Grand Prize at the İnternational Conference "Management Science and Engineering Management" at Islamabad, Pakistan.

He is the Honorary Academician of Academy of Sciences of Moldova, Foreign Member of the Mongolian National Academy of Sciences, Member of TWAS (The World Academy of Sciences), Honorary Professor of Chengdu University (China) and the elected member of the International Statistical Institute.

He has also the honor of holding the Office of the Vice-President of the Parliamentary Assembly of the Black Sea Economic Cooperation Organization and since 2015; he serves as the Secretary General of the same organization (PABSEC). He shares his talents and expertise on the boards of many international academic organizations and institutions. He has more than 135 peer reviewed scientific publications to his credit, published in the highly reputed journals.

Chapter 3
Ensuring Food Security Through Increasing Water Productivity and Cereal Yields Forecasting – A Case Study of Ouled Saleh Commune, Region Casablanca-Settat, Morocco

Abdelhadi Mouchrif, Fouad Amraoui, and Abdalah Mokssit

3.1 Introduction

Agriculture is the mainstay of the Moroccan economy. The sector accounts for around 14% of Gross Domestic Product (GDP). Cereals (soft wheat, durum wheat and barley) are produced all over the country, occupying nearly 52% of useful agricultural lands (MAPM 2014).

Due to the growing population and demand of food, and considering the persistently growing pressure on finite freshwater and soil resources, it becomes increasingly clear that increasing water productivity holds the key to future water scarcity and food security challenges, a situation that is expected to be exacerbated by a changing climate. At the 6th World Water Forum held in Marseille in 2012, improving water productivity of rainfed agriculture was one of the major targets of the thematic priority '*Contribute to Food Security by Optimal Use of Water*' (WWF 2012).

Production and crop yields estimates are of great interest to the agricultural sector. In Morocco, cereals represent a strategic crop commodity in terms of food security, and therefore early forecasting yields before the harvest is crucial. This will enable an adequate preparation to manage the consequences of any potential shortage in production through various actions aiming to reduce the vulnerability to climatic risks. It also allows planning actions like aids to farmers or cereal imports (Balaghi et al. 2013).

The objective of this research was to calculate the soft wheat water productivity (WP) for the period 2002–2012, and to test the ability of the AquaCrop model to

A. Mouchrif (✉) · F. Amraoui
Laboratory of Geosciences Applied to the Planning Engineering (GAIA), Faculty of Sciences, Hassan II University of Casablanca, Casablanca, Morocco

A. Mokssit
National Meteorological Office, Casablanca, Morocco

M. Behnassi et al. (eds.), *Climate Change, Food Security and Natural Resource Management*, https://doi.org/10.1007/978-3-319-97091-2_3

simulate the yields for cropping seasons from 2008–2009 to 2011–2012 under rainfed and semi-arid conditions at Ouled Saleh commune.

3.2 Study Area

3.2.1 Geographic Location

The region of Grand Casablanca is located on the Atlantic coast in the northwestern of Morocco (Fig. 3.1). It borders the Atlantic Ocean to the west, the region of Chaouia-Ouardigha to the east, and the south and province of Benslimane to the north. It is divided into two prefectures (Casablanca and Mohammedia) and two provinces (Mediouna and Nouaceur) (HCP 2010).

The province of Nouaceur is spread over an area of 44.583 hectares. It is subdivided into three municipalities (Bouskoura, Nouaceur, and Dar Bouazza) and into two rural communes (Ouled Saleh and Ouled Azzouz). Ouled Saleh is an important agricultural area of the province of Nouaceur having great potential mainly due to its soil fertility (about 13.499 ha hence representing 30% of total area). Useful Agricultural Surface occupies 12.260 ha and agricultural exploitations number 1307. Cereal production is the most significant agricultural resource and winter soft wheat is the major crop cultivated in the province (DRA 2010).

3.2.2 Soil Types

Major soil types encountered at the rural commune of Ouled Saleh are Tirs, Hamri and Hrach (DRA 2010). Tirs is a black to dark brown soil with high fertility, good agricultural potential and good permeability. It contains high percentage of expanding clay and low stones and is resistant to erosion. Hamri is a red soil that has particles of loam and clay diameter in mixture with stones. It has a very good agricultural potential. Hrach is a grayish soil with medium agricultural potential. It is slightly permeable, has large particles and stones (Encyclopedia of Soil Science 2006). Figure 3.2 highlights the soil fertility in Ouled Saleh since it consists of 60% of Tirs and 30% of Hamri.

3.2.3 Soft Wheat

Soft wheat (*Triticum aestivum*) is the number-one cereal crop with respect to area and production in Morocco. It is of particular economic importance. It represents 54% of the national cereal production at the end of the growing season 2011–2012 against 22% for durum wheat and 24% for barley (ONICL 2012). Soft wheat, which can be cultivated in all agricultural regions, requires healthy and well-drained soil

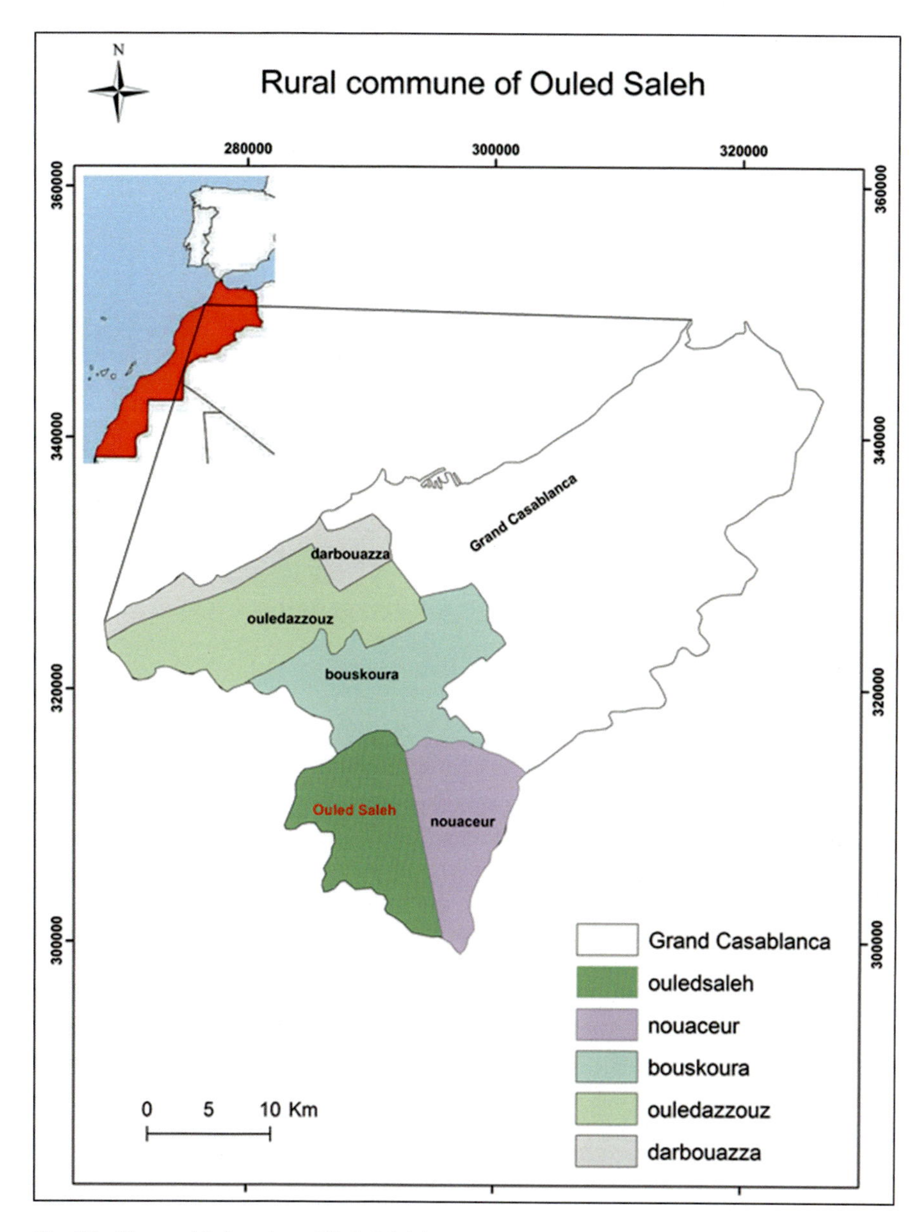

Fig. 3.1 Geographic location of Ouled Saleh

and is sensitive to water stress, especially during the period of reserve accumulation in grains (Alaoui n.d.). According to local agricultural practices, the majority of farmers sow wheat in mid-November and harvest in May. Achtar is the most important soft wheat cultivar in use in the rural commune of Ouled Saleh.

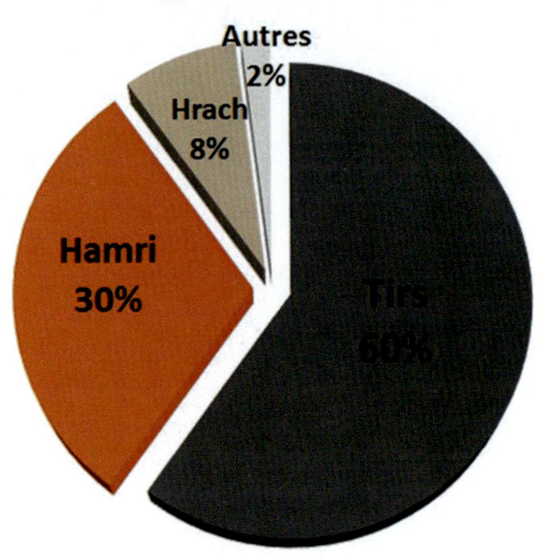

Fig. 3.2 Soil types encountered at Ouled Saleh with predominance of Tirs. (Source: DRA 2010)

3.3 Agro-meteorological Modeling

3.3.1 AquaCrop Model

AquaCrop is a crop water productivity model developed by the Food and Agricultural Organization (FAO) of the United Nations. It simulates soil water balance, water productivity, yield response to water of herbaceous crops, and is particularly suited to address conditions where water is a key limiting factor in crop production (Steduto et al. 2009). The particular feature that distinguishes AquaCrop from other crop models is its focus on water, the use of ground canopy cover instead of leaf area index, and the use of water productivity values normalized for atmospheric evaporative demand and of carbon dioxide concentration. This confers the model an extended extrapolation capacity to diverse locations and seasons, including future climate scenarios (Raes et al. 2009).

3.3.2 Input Data

The AquaCrop model uses input variables that require simple methods for their determination (FAO 2009). It structures the soil-plant-atmosphere system by including:

- *Atmosphere*: thermal regime, rainfall, reference evapotranspiration ETo and carbon dioxide concentration.

- *Plant*: processes of growth, development, and yield.
- *Soil*: water and nutrients budgets.
- Management practices: major agronomic practices such as planting dates, fertilizer application and irrigation if any.

Input climatic data for AquaCrop were obtained from the Nouaceur meteorological station and reference evapotranspiration values were calculated by the Penman-Monteith method using ETo calculator (Raes 2009). However, other data (crop, soil, and management) were obtained from Bouskoura Technician Works Center (CT).

3.4 Results and Discussion

3.4.1 Relationship Between Rainfall and Soft Wheat Yields

In Morocco, it has been demonstrated from long time that a close relationship exists between cereal yields and rainfall. Figure 3.3 clearly establishes the influence of inter-annual rainfall distribution on soft wheat yields at Ouled Saleh from 2002 to 2012. In general, yields evolve in the same direction as the rainfall. However, intra-annual rainfall distribution has been analyzed (Mouchrif et al. 2015) in order to explain that an inadequate distribution may decrease the yield in spite of a high rainfall. Therefore, rainfall, in terms of total amount and distribution during the cropping season, strongly influences soft wheat yields at Ouled Saleh commune level.

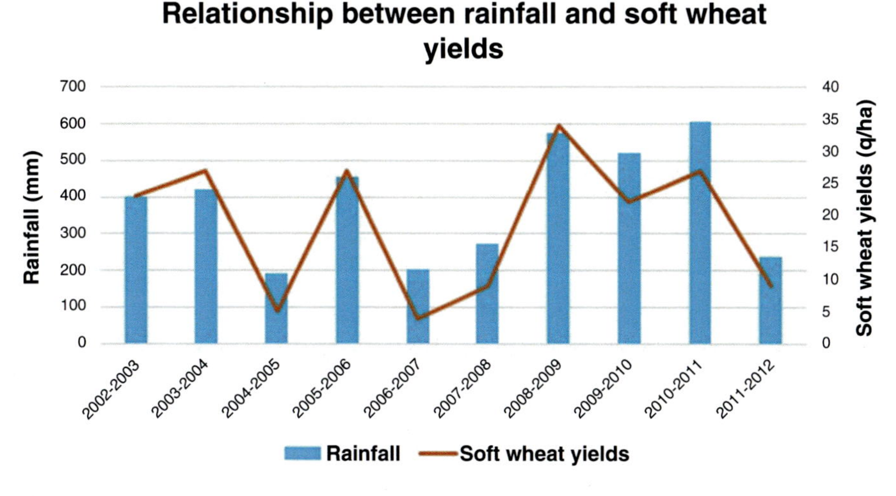

Fig. 3.3 Relationship between rainfall distribution (mm) and soft wheat yields (q/ha) for the period 2002–2012 in Ouled Saleh

3.4.2 Relationship Between Rainfall and Rainwater Productivity

Rainwater productivity (RWP) expressed by crops is defined as the quantity of grain produced per unit of rainwater of a cultivated land area. It is equal to grain yield weight per hectare or meter squared (grams/m^2) divided by the quantity of rainfall (in liters or millimeters) received during crop cycle (Balaghi et al. 2013). Figure 3.4 indicates soft wheat yield regression against cumulated rainfall during the cropping season, the slope of the regression line represents RWP, which is in average 0.65 g/l for the period 2002–2012 at Ouled Saleh commune level.

The better use of rainwater by soft wheat is related probably to the fact that Achtar, a cultivar of soft wheat developed by the National Institute for Agronomic Research (INRA), is largely adopted by farmers at Ouled Saleh commune for its high productivity and resistance to drought and diseases (Jlibene 2009). High RWP is more beneficial to the rural commune where the climate is semi-arid and the water resources are limited since it can combat drought and contribute to food security in an era of climate change.

It should be noted that it is possible to produce more with less water and take advantage of each droplet of rain to increase crop yields by using new cultivars that have been genetically improved by INRA and which can resist to drought and use the available water resources with efficiency. Their RWP has been estimated at 2.2 g/l in agro-ecological zone '*Bour Défavorable*' (unfavorable rainfed lands), which represents a saving of water equivalent to 0.77 mm per year and per hectare in comparison with previous cultivar Nasma (Jlibene 2009). In addition, increasing RWP requires more farmers' input investments, mainly concerning fertilization management and plant protection (Balaghi et al. 2013).

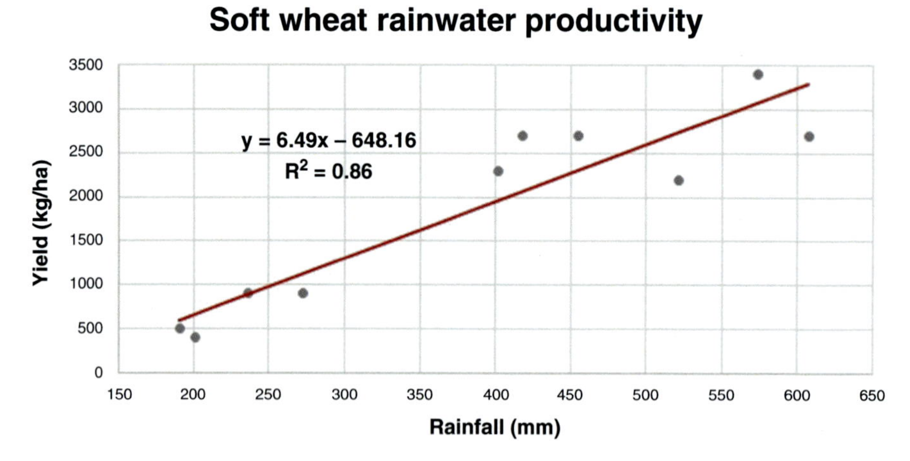

Fig. 3.4 Relationship between soft wheat yield (kg/ha) and cumulated rainfall during the cropping season (mm) for the period 2002–2012 at Ouled Saleh

Table 3.1 Comparison between observed and modelled yields simulated by AquaCrop

Cropping season	Observed yield (q/ha)	Modelled yield (q/ha)
2007–2008	09	09
2008–2009	36	42
2009–2010	24	47
2010–2011	27	21
2011–2012	09	07

3.4.3 Soft Wheat Yields Forecasting

3.4.3.1 Model Calibration and Validation

AquaCrop model version 4.0 (2012) was applied with daily time steps to estimate soft wheat grain yields under rainfed conditions between 2007 and 2012. To calibrate and validate the model, we tried to match the model yields with the observed yields by adjusting parameters that are dependent on location, crop cultivar, management practices, and keeping constant conservative parameters. Cropping season of 2007–2008 was used for calibration, while observed yields of 2008–2009 to 2011–2012 were used for validation. The results of the model simulations are shown in Table 3.1.

The forecasts were close to official yields (differences of 2 to 6 q/ha). However, it should be noted that pests, diseases and weeds are not yet integrated in AquaCrop (FAO 2009). Thus, the model tends to overestimate grain yield for cropping season of 2009–2010, which was particular due to the existence of massive weed infestations observed in most cereal fields and diseases that have affected the crop production (El Aydam et al. 2010).

3.4.3.2 Model Performance

To assess the 'goodness-of-fit' of the model, Coefficient of determination (r^2), Root Mean Square Error (RMSE), and Nash-Sutcliffe efficiency (EF) were used as Criterions. These statistical indicators are computed from the pairs of observed and predicted values Oi and Pi (i = 1, 2,... n) whose means are respectively $\bar{O}$ and $\bar{P}$. n is the number of observations.

Coefficient of Determination (r^2)

The coefficient of determination r^2 is defined as the squared value of the coefficient of correlation (r) according to Bravais-Pearson. It describes the proportion of the total variance in the observed data that can be explained by the model. It ranges from 0 to 1, where the higher values indicate better agreement between the observed and predicted data. r^2 is calculated as follows:

$$r^2 = \left(\frac{\sum_{n}^{i=1} \left(O_i - \bar{O}\right)\left(P_i - \bar{P}\right)}{\sqrt{\sum_{n}^{i=1} \left(O_i - \bar{O}\right)^2} \sqrt{\sum_{n}^{i=1} \left(P_i - \bar{P}\right)^2}} \right)$$

The fact that only the dispersion is quantified is one of the major drawbacks of r^2 if it is considered alone. A model which systematically over- or underpredicts all the time will still result in good r^2 values close to 1 even if all predictions were wrong (Krause et al. 2005). Willmott (1984) and Legates and McCabe (1999) indicate that r and r^2 values are inappropriate statistics for quantitative comparisons of model performance because they are highly sensitive to outliers. However, because the coefficient of determination is commonly used, we have chosen to include it along with two measures of RMSE and EF to evaluate model performance.

Root Mean Square Error (RMSE)

RMSE is one of the most widely used statistical indicators in environmental estimation models (Jacovides and Kontoyiannis 1995). It represents a measure of the overall, or mean, deviation between observed and simulated values, that is, a synthetic indicator of the absolute model uncertainty. Small RMSE values indicate good performance. It is calculated as follows (Loague and Green 1991):

$$RMSE = \left[\frac{\sum_{n}^{i=1} \left(P_i - O_i\right)^2}{n} \right]^{0.5}$$

RMSE does not provide indication about over and under estimation of a model (Jacovides and Kontoyiannis 1995).

Nash-Sutcliffe Efficiency (EF)

The efficiency EF proposed by Nash and Sutcliffe (1970) is a normalized statistic that determines the relative magnitude of the residual variance compared to the measured data variance (Moriasi et al. 2007) and is defined as:

$$EF = 1.0 - \frac{\sum_{n}^{i=1} \left(O_i - P_i\right)^2}{\sum_{n}^{i=1} \left(O_i - \bar{O}\right)^2}$$

The range of EF (non-dimensional) lies between 1 (perfect fit) and $-\infty$. It approaches 1 when the residual variance is much smaller than the measured data variance, while

Table 3.2 Statistical indices derived for evaluating the performance of AquaCrop model in predicting grain yield

Statistical index	Calculated value
r	0.95
r^2	0.91
RMSE (q/ha)	5.03
EF	0.8

negative EF values indicate that the mean is a better estimator than the model (Moriasi et al. 2007).

Results

The values of r^2, RMSE and EF for soft wheat yield simulation are presented in Table 3.2.

Overall, the agreement between simulated and observed wheat grain yield was satisfactory with $r = 0.95$, $r^2=0.91$, RMSE=5.03 (q/ha) and EF=0.8. Therefore, the AquaCrop model has a good simulation for soft wheat grain yield.

3.5 Conclusion and Recommendations

In this study, soft wheat RWP was computed for the period 2002–2012 and AquaCrop model was used to simulate yields from 2008 to 2012 under rainfed conditions and semi-arid climate of Ouled Saleh rural commune, Region Casablanca-Settat.

The results showed that RWP is not higher and still needs improvement in order to produce more 'crop per drop' of rain, and that AquaCrop is capable of simulating winter soft wheat under rainfed conditions at Ouled Saleh commune level with a reasonable accuracy. Therefore, we concluded that it is a useful tool for relevant planning and policy making processes. Finally, we present some suggestions for further efforts to increase the RWP and improve the crop simulation, which include:

- Taking advantage of technological progress concerning varieties with increased productivity, resistance to pests, and drought;
- Encouraging farmers to convert to new practices like agro-ecology which is capable of delivering productivity goals without depleting the environment by using ecological concepts and principles for the design and management of sustainable agricultural systems;
- More observed climate and crop data covering a longer period and other data used in modeling, such as soil information and nitrogen applications in order to enhance AquaCrop predictive capacity under local conditions; and
- Spatializing Aquacrop model to cover larger areas with very heterogeneous climate, soil, and management practices.

References

Alaoui S.B. (n.d.). *Référentiel pour la conduite technique de la culture du blé tendre (Triticum aestivum)*, 14 p. available at: www.fellah-trade.com/ressources/pdf/ble_tendre.pdf (visited January 23, 2017).

Balaghi R., Jlibene M., Tychon B. & Eerens H. (2013). *Agrometeorological cereal yield forecasting in Morocco*. INRA, Rabat, 157 p.

Direction Régionale de l'Agriculture de Casablanca (DRA). (Décembre 2010). *Monographie agricole de la province de Nouaceur*.

El Aydam M., Baruth B. & Balaghi R. (2010). MARS Agrometeorological Vol. 18 No. 7 – Crop Monitoring in Morocco.

Encyclopedia of Soil Science. (2006). Vol. 1. R. Lal. CRC Press, 1923 pages. Pages: 121–126.

Food and Agriculture Organization of the United Nations (FAO). (2009). Software: AquaCrop. (http://www.fao.org/nr/water/index.html)

Haut Commissariat au Plan (HCP), Direction régionale du grand Casablanca. (Juillet 2010). Monographie de la région du Grand Casablanca.

Jacovides, C. P., & Kontoyiannis, H. (1995). Statistical procedures for the evaluation of evapotranspiration computing models. *Agricultural Water Management, 27*, 365–371.

Jlibene M. (2009). *Amélioration génétique du blé tendre au Maroc à l'aube du 21ème siècle*. 80 pages, Editions INRA.

Krause, P., Boyle, D.P. & Bäse, F. (2005). Comparison of different efficiency criteria for hydrological model assessment. *Advances in Geosciences*. pp. 89–97.

Legates, D. R., & McCabe, G. J. (1999). Evaluating the use of 'goodness-of-fit' measures in hydrologic and hydroclimatic model evaluation. *Water Resources Research, 35*, 233–241.

Loague, K., & Green, R. E. (1991). Statistical and graphical methods for evaluating solute transport models: overview and application. *Journal of Contaminant Hydrology, 7*, 51–73.

Ministère de l'Agriculture et de la Pêche Maritime (MAPM). (2014). Agriculture marocaine en chiffres. Rabat, Maroc.

Moriasi, D. N., Arnold, J. G., Van Liew, M. W., Bingner, R. L., Harmel, R. D., & Veith, T. L. (2007). Model evaluation guidelines for systematic quantification of accuracy in watershed simulations. *Transactions of ASABE, 50*(3), 885–900.

Mouchrif A., Amraoui F. et Mokssit A. (2015). Caractérisation agro-climatique et liens avec les rendements agricoles : cas du blé tendre dans la région du Grand Casablanca. *Afrique Science*, Vol.11, n°4. http://www.afriquescience.info/document.php?id=5017

Nash, J. E., & Sutcliffe, J. V. (1970). River flow forecasting through conceptual models, Part I -A discussion of principles. *Journal of Hydrology, 10*, 282–290.

Office National Interprofessionnel des Céréales et des Légumineuses (ONICL) (2012). Production des céréales, campagne agricole 2011–2012.

Raes D. (2009), "ETo Calculator v3.1," Land and Water Digital Media Series No. 36, Food and Agriculture Organization of United Nations, Rome.

Raes, D., Steduto, P., Hsiao, T. C., & Fereres, E. (2009). AquaCrop – The FAO crop model to simulate yield response to water: II. Main algorithms and soft ware description. *Agronomy Journal, 101*, 488–498.

Steduto, P., Hsiao, T. C., Raes, D., & Fereres, E. (2009). AquaCrop – The FAO crop model to simulate yield response to water: I. Concepts and underlying Principles. *Agronomy Journal, 101*, 438–447.

Willmott, C. J. (1984). On the evaluation of model performance in physical geography. In G. L. Gaile & C. J. Willmottt (Eds.), *Spatial statistics and models* (pp. 443–460). Boston: D. Reidel.

World Water Forum (WWF). (2012). *Contribute to food security by optimal use of water*. http://www.worldwaterforum6.org/commissions/thematic/priorities-for-action-and-conditions-for-success/priority-for-action-22/

Dr. Abdelhadi Mouchrif holds a PhD in Geomatics and a master's degree in Geographic Information Systems (GIS) and Land Management from Hassan II University of Casablanca, Morocco.

Dr. Fouad Amraoui is specialist in Hydrogeology and Consultant on water sciences. He is also Professor at the Faculty of Sciences, Hassan II University of Casablanca, Morocco.

Mr. Abdalah Mokssit is a Meteorologist and an International expert on climate change.

Chapter 4
Food Waste in the Kingdom of Saudi Arabia: Implications for Extension Education

Khodran H. Al-Zahrani, Mirza Barjees Baig, and Gary S. Straquadine

4.1 Introduction

Saudi Arabia, with its scarce water resources, harsh climatic features like high temperatures and low rainfall, is one of the driest countries of the Arabian Peninsula, which suffers from many restrictions to their agricultural growth. Before 1970, agriculture was practiced on a small scale. In 1980, the Kingdom began placing greater emphasis on technically sound agriculture to realize self-sufficiency. Farmer-friendly policies were framed. Among the best programs were soft and interested free loans, free lands for potential growers, and modern irrigation systems were installed. By employing scientific and advanced technologies in agriculture was practiced the Kingdom was able to achieve the goal of food security. The Kingdom also did realize its self-sufficiency in many agricultural commodities and became the sixth largest exporter of wheat. The farming systems caused a serious drain on the Kingdom's water resources, drawing mainly from non-renewable aquifers (Al-Shayaa et al. 2012; Baig and Straquadine 2014). However, many scientific studies indicated that the agricultural sector had damaged the limited water resources, and were unable to sustain the present levels of agricultural production at the expense of luxurious supplies of fossil water (Al-Shayaa et al. 2012; Baig and Straquadine 2014).

In the year 2008, Saudi Arabia adopted an annual reduction of wheat production (12.5% annually) since 2009, by off-setting such reductions with reasonable imports. The government has firmly followed the complete phase-out wheat

K. H. Al-Zahrani · M. B. Baig (✉)
Department of Agricultural Extension and Rural Society, College of Food and Agriculture Sciences, King Saud University, Riyadh, Saudi Arabia
e-mail: khodran@ksu.edu.sa; mbbaig@ksu.edu.sa

G. S. Straquadine
Utah State University – Eastern, Price, UT, USA
e-mail: gary.straquadine@usu.edu

M. Behnassi et al. (eds.), *Climate Change, Food Security and Natural Resource Management*, https://doi.org/10.1007/978-3-319-97091-2_4

production program by 2016 and has placed a ban on planting crops with high water requirements. An implementation of the phase-out program (as depicted in Fig. 4.2) has also reduced the country's food self-sufficiency, making it essential to explore other sustainable development avenues that could ensure the continued supplies of food commodities in domestic markets. Obviously such practice will lead to rely upon continued and increasing imports in the future.

In the situation, the suitable strategy adopted to meet the food requirements of its citizens was to import higher volumes food and agricultural commodities from food surplus countries. Though prices for the food commodities are on the rise in the international markets yet the Kingdom provides food items to the consumers at the highly subsidized rates. Since the prices for the food items are not very high, therefore, Saudi citizens take food items as granted. Due to low prices of the food commodities, the Kingdom is known as the prime food waster in the Gulf Cooperation Council (GCC) countries. Generally Saudi citizens have welcoming attitudes towards guests and are hospitable and food lovers. Also, as the culture in the KSA is based on festivals and celebrations, huge quantities of food are served in routines. Substantial portion of food commodities that are wasted and end up in the landfills, threatening the Kingdom's national food security goal; damaging natural resources; and creating environmental issues and economic losses.

At the moment, food waste is the major contributor to the landfills, establishing the need to reduce food waste in the KSA by adopting strategies and processes that are economically viable, socially acceptable and environmentally friendly. Although complete prevention seems difficult yet food waste can be reduced significantly by educating the public, creating awareness, and changing their attitudes towards natural resources including food and water in the Kingdom. The purpose of this chapter is to examine the reasons causing food waste in the Kingdom and establishing the need to reduce and stop food waste to the possible level. The impacts of food waste in realtion to achieving food security has also been discussed. Based on the discussions and arguments made in the chapter, it is anticipated that it would not be possible to achieve the goal of food security without reducing food wastes in the Kingdom. In the light of these facts and the present scenario on food waste, a national, comprehensive campaign and sound extension education programs would be required. In this chapter, an effort has been made to highlight the problem of food waste and outline the suitable strategies to minimize waste.

4.2 Why the Kingdom Cannot Afford to Waste Food: Possible Reasons

4.2.1 Availability of Low Arable Lands

The diminishing water resources, low arable land, and harsh climatic conditions prevent the development of large-scale agriculture Agricultural lands cover only 1,733,550 km^2, and similarly arable lands cover as low as 1.44–1.5% of the land

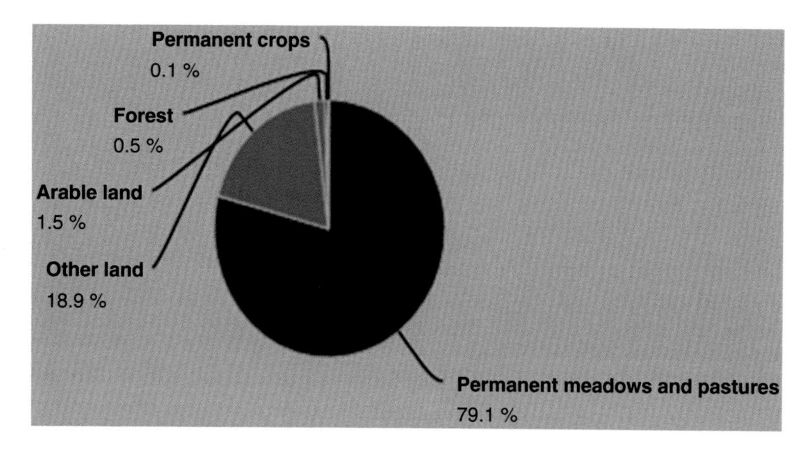

Fig. 4.1 Land use Area under crops, forests and arable lands in 2011. (Source: http://faostat3.fao.org/browse/area/194/E)

area (World Bank 2011a, b, c; FAO 2015a, b) as depicted in Fig. 4.1. With the lowest per capita arable land i.e. 0.11 (see the Table 4.1), agriculture is now practiced on the limited and small scale in the Kingdom.

4.2.2 Location and Limited and Diminishing Water Resources

The kingdom falls under arid, semi-arid and desert climate with limited arable lands and dwindling water resources. Many experts reported that low rainfalls and less precipitations were unable to recharge deep surface wells and aquifers causing substantial variability and in cases serious declines in crops yields (Darfaoui and Al-Assiri 2010). The scarce fossil water resources remained of paramount concern for the last many decades and were not in a position to produce agricultural crops at the present levels. The most recent National Development Plans place greater emphasis on growing crops with less water requirements. Therefore, Saudi Arabia made an important strategy shift, aiming to favor sustainable agriculture while abandoning conventional production systems and banning the fodder crops with high water-requirements.

4.2.3 Complete Phase Out of Wheat by 2016

Today the agricultural sector in the Kingdom of Saudi Arabia is facing two crucial challenges: the scarcity of water resources and ensuring food security. The Arabian Peninsula is one of the world's driest regions, with rainfall averaging less than

Table 4.1 Key indicators for social and National Development of the Kingdom of Saudi Arabia

Indicator	Year	Unit	Value
Total population	2014	People	30,770,375[a]
Population growth rate	2014	Percent	2.55
Urban population growth (annual %)	2014	Percent	2.5
Saudi population growth rate	2014		2.1
Population density (person/sq. km)	2014	(Person/sq. km)	15.3
Rural population	2014	Percent	17
Rural population (% of total population)	2012	Percent	17.50
Agricultural population (% of total population)[a]	2011	Percent	1.82
Total economically active population	2012		10,382,733
Total economically active population in agriculture[a]	2010		515,000
Total economically active population in agriculture (in % of total economically active population)	2011	Percent	4.96
Female economically active population in agriculture (% of total economically active population in agriculture)[a]	2010	Percent	0.27
Agricultural land (sq. km)	2011		1,733,550
Agricultural land (% of land area)	2011	Percent	80.64
Arable land (hectares)	2011	Hectares	3,110,000
Arable land (% of land area)	2011	Percent	1.44
Arable land (hectares per person)	2011	Hectares	0.11
Saudi population	2014	(People)	20,702,536[a]
Saudi population growth rate	2014		2.1
GDP growth at constant prices	2015	Percent	3.35
Per capita GDP at current prices in	2015	(SAR)	77,711
Economic diversification indicators			
Private sector's contribution to GDP at constant prices for 2015	2015	Percent	39.46
Proportion of private sector growth at constant prices for 2015	2015	Percent	3.74
Proportion of non-oil exports to imports	2015	Percent	28.35
Growth of exports of non-oil goods	2015	Percent	−17.99
Growth of imports of goods	2015	Percent	−3.65
Exports contribution to GDP for 2015 at current prices	2015	Percent	33.33
Exports of goods and services (% of GDP)[b]	2014	Percent	47.5
Imports of goods and services (% of GDP)[b]	2014	Percent	34.2
Agriculture, Value added (% of GDP)[b]	2014	percent	1.9
The cost-of-living index	2014		130.1
Change in the cost-of-living index (inflation) for 2014		Percent	2.7
Unemployment rate (>15 years)	2014	Percent	5.7
Saudi's unemployment rate (>15 years)	2014	Percent	11.7
Employment as percentage of population	2014	Percent	35.7
Revised economic participation rate	2014	Percent	41.1
Infants mortality rate (per thousand live births)	2014		15.5

(continued)

Table 4.1 (continued)

Indicator	Year	Unit	Value
Literacy rate, adult total (% of people ages 15 and above)	2011	Percent	87.15
Literacy rate, youth female (% of females ages 15–24)	2011	Percent	97.00
Literacy rate, youth male (% of males ages 15–24)	2011		98.99
Ratio of young literate females to males (% ages 15–24)	2011	Percent	97.99
Ratio of female to male secondary enrollment (%)	2012	Percent	97.55
Gross enrollment rate in primary education	2014	Percent	108.17
Net enrollment rate in primary education	2014	Percent	97.39
Primary completion rate, both sexes (%)	2014	Percent	116.6
Gross enrolment ratio, primary, both sexes (%)	2014	Percent	109.8
Gross enrolment ratio, secondary, both sexes (%)	2014	Percent	124.3
Gross enrolment ratio, primary and secondary, gender parity index (GPI)	2014		1.0
Annual freshwater withdrawals, total (% of internal resources)	2014	Percent	986.3
Improved water source (% of population with access)	2014		97.0
Improved sanitation facilities (% of population with access)	2014	Percent	100.0

[a]According to Population Census 2010

Source: http://www.cdsi.gov.sa/english/; World Bank (2015); FAO: http://faostat.fao.org; http://data.worldbank.org/indicator/SP.RUR.TOTL.ZS/countries/SA?display=default

130 mm per year, making water the main determinant and the biggest challenge for the agricultural sector in the Kingdom of Saudi Arabia. Irrigations to grow agricultural crops rely heavily on limited and non-renewable groundwater resources. Agricultural sector accounts for the largest share of water consumption, amounting to 86.5% of total water consumed in the Kingdom (KACST 2009). Consequently, the absence of alternative water resources for growing crops, ground waters are depleting very fast. Also the water extraction rate is higher than the recharge making the aquifers dry at an alarming rate. To be the sustainable water user, the Kingdom made a strong policy shift in 2008 and fully implemented wheat phase out by 2016 in order to conserve its drying fossil water resources. To ensure food security, the Kingdom fully relies on imports of wheat, rice and other food commodities (WTO 2016; FAO 2016). The reduction in wheat cultivation is presented in Figs. 4.2, 4.3, 4.4, 4.5, 4.6, and 4.7.

4.2.4 *Climate Change*

The Kingdom like many other countries is experiencing negative outcomes of the climate change. For example, recently it has witnessed heavy floods and continued dry spells at frequent intervals that could be primarily associated with

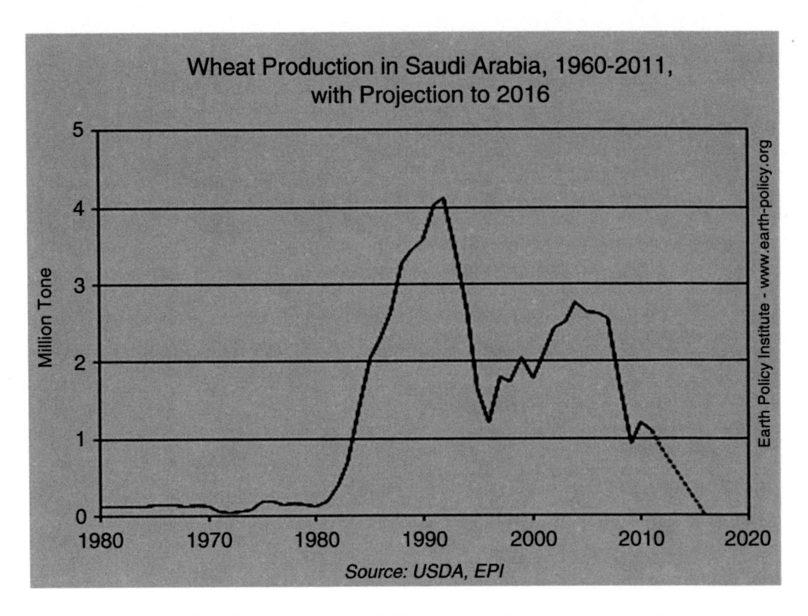

Fig. 4.2 Wheat production decreases after 2009 and complete phase-out wheat production program was implemented by 2016. (Source: Spackman 2015)

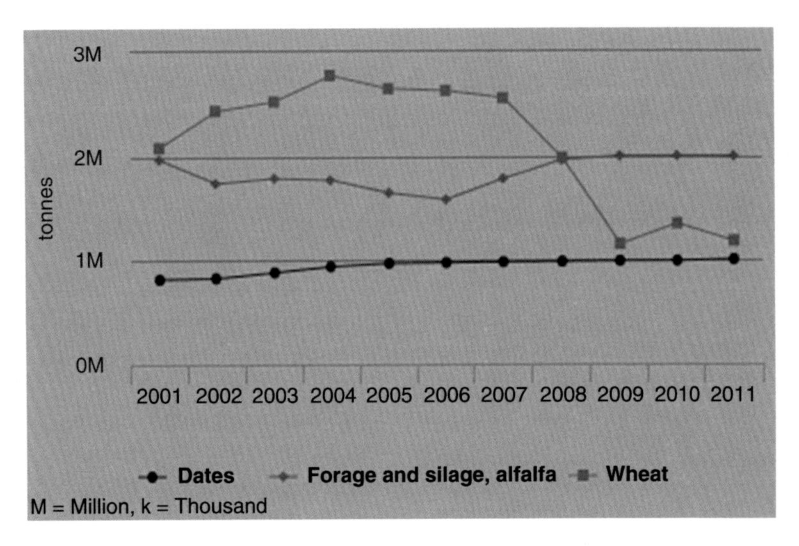

Fig. 4.3 Wheat production decreases after 2004. (Source: http://faostat3.fao.org/browse/area/194/E)

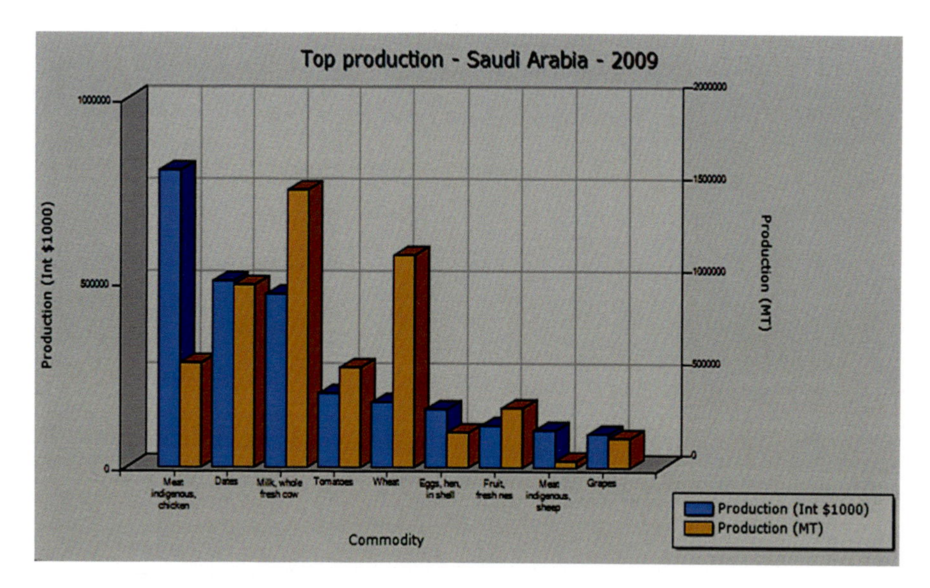

Fig. 4.4 Production of crops with high water requirements decreases after 2009. (Source: http://faostat.fao.org/default.aspx)

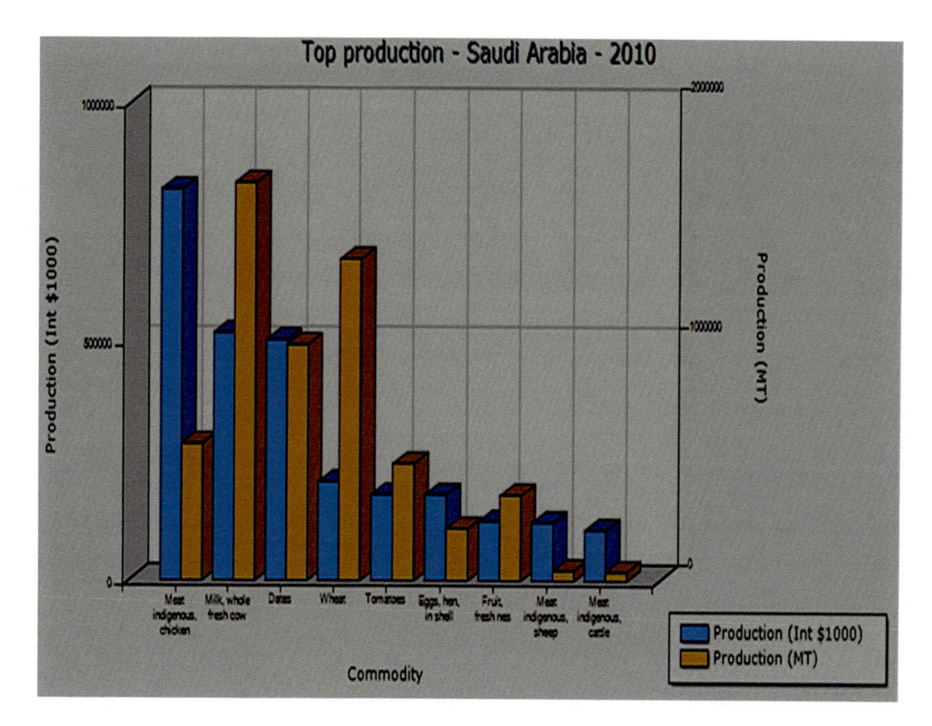

Fig. 4.5 Production of crops with high water requirements decreases after 2009. (Source: http://faostat.fao.org/default.aspx)

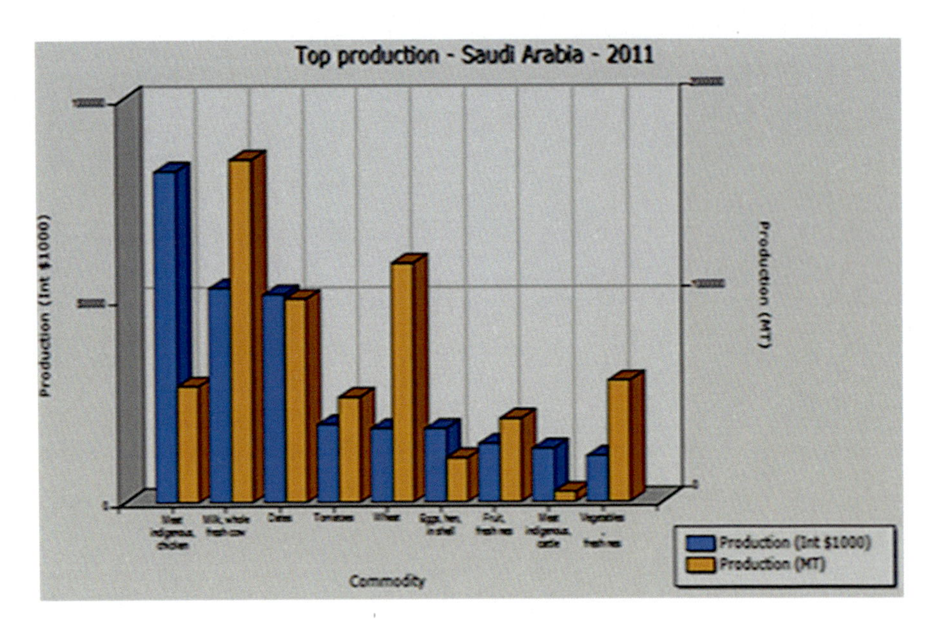

Fig. 4.6 Production of crops with high water requirements decreases after 2009. (Source http://faostat.fao.org/default.aspx)

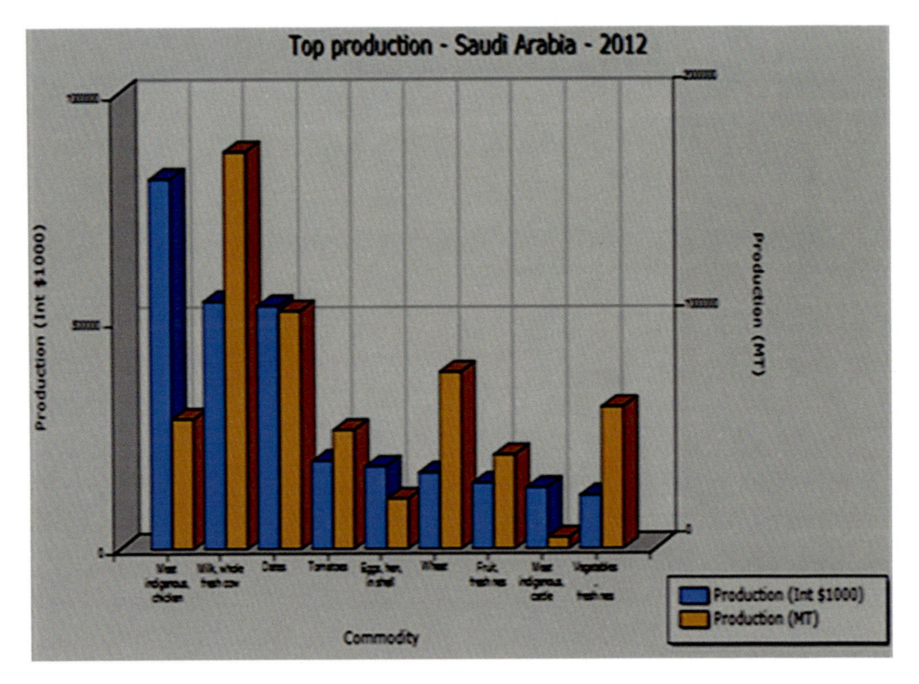

Fig. 4.7 Production of crops with high water requirements decreases after 2009. (Source: http://faostat.fao.org/default.aspx)

climate change. Researchers like Darfaoui and Al-Assiri (2010) warned that climate change in the Kingdom of Saudi Arabia could have major impacts on its agriculture and food production systems primarily due to the reduced availability of water. They reported that the expected impact of climate change could be through an increase in temperatures, up to 3 °C higher by 2040, greater rainfall variability and rising sea levels. These changes in climate could have an intense and marked influence on agriculture and food production, which are already under serious stress due to the lowering of groundwater tables and diminishing aquifers.

4.2.5 An Increase in the Population Means More Mouths to Feed in the Kingdom

The country's population was estimated at about 30 million in 2014 as depicted in Fig. 4.8 (CSDI 2015) and expected to reach 40 million by 2025 (USDA 2014). As populations are expected to increase to 53.4 million by 2020, it can be predicted that at least one fifth of the people in Bahrain, Oman, Qatar and Saudi Arabia will become food insecure (Fig. 4.9). The rising trends in populations as depicted in Fig. 4.9 will boost demand for more food imports, as Saudi Arabia relies mostly on foreign imports to satisfy about 80% of its food consumption needs (USDA 2014). According to Intini et al. (2012), about 60–90% of food products consumed in the GCC countries at present come through imports. Pradhan (2010) also anticipates a larger food gap between indigenous low production and higher imports due to growing populations in the region. Similarly, Sadik (2014) maintains that an increase in

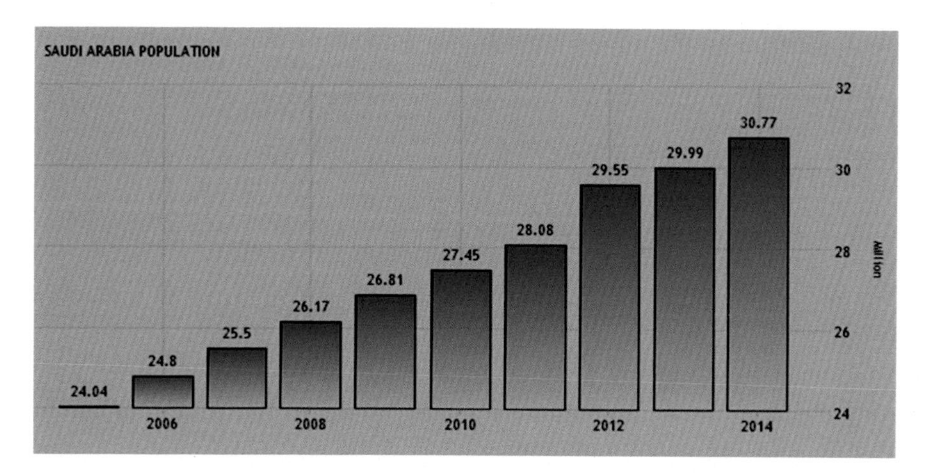

Fig. 4.8 Population increases from 2006 to year 2014. (Source: Central Department of Statistics and Information 2015)

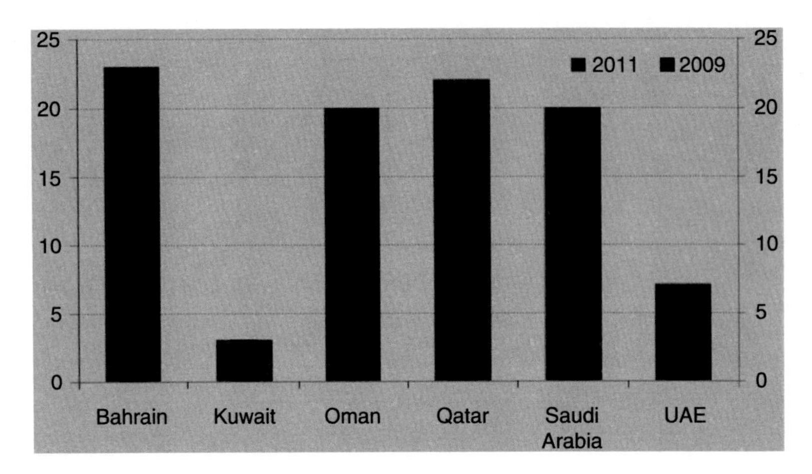

Fig. 4.9 One fifth of the people in Saudi Arabia food insecure in 2020. (Source: Intini et al. 2012)

population, especially in those Arab countries with high population growth rates, not only exerts immense pressures on limited agricultural resources but it also calls for higher food imports.

4.2.6 The Kingdom Meets its Food Requirements Through Heavy Imports

USDA (2014) considers Saudi Arabia as the largest importer of food and agricultural products among the GCC countries, with a population more than double that of the five GCC states (UAE, Kuwait, Qatar, Oman, and Bahrain). Saudi Arabia on an average has been spending $12 billion/annually on importing food and agricultural commodities and ranked 19th among the world's largest importers in monetary terms. To meet almost 80% agri-food requirements, only 6% of its GDP gets consumed through the imports of agricultural commodities. The top ranking imports, namely barley, sheep, rice, chicken and wheat, account for 40% of total imports as depicted in Fig. 4.10. Saudi Arabia and UAE are the major importers of agricultural products due to their highest populations among the GCC countries. Both of these countries account for about 80% of the total agricultural imports of the region while the other five nations import only about 20% of the total food and agricultural commodities (Boughanmi et al. 2014). However, due to limited agricultural production within the Gulf Cooperation Countries (GCC) and continued declines in domestic production many authors like Pradhan (2010), Woertz (2010), and Intini et al. (2012) anticipate that an increased dependence on food and agricultural imports to meet their domestic food requirements would be higher in the future.

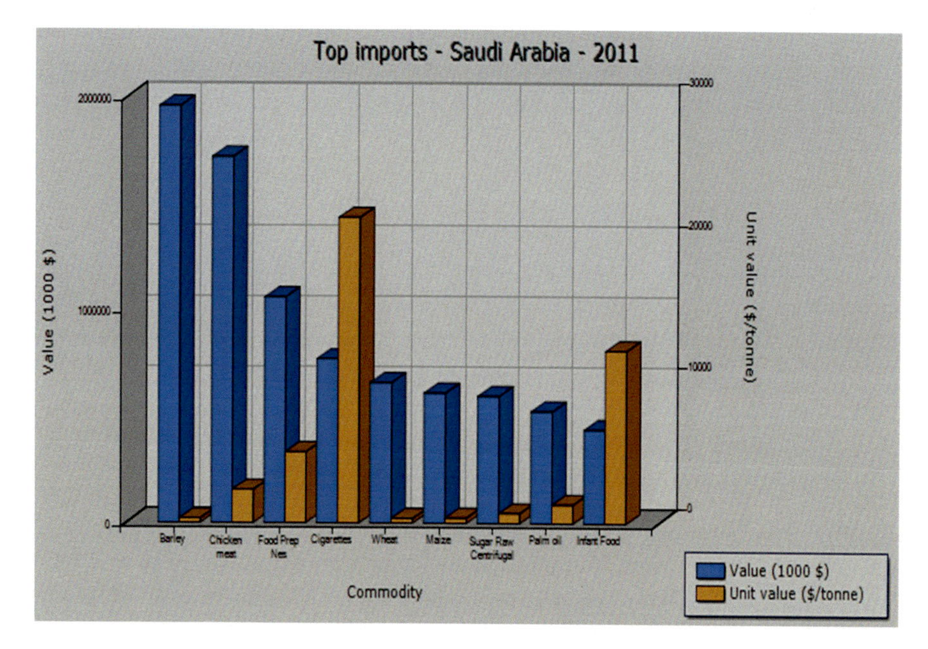

Fig. 4.10 Production of crops in the Kingdom decreases and imports of agricultural commodities decreases after 2009. (Source: http://faostat.fao.org/default.aspx)

4.2.7 Rising Populations, High Reliance on Imports and Price Hikes in the International Markets

The food commodities have witnessed high hikes in prices in the international markets as revealed by the Fig. 4.11. Commodity prices have been increasing since early 2009, following the sharp drop in late 2008 as the financial crisis unfolded. In the second half of 2010, commodity prices began rapidly rising, particularly for food and oil as depicted in the Fig. 4.11 (World Bank 2011a, b, c). Heavy dependence of Arabian countries on imported food and rising international prices can have substantial growing pressure on national and household budgets, depending on the level of subsidies offered in different Arabian countries.

4.2.8 High Subsidies Are Offered on the Food Items to Make Them Affordable

In oil rich Middle Eastern countries, such prices hikes and the cost of importing grain may not have much impact on their citizens because governments often absorb the part of the food-price increases at the country-level and regulate prices as seen in the KSA. According to Boughanmi et al. (2014), the Saudi government uses price

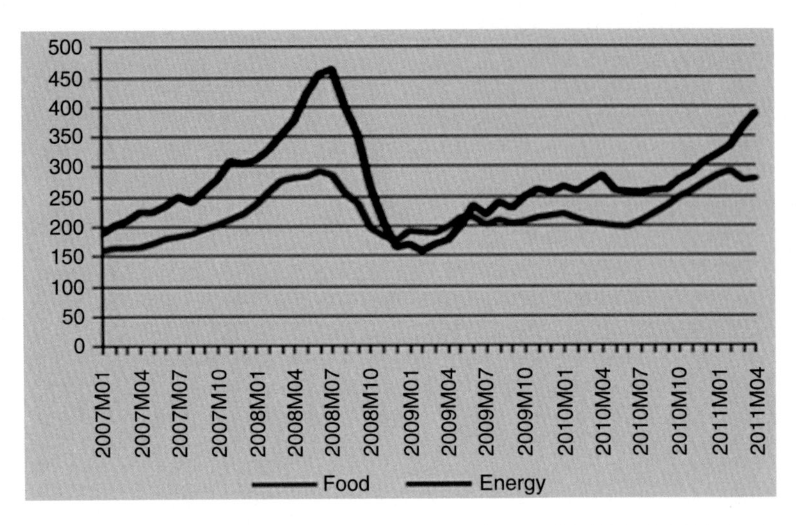

Fig. 4.11 Prices of food and energy are on the increase. (Source: World Bank 2011a, b, c)

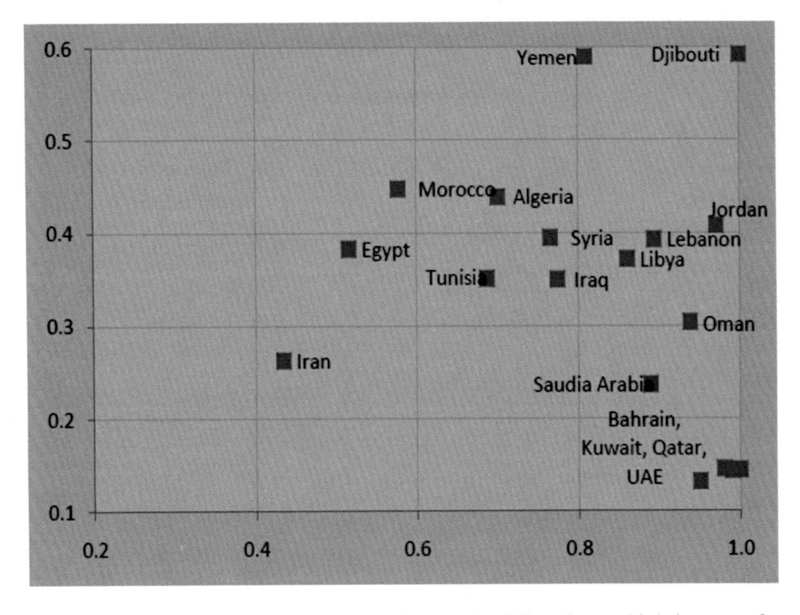

Fig. 4.12 Saudi Arabia is among less vulnerable to price hikes due to high incomes from oil. (Source: Adapted from World Bank 2011c)

caps and provides generous subsidies to the food producers and retailers to enable them to provide food to the consumers at the low prices. Particularly, the government regulates the prices of staple foods, such as milk, and provides supports to food producers and retailers by compensating the private sector for their slim profit margins. Saudi Arabia falls among the countries, less vulnerable to price hikes due to high incomes from oil as depicted in Fig. 4.12.

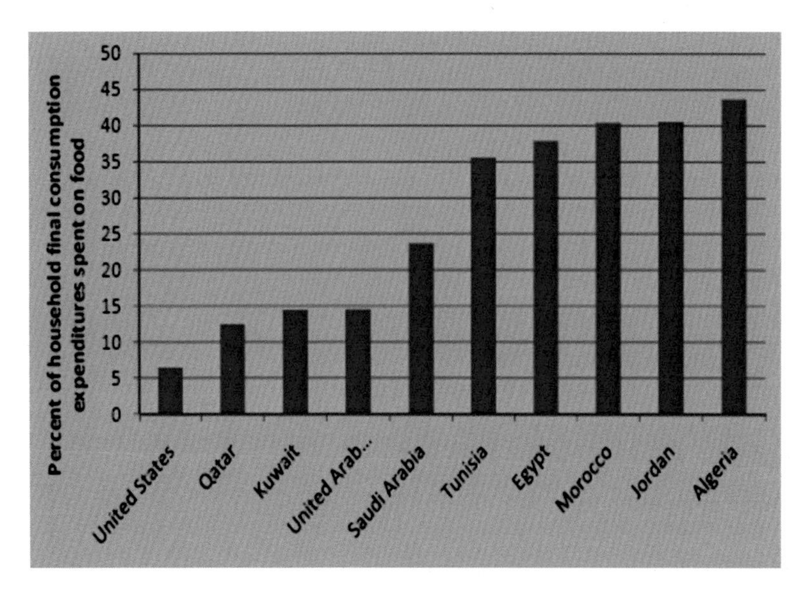

Fig. 4.13 Percentage of household final consumption expenditures spent on food in the selected Arab countries. (Source: USDA 2014)

The fiscal pressures vary from country to country as some governments have been more successful than others in regulating subsidies, and extending assistance to the poor. However, substantial increases in international prices of a broad range of foods (World Bank 2011b) and fast-growing domestic food demand could have an impact on people as typically they spend from one-third to two-thirds of their incomes on food items. Despite ample and generous subsidies offered by the Kingdom to keep the consumers comfortable, Saudis still spend about 27% of their incomes on food (USDA 2014). Yet, some countries in the Middle East and North Africa (MENA) have been facing fiscal as well as domestic inflationary pressures, as shown in Fig. 4.13 (Crowley 2010). Most recently the prices of oil in the international market have dropped down from $100 to $26, the prices for the Benzene, electricity and the food commodities in the KSA are steadily going up to cope with the budget deficit.

4.3 The Definitions of Food Waste and Global Significance

Lipinski et al. (2013:1) define food waste as the *"food that is of good quality and still fit for human consumption but that does not get consumed because it is discarded – either before or after it spoils"*. Food waste is the result of negligence or a conscious decision of not using the food and is thrown it away. The Food and Agriculture Organization (FAO 2013) *defines food waste as the "food losses resulting from decisions to discard food that still has value"*.

4.3.1 Food Waste Is a Global Issue Including the Kingdom of Saudi Arabia

By 2075, the United Nation's mid-range projection for global growth predicts that the population will peak at about 9.5 billion people (IME 2013; Lipinski et al. 2013). This means that there could be an extra three billion mouths to feed by the end of the century, implying to grow more food and reduce its waste. Approximately 4 billion metric tons of food is produced at the global level per year. Yet because of poor harvesting, careless storage, and inefficient transport, at least 1.3 billion tons of this produce never reaches the table as revealed in Fig. 4.14 (FAO 2011, 2012, 2016; Arab New 2015). At the same time, more than 870 million people worldwide do not have sufficient access to the food they need to feed themselves and their families (Arab News 2015; FAO 2015b) as depicted in Fig. 4.14.

The FAO (2016) reports that the amount of food wasted by the consumers of industrialized countries is more or less equal to the entire food produced by the sub-Saharan Africa. Food waste remains the third largest component of generated waste by weight globally. According to a United Kingdom-based Waste and Resources Action Program, UK produces over 11 million tons of food wastes annually, causing an environmental impact more or less equal to one fifth of all emissions produced by cars in the country. In the European Union, 89 million tons of food is wasted annually, and it may rise to 126 million tons by 2020 if improvement measures are not implemented (WRAP 2011; BQ Magazine 2014). In 2010, Americans wasted 33.79 million tons of food, and an average American throws away up to 115 kg of food every year. These extraordinary figures of food waste illustrate an increase of 16% as compared to the year 2000 (Myers 2015). Such food wasting behaviors continue to increase, despite the rising US passion for healthy living and regular exercise.

Fig. 4.14 Food waste in the world. (Source: Arab News 2015) Available at: http://direct.arabnews.com/saudi-arabia/news/722026

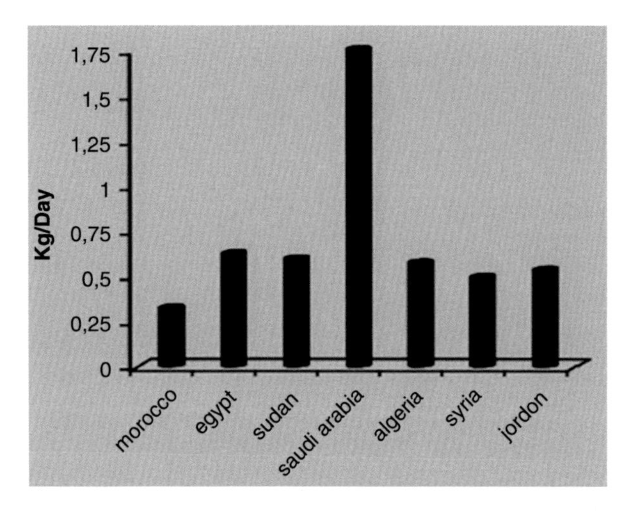

Fig. 4.15 Per capita solid waste generation in the Kingdom of Saudi Arabia as compared to the other Arabian countries. (Source: Khan and Kaneesamkandi 2013)

Food waste remains a global issue and the Middle East is not an exception. Data on food waste from the Middle East are similar to those reported by the global statistics. The GCC countries, being rich are among the top food wasters of the world, particularly food in bulk are wasted during Ramadan (Environment Agency-Abu Dhabi 2013). Food waste in the UAE in a normal month comprises 39% of a household's organic waste (Environment Agency-Abu Dhabi 2013). However, in the holy month of Ramadan the waste increased to 55% of household waste or 1850 tons is thrown away every day (BQ Magazine 2014).

In an estimate made by the Bahrain's Supreme Council for Environment Waste Disposal during the holy month of Ramadan, food waste in Bahrain reaches above 400 tons per day. Similarly, each household in Bahrain throws away about 25% of food bought during Ramadan with an approximate cost of $395 per year (BQ Magazine 2014). KSA generates more than 15 million tons of solid waste per year whereas per capita waste generation ranges from 1.5 to 1.8 kg person/day as depicted in Fig. 4.15.

According to a survey conducted in 2012 by an internet-based market research firm YouGov found that some 78% of respondents (in Saudi Arabia and the UAE) throw food away every week to make room for new supply of groceries in their refrigerators, and two out of three UAE residents do not view food waste as an issue of global environmental concern as evident from Fig. 4.16.

The per capita production of solid waste in cities of Riyadh (Saudi Arbia), Doha (Qatar) and Abu Dhabi (UAE) is over 1.5 kg per day, placing these population centers among the highest per capita waste producers in the world. Khan and Kaneesamkandi (2013) found that food waste constitutes a significant portion of the organic waste generated by the major cities in the KSA (Fig. 4.17) due to an increasing population and urbanization. Solid waste mismanagement is one of the main reasons causing pollution and resulting environmental challenges and issues (Nitivattananon et al. 2012).

Fig. 4.16 The state of food waste in the Middle East. (Source: http://www.bq-magazine.com/industries/2014/06/gcc-among-top-food-wasters)

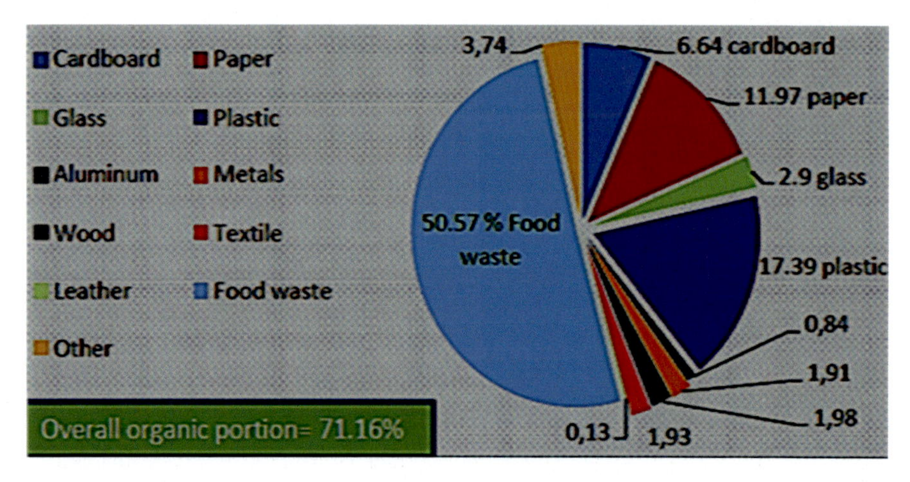

Fig. 4.17 Solid waste in the Kingdom of Saudi Arabia has high organic contents. (Source: Khan and Kaneesamkandi 2013

4.3.2 An Overview of Food Availability and Its Waste in the Kingdom of Saudi Arabia

The country ranks among the world's top food wasters. Food items make the chief constituents of the waste dumping sites. Food waste includes un-eaten and unfinished food and food preparation scraps from leftovers coming from residences or households, commercial establishments like restaurants, schools, institutions, and cafeterias. People in the KSA prefer to throw away the leftover food as they think it's unhealthy to eat or they don't care to eat the same foods – variety becomes supreme. The teens, younger generation and children are one of major food waster groups. They do not hesitate throwing away their half empty and partially eaten packets of chips, chocolates, burgers, sandwiches, and soft drink bottles. Not only do such items litter the roadways, but also demonstrate evidence of wasted food.

Due to such huge volumes of food wastes and several other reasons, Saudi Arabia has to depend on imports food commodities to feed its population. After observing the roll-back policy regarding mass-scale agriculture of 2008, the KSA had to spend $20-billion worth of food in 2010 (Chatham House 2013). The promising oil-based economy, strong fiscal balance and large oil reserves make the Kingdom a food-secure country. The Kingdom is quite capable of importing large volumes of food commodities to feed its population and making them available to the people at the subsidized rates, being the welfare state. The consumption, production and the imports levels have been presented in the Figs. 4.18, 4.19, 4.20, and 4.21, indicating that import dependency of Saud Arabia, will remain high, particularly for strategic commodities such as cereal grains as it is almost impossible to produce domestically.

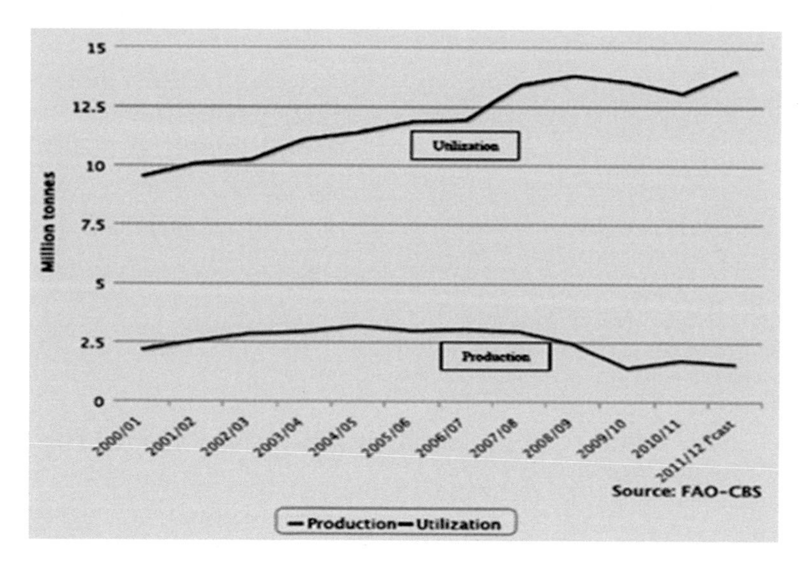

Fig. 4.18 Trends in cereal production, utilization (consumption) in Saudi Arabia. (Source: Intini et al. 2012)

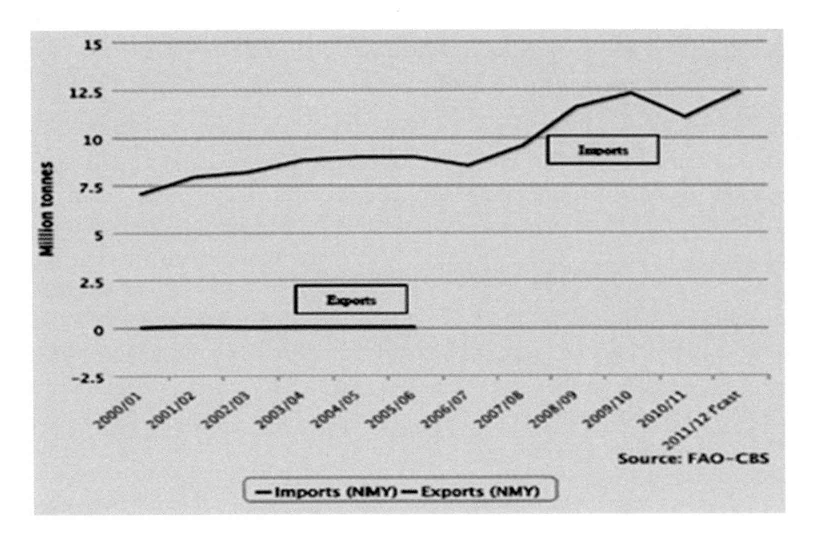

Fig. 4.19 Trends in cereal import, and export in Saudi Arabia. (Source: Intini et al. 2012)

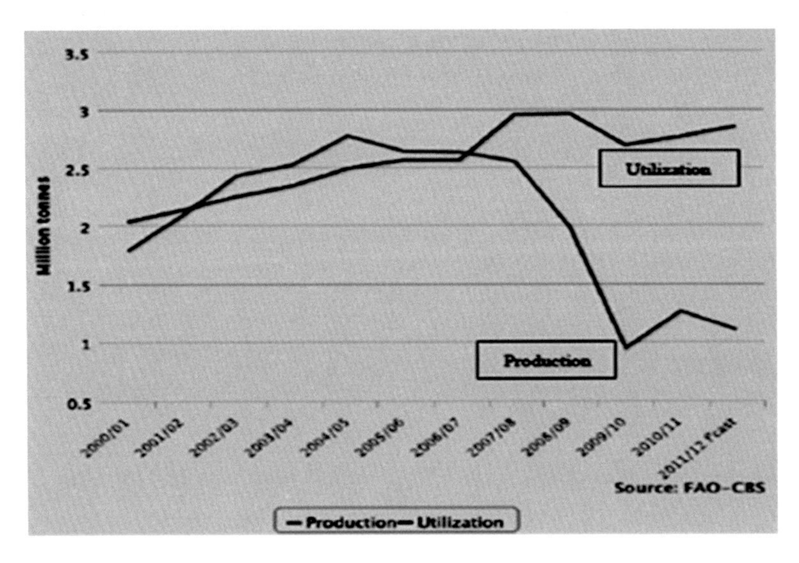

Fig. 4.20 Trends in wheat production, utilization (consumption) in Saudi Arabia. (Source: Intini et al. 2012)

4.3.3 Food Waste Is Unavoidable as Food Is an Integral Part of Saudi Arabian Culture

Arabs are generous in hospitality; the provision of food is a gesture to welcome the guests. Even when a family might have little food to spare, the visitors are often fed with the surplus. People in the Kingdom usually buy food in bulk and prepare more

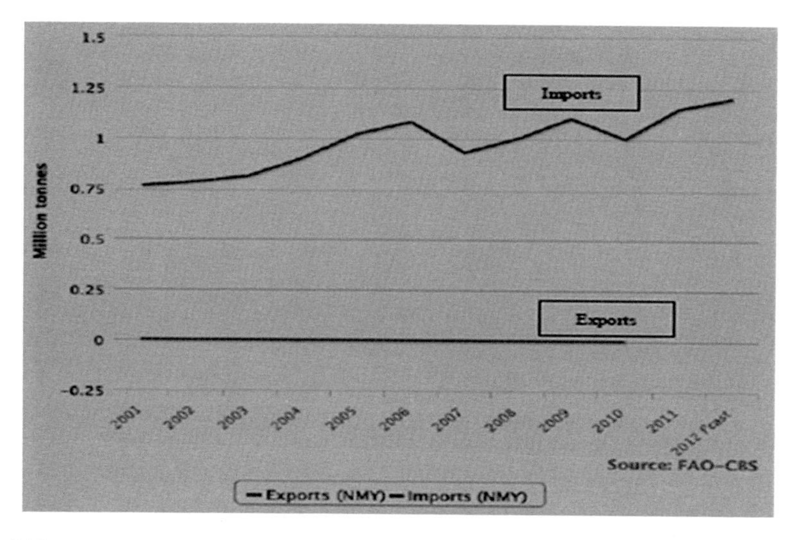

Fig. 4.21 Trends in wheat import, and export in Saudi Arabia. (Source: Intini et al. 2012)

food than they could consume, resulting significant quantities go in waste. According to the country's cultural values, Saudis love setting up lavish food tables during Eid festivals, weddings, parties or informal get-togethers. The culture in the KSA is based on festivals, designating availability of huge quantities of food and yet a great portion from that food ends-up in the landfills. They love to organize innumerable banquets where the wasting of food is an indispensable feature. Food waste in the Kingdom has emerged as a serious societal behavioral issue. The high levels of food waste could be due to the people's attitudes towards food. The availability of the abundant supplies of food commodities at the extremely subsidized rates makes them take food for granted.

Saudi Arabia being the biggest consumer of food commodities in the GCC, accounts for 60% of the total consumption in the region. An ordinary Saudi citizen consumes more food than his needs, causing more than 50% of waste and in the public events this level of waste goes beyond 70%. Approximately 13 million tons of discarded food and its products make about 35% of the domestic garbage thrown in the landfills in the Kingdom. It is also estimated that almost four million meals are not consumed and they just end up in the landfills in the Eastern Province daily. Such behaviors cause a serious drain on the budget of ordinary households that can be spared if they rationalize their food consumption. Economists also estimate that if the citizens reduce this waste by 30%, food prices would go down at least by 15%. Therefore, in 2014 the Kingdom designed and implemented a food security strategy to ensure food availabilty.

Rice is one of the key commodities, consumed in the KSA and is associated with food security. Indeed, the Kingdom imports about 1.3 million tons of rice annually at an estimated cost of SR 4 billion, of which nearly 440 tons are wasted every year. Saudis throw away between 35% and 40% of cooked rice a year, the waste is worth nearly SR1.6 billion (Saudi Gazette 2014). This wasteful practice could be attributed to the lack of understanding of the Saudi citizens on the importance of conserving food and its sustainable consumption.

4.3.4 Impacts of Food Waste in the Kingdom of Saudi Arabia

In addition to the impact of food waste on financial resources, it has got social, environmental and religious implications too. Food waste results in the waste of resources, including water, land, energy, labor, and capital etc. Environmentally, food waste remains the wasteful use of chemicals such as fertilizers and pesticides. Fuel used for transportation of food goes to waste; and rotting food on the landfills releases more methane – one of the most harmful greenhouse gases contributing to climate change. Methane is known to be 23 times more potent and toxic than CO_2 in our atmosphere and more deleterious to the environment than the widely feared CO_2 as a greenhouse gas. Despite its potency, methane is typically receives low attention because of its much smaller percentage of total emissions. But this particular gas has a significantly higher potential to cause warming than carbon dioxide. Ruthless dumping of huge volume of food as waste into the landfills significantly contributes to global warming (Psomopoulos et al. 2009; Al Ansari 2012). Food waste not only causes problems in its disposal but also imposes extra costs on the consumer. If some restrictions were placed on the purchase of perishables items such as fruits and vegetables (including tomatoes) for the daily needs, the price of such items would go down by 50%, in response to the law of supply and demand. Such behaviors of luxurious and wasteful consumption of food cannot be justified as they adversely affect the economy, household budgets and personal savings.

4.3.5 Measures Taken to Reduce Food Waste in the Kingdom

The discussions made in the previous sections indicate that several factors are responsible for the food losses and waste. Experts categorize these factors as: lack of awareness; lack of shopping planning; reluctance in consuming left-overs; waste at the household levels, restaurants, parties and occasions. Saudis believe that fresh food is hygienic, standardized and safe to be consumed. The members of society throws leftover food in the garbage, thus not wasting food remains a tough challenge in a culture where food is readily discarded. Food waste is not just an ethical issue (since many people view it immoral to throw food away) but affects the environment as well. Excessive food rottening and its decomposition in the landfills release harmful greenhouse gasses; this food is an untapped energy source yet no significant efforts are made to realize its potential. Keeping all these facts in view, several measures to combat food waste and bring the concepts of sustainable consumption of food in the society have been taken which include:

4.3.5.1 Creation of Awareness

Society must learn the significance and importance of food and water – including the reduction of waste. There is a great need to reduce the food waste in the KSA by employing different strategies. One of the most viable options is to increase the awareness among all the segments of society on the food and water situation in the Kingdom through a national, comprehensive campaign with the goal to save food and use it wisely since a significant proportion of food is imported. In the Kingdom, a well-organized campaign is needed to stop the wasting of food, not only in the residential areas and households but also in restaurants, hotels, schools, institutions, and universities etc.

4.3.5.2 Setting-Up a Committee to Study Ways and Means to Reduce Food Waste

In considering food waste a serious issue, the Kingdom allocates luxurious budgets for the municipal services sector to look after the water drainage and waste disposal. Realizing the gravity of food waste issue, the Ministry of Food and Agriculture in Riyadh, has set-up a committee to study ways and means to reduce food waste and address the issue in a sustainable manner. The committee has been entrusted with the mission to devise mechanisms for reducing food wastage in the society.

4.3.5.3 Setting Up of the Food Charitable Society

An organization namely 'The It'aam Food Charitable Society' has been established to collect surplus quantities of food. Extension programs have also been accessed and aligned with It'aam Food Charitable Society. It'aam has also signed a number of memorandums of understanding with several organizations, such as the Saudi Press Agency and Saudi Aramco. In addition to donating food to the needy, It'aam is planning to publish a magazine namely 'Meerah', to convey its message 'Not to waste food' to the readers and the general public. Since young children make the principal segment of the society, therefore the society is also striving to produce and release animated short movies for them to bring a behavioral change and to give a new vision to the teachers and children. The organization It'aam, through its Extension Education Programs, is creating awareness on the importance of food and taking essential measures to reduce its wastage. Its organizers believe that such awareness campaigns may have a positive impact on both society and national economy. In addition, the organization, through its food bank, collects food and re-distributes it among the needy families in special packages. The organization has set-up another branch in Riyadh providing 12,000 meals to the needy every day. Shortly branches of such food banks will be functioning in Qatif and its suburbs. In Saudi Arabia wastage of food can be minimized by redistributing the surplus among the needy families.

Fig. 4.22 A Saudi man placed a fridge in front of his house in Hail and invited others to donate food (Image Credit: Mezmez). (Source: htpp//gulfnews.com/news/gulf/saudi-arabia/man-installs-charity-fridge-outside-hishouse-1.1328931)

Realizing the importance of food and to show their concerns regarding the waste of food in huge volumes, some individuals have also taken initiatives on their own. For example, in order to reduce food waste and make the food available to the needy, a rich person in the northern city of Hail has put a big refrigerator in front of his house (Fig. 4.22) and invited other people of the neighborhood to donate food to help the needy. The open-air donation and ready availability of food to the needy saves them from embarrassments while meeting their needs for food (Toumi 2014).

4.4 Implications for the Extension Education and Capacity Building Programs

A large volume of food waste could be attributed to a culture that fails to emphasize the importance of conserving food and the resulting effect on the environment. However, the solution begins first of all with the family and then society as there should be awareness about wasting food, especially food products and commodities that are imported in huge volumes, such as rice. The deteriorating situation has created severe and serious implications for the National Extension Service of the Kingdom. Therefore, the Kingdom badly needs the very sound and appropriate extension education programs to bring changes in the behavior and attitudes of Saudis, usually over-buying groceries and cooking more and extra food. Though it seems difficult to control waste yet by involving and educating each and every

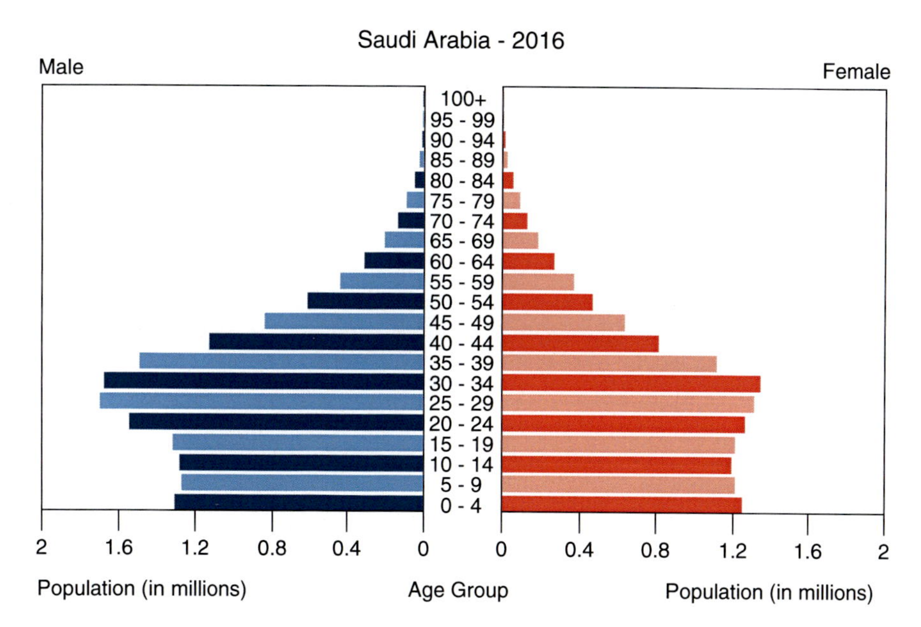

Fig. 4.23 Women and youth are the largest demographic group in the Kingdom of Saudi Arabia. (Source: http://www.indexmundi.com/saudi_arabia/age_structure.html)

member of the society, the issue food waste can be reduced. In addition, there are some simple practices to reduce waste which include:

National Extension Service needs to design educational programs for the youth and women

It is possible for families to reduce food waste by avoiding over-buying and – cooking more than what's needed to feed the members within a particular family. However, the government alone cannot take any measures, unless these measures are accepted by the public. Civil society in the Kingdom generally happens to be traditional, collective, and conservative. Youth are the largest demographic segment of the society. About 51% of the population is below the age of 25, and two-thirds below the age of 29. Women make up about 49% of the total population as indicated in Fig. 4.23.

To address the issue and the challenge of food waste faced by the nation, it is of paramount imporatnace that Extension Department must focus on the youth and women. National Extension Service must organize workshops and conferences for the youth and women groups, planners, policy makers, opinion makers such as civil society organizations to address the issue and catort he needs of youth and women. It is really important to target youth (as the major group) by organizing programs on how to save food and stop food waste; Special TV shows on cooking and recipes catering the interests of women and house-wives as they are main agents for food waste.

Extension Department needs to focus on all possible options that could address the issue. It should:

- Adopt all possible measures and find ways to minimize waste of food and loss of agricultural products; establish food collection organization near the farmers' fields and farms that could collect, reuse or recycle surplus in crops and products as one of means of mitigating the problem; Establish charity organizations to collect leftover food to distribute among the poor and the needy consumers;
- Plan to buy food in reasonable quantities that families could realistically consume and avoid buying groceries in bulk; develop the habit of re-using left-overs by consuming them as next meal; promote better recipes for using left-over foods at the household level;
- Introduce an appropriate food waste management strategy for the municipalities so that food waste could be reduced; rethink and build a model that could result in a better and smarter system in Saudi cities to stop food waste and reduce pressures on landfills;
- Frame laws and draft legislation and develop regulations that prohibit wasting of all types of foods including fruits and vegetables; educate the community to turn the food that turns un-consumed food into the compost to avoid sending it to the landfill;
- Initiate rigorous campaigns to disseminate information on subject of food waste, targeting all the segments of the society on the extent and causes of the food losses. To realize success, it would be important to employ all possible means of communications and using media including the print media (i.e. brochures, newspapers, and magazines), and electronic media (i.e. TV, Radio, messages through cell phones); and social networking websites including the use of whats up groups, Apps, blogs, Facebook, Twitter and YouTube etc.

In Saudi Arabia, Islamic teachings exert a substantial influence on society. Islam demands the rich to feel the pain of poor and hungry and expects the Muslims to develop a feeling of solidarity towards hungry not having a single meal each day; Islam strictly forbids the waste including water and food. It would be of great help if the following verse of Holy Quran (Fig. 4.24) be brought mainstreamed in public consciousness:

4.5 Conclusions and Recommendations

The KSA is not in a position to feed its citizens through the domestic agricultural production; therefore it will have to depend on food imports. Therefore dependency on importing food and agricultural commodities is expected to rise further due to the rapidly growing population, improved living conditions, sustained economic/industrial development, and depleting natural resources (mainly water and land resources). Moreover, climate change is also expected to have a major effect on water resources in the country. An overview of the resources making food avaialable in the Kingdom are presented in Table 4.1.

Fig. 4.24 Allah says: "O children of Adam! Beautify yourselves for every act of worship, and eat and drink (freely), but do not waste: verily, he does not love the wasteful!" (7:31). (Source: http://prophetic-path.com/purification-of-the-stomach-using-prophetic-path/)

Despite the Kingdom meets the 80–90% food requirements of its citizens yet successfully manages to make them available in abundance at the highly subsidized rates. Therefore, people in the Kingdom, take food as granted resulting huge volumes of food go in waste and end-up in the landfills. Consequently, the Kingdom has shown its serious concerns on the enormous levels of food waste. The consumers can address food waste by reducing its production by keeping their food consumption reasonable.

Households can help by adopting few simple measures such as restricting their shopping to their needs. They need to make an appropriate plan for shopping food and groceries and must seek proper knowledge about the storage of food items to reduce their waste.

Food waste is a complex issue. There would not be a single solution to combat waste and to ensure food security in the Kingdom requiring an integrated approach. There is a critical need to educate the local population on how precious food is and how much it hurts and harms the economy, contributes to inflation, damages the environment and causing many other problems invisible to an ordinary citizen.

However, Extension education has greater and more implications to address the issue in the Kingdom. Some important measures include: awareness creation among the communities, households, businesses, food industries, educational programs and behavioral change. To stop this phenomenon in the Kingdom, a viable option would be to educate the society to change their attitudes towards food waste. It is really important to foster a behavioral change among youth and women. Extension Education and capacity building programs can help creating awareness and bring behavioral change by holding workshops, conferences, seminars and launching anti-food waste campaigns for all the stake-holders keeping in view the type and nature of the targeted population.

To boost awareness among youth extension messages be released on cell phones; Twitters, face book etc. Since Youth groups are regular users and have permanent access to Internet and media, therefore Extension needs to further focus on youth and women through these means. Billboards on roads and signs in the parks could be effective extension tools to boost awareness, educate the society and bring behavioral change in the overall society. Morning shows on TV could be an effective medium to educate the women groups on the importance food, negatives of food waste causing economic losses and environmental damages. Above all, food waste is against the religious teachings of Islam. Women are also to be taught how to convert leftovers into the delicious dishes.

The eating places serving buffets in the cafeterias, restaurants can help reducing food waste by displaying courteous posters on using the right and the appropriate sizes of food in their plates for their consumption. Food waste is one of the prime factors influencing financial, environmental and social issues. However, there is a pressing need to develop more scientific studies to reveal accurate estimates about the volume of food wasted in the Kingdom. Food wastage in the Kingdom is alarming, making it necessary to adopt an approach to build resilience and improve sustainability. This will require inter-disciplinary and multidisciplinary collaboration by involving all the stakeholders.

Acknowledgements The authors are extremely grateful and express their gratitude to the Saudi Society of Agricultural Sciences for extending us all the possible help and sincere cooperation toward the completion of this piece of work and research.

References

Agri-food Canada. (2013). *Agri-food sector profile – Riyadh Saudi Arabia*. Produced by the Canadian Trade Commissioner. Available at: http://www.enterprisecanadanetwork.ca/_uploads/resources/Agri-Food-Sector-Profile-Saudi-Arabia.pdf

Al Ansari, M. S. (2012). Improving solid waste management in gulf co-operation council states: Developing integrated plans to achieve reduction in greenhouse gases. *Modern Applied Science, 6*(2). doi:https://doi.org/10.5539/mas.v6n6p60.

Al-Fawaz, N. (2015). *Need to reduce food waste: Experts*. Published – Monday 23 March 2015. Available at: http://www.arabnews.com/saudi-arabia/news/722026

Al-Shayaa, M. S., Baig, M. B., & Straquadine, G. S. (2012). Agricultural extension in the Kingdom of Saudi Arabia: Difficult present and demanding future. *Journal of Animal Plant Science, 22*(1), 239–246.

Al-Zahrani, K. H. (2014a). Food wastage remains a serious concern in KSA. An interview published by *The Daily Arab News*—Sunday 13 April 2014.

Al-Zahrani, K. H. (2014b). Environmentalists alarmed at food wastage. An interview published by *The Daily Arab News*—Monday 21 April 2014.

Baig, M. B., & Straquadine, G. S. (2014). Sustainable agriculture and rural development in the Kingdom of Saudi Arabia: Implications for agricultural extension and education. In M. Behnassi, M. Syomiti, R. Gopichandran, & K. Shelat (Eds.), *Chapter 7. Climate change: Toward sustainable adaptation strategies* (pp. 101–116). Dordrecht: Springer Science +Business Media B.V.. https://doi.org/10.1007/978-94-017-8962-2_7.

Boughanmi, Houcine, Kodithuwakku, Sarath, Weerahewa, Jeevika. (2014). Food and agricultural trade in the GCC: An opportunity for South Asia? Available at Arab New (2015).

Need to reduce food waste: Experts. Available at: http://www.arabnews.com/saudiarabia/news/722026

BQ Magazine. (2014, June 17). *Half of Ramadan feast will end up in the trash.* http://www.bq-magazine.com/industries/2014/06/gcc-among-top-food-wasters

Chatham House. (2013). *Global food insecurity and implications for Saudi Arabia. Energy, environment and resources (summary)* (pp. 1–10). London: Chatham House.

Crowley, J. (2010). *Commodity prices and inflation in the Middle East, North Africa, and Central Asia, IMF Working Paper WP/10/135.* Washington, DC: International Monetary Fund.

CSDI. (2015). Central department of statistics and information. Population in Saudi Arabia. Available at: https://www.stats.gov.sa/en/43; https://tradingeconomics.com/saudi-arabia/population

Darfaoui, E. M., & Al-Assiri, A. (2010). *Response to climate change in the Kingdom of Saudi Arabia.* Cairo: FAO-RNE.

Environment Agency Abu Dhabi. (2013). *Think, eat save reduce your food print!* https://envirocompetition.ead.ae/_layouts/15/IMAGES/EAD.EnvironmentalCompetition/_data/global/facts/FactSheet.pdf

FAO. (2011). *Global food losses and food waste – extent, causes and prevention.* Rome: FAO.

FAO. (2012). *Global initiative on food losses and waste reduction.* Rome: FAO.

FAO. (2013). *Food loss and waste: Definition and scope.* Unpublished.

FAO. (2015a). *UN Food and Agriculture Organization (FAO).* FAOSTAT. http://faostat.fao.org/default.aspx

FAO. (2015b). *New knowledge-sharing initiative to measure and reduce food loss and waste.* Available at: http://www.fao.org/news/story/en/item/357085/icode/

FAO. (2016, September 6). *Food security – Country brief- Saudi Arabia.* Global Information and Early Warning System (GIEWS).

Gulf News, May 06, 2014

IME. (2013). *Global food: Waste not want not* (pp. 1–35). Institution of Mechanical Engineers. Available online at http://www.wanttoknow.nl/wp-content/uploads/IMechE+Global+Food+Report.pdf

Intini, V., Breisinger, C., Brnovic, I., Byringiro, F., Ecker, O., Iversen, K. (2012). *Food security strategies in the GCC countries. Economic and Social Commission for Western Asia (ESCWA).* Strengthening Development Coordination among Regional Actors in ESCWA Region. E/ESCWA/ECRI/2012/Technical Paper. 2. This Paper was presented by Mr. Nadim Khouri (Deputy Executive Secretary, United Nations Economic and Social Commission for Western Asia UN-ESCWA) at the Emirates Center for Strategic Studies and Research 17th Annual Meeting "Watter & Food Security in the Arabian Gulf" – Abu Dhabi, United Arab Emirates on March 27, 2012.

KACST. (2009). *Utilizing high technical extrusion technology in food waste recycling as a potential animal feed.* Riyadh: KACST.

Khan, M. S. M., & Kaneesamkandi, Z. (2013). Biodegradable waste to biogas: Renewable energy option for the Kingdom of Saudi Arabia. *International Journal of Innovation and Applied Studies, 4*(1), 101–113 Available at: http://www.ijias.issrjournals.org/abstract.php?article=IJIAS-13-142-18

Lipinski, B., et al. (2013). *Reducing food loss and waste, Working Paper, Installment 2 of Creating a Sustainable Food Future.* Washington, DC: World Resources Institute Available online at http://www.worldresourcesreport.org

Myers, D. (2015, November 17). *10 mind-blowing facts about food waste in America.* Available at: http://www.thedailymeal.com/10-mind-blowing-facts-about-food-waste-america

Nitivattananon, V., et al. (2012). Renewable energy possibilities for Thailand's green markets. *Renewable and Sustainable Energy Reviews, 16*, 5423–5429.

Pradhan, S. (2010). *Gulf-South Asia economics relations; realities and prospects.* Centre for economic policy research. Available at: http://www.globaltradealert.org/sites/default/files/GTA4.pdf. Accessed 10 Jan 2014.

Psomopoulos, C. S., Bourka, A., & Themelis, N. J. (2009). Waste-to-energy: A review of the status and benefits in USA. *Waste Management, 29*, 1718–1724.

Sadik, A. K. (2014). The state of food security and agricultural resources. Chapter 1. In A. Sadik, M. El-Solh, & N. Saab (Eds.), *Arab environment: Food security*. Annual Report of the Arab Forum for Environment and Development, 2014. Beirut: Technical Publications Available at: http://www.afedonline.org/Report2014/E/p12-43%20chp1eng.pdf

Saudi Gazette. (2014). Saudis throw away 440 tons of rice annually. *Saudi Gazette*. Friday, 4 July 2014. Available at: http://english.alarabiya.net/en/business/economy/2014/07/04/Saudis-throw-away-440-tons-of-rice-annually.html

Spackman, N. (2015). *Saudi Arabia's food security double whammy*. Available at: http://www.twovisionspermaculture.com/saudi-arabias-food-security-double-whammy/

Toumi, H. (2014). Man installs 'charity fridge' outside his house. *Gulf News – Saudi Arabia*. May 6, 2014. Available at: http://gulfnews.com/news/gulf/saudi-arabia/man-installs-charity-fridge-outside-his-house-1.1328931

United States Development Agency (USDA). (2014). *USDA foreign agricultural service*. Global Agricultural Information Network GAIN Report Number: SA1417 12/15/2014.

Woertz, E. (2010). *The Gulf food import dependence and trade restrictions of agro exporters in 2008*. Available at: http://www.globaltradealert.org/sites/default/files/GTA4.pdf. Accessed 10 Jan 2014.

World Bank. (2011a, May). *MENA facing challenges and opportunities. Middle East and North Africa Region Regional Economic Update*.

World Bank. (2011b). *Global economic prospects*.

World Bank. (2011c). *Regional economic update: MENA facing challenges and opportunities. Middle East and North Africa Region*.

World Bank. (2015). Country profile–Saudi Arabia. Available at: http://databank.worldbank.org/data/views/reports/reportwidget.aspx?Report_Name=CountryProfile&Id=b450fd57&tbar=y&dd=y&inf=n&zm=n&country=SAU

WRAP. (2011). *New estimates for household food and drink waste in the UK*.

WTO. (2016, February 29). *rade policy review body on the Kingdom of Saudi Arabia, Report # WT/TPR/G/333*. The Kingdom of Saudi Arabia: World Trade Organization. Available at: https://www.wto.org/english/tratop_e/tpr_e/g333_e.pdf

Dr. Khodran H. Al-Zahrani is the Professor at the Department of Agricultural Extension and Rural Society, College of Food and Agriculture Sciences, King Saud University, Riyadh, Saudi Arabia. He did his B. Sc and M.Sc College of Food and Agriculture Sciences, King Saud University and Ph.D. from the Ohio State University, USA. He has conducted his research on the major issues like water and climate change faced by the Kingdom. He has done several studies to explore the shortcomings and challenges faced by the National Extension Service of Kingdom and did produce recommendations for improving the Extension System and enhancing the efficiency of the Extension Staff. His professional interests include: sustainable use of water, Conservation of natural resources, combating food waste and innovations in Extension Education. He is an author of 2 books and has published extensively in his areas of interests. He has represented the Kingdom of Saudi Arabia at many international fora and has delivered many scholarly talks.

Dr. Mirza Barjees Baig is working as a Professor of Agricultural Extension and Rural Development at the King Saud University, Riyadh, Saudi Arabia. He has received his education in both social and natural sciences from USA. He completed his Ph.D. in Extension Education for Natural Resource Management from the University of Idaho, Moscow, Idaho, USA. He earned his MS degree in International Agricultural Extension in 1992 from the Utah State University, Logan, Utah, USA and was placed on the "Roll of Honour". During his doctoral program, he was honored with "1995 Outstanding Graduate Student Award". Dr. Baig has published extensively in the national and international journals. He has also presented extension education and natural resource management extensively at various international conferences. Particularly issues like degradation of natural resources, deteriorating environment and their relationship with society/community are his areas of interest. He has attempted to develop strategies for conserving natural resources, promoting environment and developing sustainable communities through rural development programs. Dr. Baig started his scientific career in 1983 as a researcher at the Pakistan Agricultural Research Council, Islamabad, Pakistan. He has been associated with the University of Guelph, Ontario, Canada as the Special Graduate Faculty in the School of Environmental Design and Rural Planning from 2000-2005. He served as a Foreign Professor at the Allama Iqbal Open University (AIOU) through Higher Education Commission of Pakistan, from 2005-2009. Dr. Baig the member of IUCN – Commission on Environmental, Economic and Social Policy (CEESP). He is also the member of the Assessment Committee of the Intergovernmental Education Organization, United Nations, EDU Administrative Office, Brussels, Belgium. He serves on the editorial Boards of many International Journals Dr. Baig is the member of many national and international professional organizations.

Prof. Dr. Gary S. Straquadine serves as the Interim Chancellor, Vice Chancellor for Academic Programs and Vice Provost at the Utah State University – Eastern, USA. He did Ph.D. from the Ohio State University, USA. Presently he also leads the applied sciences division of the USU-Eastern campus. He is responsible for faculty development and evaluation, program enhancement, and accreditation. In addition to his heavy administrative assignments, he manages to find time to teach some undergraduate and graduate courses and supervise graduate student research. Being an extension educator, he has passion for the economic development of the community through education and has also successfully developed significant relations with agricultural leadership in the private and public sector. He has also served as the Chair, Agricultural Comm, Educ, and Leadership, at The Ohio State University, USA. Before accepting the present position as the Vice Provost, he served on many positions as the Department Head; Associate Dean; Dean and Executive Director, USU-Tooele Regional Camp and the Vice Provost (Academic). His professional interests include extension education, sustainable agriculture, food security, statistics in education, community development, motivation of youth and outreach educational programs. He has also helped several under developing countries improving their agriculture and educational programs.

Chapter 5
Sustainable Agriculture and Food Security in Egypt: Implications for Innovations in Agricultural Extension

Mirza Barjees Baig, Gary S. Straquadine, Ajmal Mahmood Qureshi, Asaf Hajiyev, and Aymn F. Abou Hadid

5.1 Introduction

Situated at the northeast corner of the continent of Africa and covering total land area of about one million square km, Egypt sustains about 94 million of population (Worldometers 2017). It occupies the most prominent and prestigious position in the Arab World due to its sound and productive agriculture. It is the main source of food and significant raw materials for industrial development. With only 3% arable land, agriculture is the most critical component of the Egyptian economy, providing livelihoods for 55% of the population (IFAD 2012), directly employing about 30% of the labor force and provides employment opportunties to 45% of all women in the workforce (USAID 2017). It also contributes 17–19% towards export earnings (CAPMAS 2016). Agriculture-related industries such as input supplies, processing and marketing of agri products account for a further 20% of GDP (IFAD 2012).

M. B. Baig (✉)
Department of Agricultural Extension and Rural Society, College of Food and Agriculture Sciences, King Saud University, Riyadh, Kingdom of Saudi Arabia
e-mail: mbbaig@ksu.edu.sa

G. S. Straquadine
Utah State University – Eastern, Price, UT, USA
e-mail: gary.straquadine@usu.edu

A. M. Qureshi
Faculty of Arts and Sciences, Harvard University, Cambridge, MA, USA

A. Hajiyev
Parliamentary Assembly of Black Sea Economic Cooperation (PABSEC), Academician of Azerbaijan National Academy of Sciences, Baku, Azerbaijan

A. F. Abou Hadid
Arid Land Agricultural Studies and Research Institute (ALARI), Faculty of Agriculture, Ain Shams University (ASU), Cairo, Egypt
e-mail: ayman_abouhadeed@agr.asu.edu.eg

M. Behnassi et al. (eds.), *Climate Change, Food Security and Natural Resource Management*, https://doi.org/10.1007/978-3-319-97091-2_5

Despite its significant contribution, for the last two decades, its contributions to GDP have declined and has not been able to maintain its production momentum. Levels due to the issues and constraints like water shortages, poverty, climate change, malnutrition food insecurity, change in rural fabric and ineffective extension service. Population and poverty in the country are increasing and more people have experienced food insecure. In an estimate about 17% of the population (13.7 million) faces food insecurity (IFPRI-WFP 2013). According to WB statistics in 2015, the the share of agriculture towards GDP was 3.92% at the global level, whereas in Egypt it was esimated around 11.2%. In addition, about 70 million Egyptians (out of a total population of 92 million) avail the benefits of food subsidy system by using smart cards. The citizens who used be well-off in the past are now struggling to suvive in the collapsing economy of the country. Food Banks are feeding the Middle-class families, thay have now fallen into the poverty (FAO 2017). Today agriculture in Egypt faces many new and complex challenges. Newly emerging changes make agriculture quite competitive, comparative and demanding. In addition, agricultural development in Egypt is gradually becoming more driven by commercialization, urbanization, and markets rather than by production of agricultural commodities. This is increasingly dominating the national agenda.

Food requirements in the country are increasing due to a growth in population, putting more pressures on present production systems. Good agricultural lands are being swallowed by the construction of new houses, settlements and water resources are under severe stress. Vella (2012) maintains that shrinking cultivated area and limited water resources has increased the risk of food insecurity in Egypt. The area under crops is decreasing everyday due to extreme land deterioration, loss of vegetation and soil moisture. Desertification is prevalent on both sides of the Nile River, a leading agricultural zone, as the Eastern and Western Deserts intrude inwards. Additionally, ongoing rise in temperatures and increasing evaporation rates are presenting different challenges for food production (Vella 2012).

In this situation, it seems imperative to critically analyse the changing scienario, identify the constraints, and outline the strategies to addrees the issues faced by the agricultural sector. Agricultural extension has great potential to create awareness on innovations, help address the issues faced by the farmers in order to realize sustainability and ensure food security. The chapter provides information on the sustainable use of natural and agricultural resources of the country. It also discusses that improvement measures adopted to increase land and water productivity would in turn enhance crop yields and help achieving achieving food security, elevating living standards and reducing the poverty among the rural inhabitants.

5.2 An Overview of Agriculture Sector in Egypt

Agricultural growth is not only important to enhance national incomes and food security, it is also vital to create employment opportunities and combat poverty in Egypt. The country has also been able to export its high value horticulture and livestock products, herbs and medicinal plants due to its favorable agro-climatic conditions, counter-seasonal production capabilities, and physical proximity to important markets. There is scope and potential for agricultural growth and food security can be enhanced through the sub-sectors of agriculture. In addition, the non-farm sector can absorb landless labor and youth, create opportunities for women and help reducing poverty and maintaining gender equity (IFAD 2012).

For the last many decades, agriculture has been able to support and strengthen the economy. Although industrialization received greater attention in the recent years, the country continues to depend largely on agricultural production. The agricultural sector absorbs about 31% of labor and it contributes almost 15% towards the GDP (El-Din 2007; Shalaby et al. 2011). The country is making concerted efforts to increase its agricultural production by adding more areas through reclamation of new lands. This practice would not only increase cultivated areas but would also attract an influx of population, meet the food requirements of growing population, and create jobs for young graduates, especially when the per capita cultivated area has decreased. In addition, the country plans to enhance vertical expansion by raising the average production per acre to narrow a constantly increasing food gap. Since farmers are either not buying or are unable to buy fertilizers, most of the production is organic and is directed to export markets. The country's main crops include cotton, wheat, rice, sugarcane, beet, fodders, clover, vegetables, peanut, sesame, sunflower, lentils, beans and onion, and fruits such as citrus and dates (Ministry of Agriculture 2011). The country produces many agricultural crops, appreciable quantities of vegetables and fruits such as tomatoes, melons, citrus, guava, and date palm (Ministry of Agriculture 2011; Shalaby et al. 2011) both for the consumption of its citizens and for export as well. Alfalfa, onions, beans, wheat, barley, sugarcane are the most important winter crops, whereas rice, sugar beet, cotton and maize are the most important summer crops. Though the country produces a variety of agricultural crops, vegetables and fruits yet its 40% food requirements are met through the imports of agricultural commodities. Cereal yields, area harvested and total productions were on the increase until 2006. After 2012, they seem stagnant as depicted in Fig. 5.1. Cereal production in Egypt and their imports from 2011 to 2016 are shown in Fig. 5.2. It clearly shows that Egypt still heavily depends on imports to meet the cereal requirements of the citizens.

While yields in the "old lands" are among the highest in the world for several staple crops, such as wheat, rice and sugar beet, yield improvements have slowed down markedly in recent years. However, yields for non-traditional high-value crops are still much below than their potential due to several reasons like traditional non-scientific farming practices, small and fragmented land-holdings etc. Experts explain the slow growth to slow adoption and reduced risk taking. An account of important agricultural crops is presented in the following sub-sections:

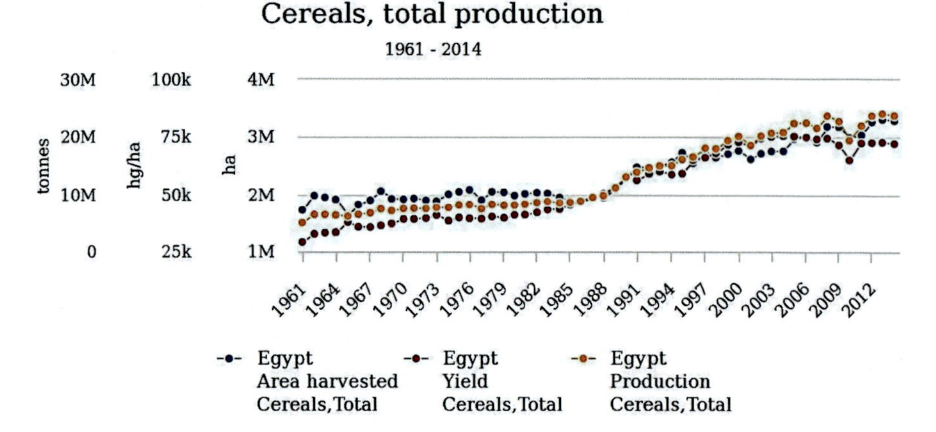

Fig. 5.1 Cereal production, area harvested and total production from 1961 to 2015 (Source: FAOSTAT Jan 03, 2017)

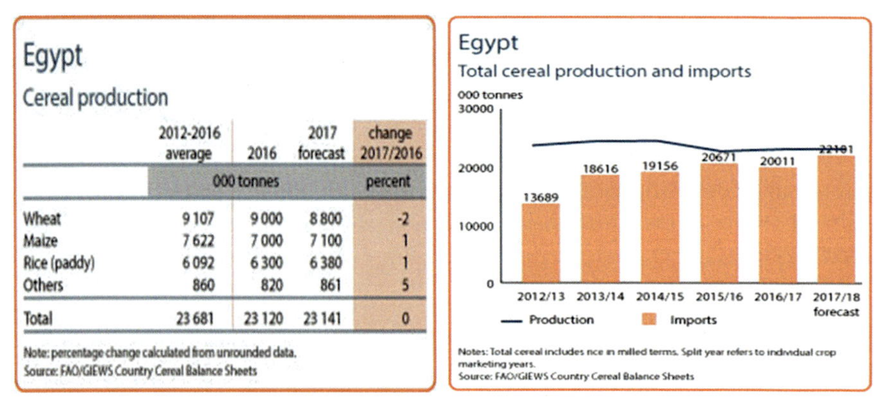

Egypt

Cereal production

	2012-2016 average	2016	2017 forecast	change 2017/2016
	000 tonnes			percent
Wheat	9 107	9 000	8 800	-2
Maize	7 622	7 000	7 100	1
Rice (paddy)	6 092	6 300	6 380	1
Others	860	820	861	5
Total	23 681	23 120	23 141	0

Note: percentage change calculated from unrounded data.
Source: FAO/GIEWS Country Cereal Balance Sheets

Fig. 5.2 Cereal production in Egypt and their imports from 2011 to 2017. (Source: http://www.fao.org/giews/countrybrief/country.jsp?code=EGY&lang=en)

5.2.1 Wheat

Egypt is the largest wheat importer in the world (Al-Ahram Weekly 2016). Among the strategic agricultural crops, wheat ranks number one for its highest consumption (Global Arab Network 2009; Shalaby et al. 2011) in the country. Many farmers chose to plant wheat due to the price hike. The cultivated area increased by 7.5%, spreading over about 2.9 million feddans. However, area under wheat cultivation within the country also decreases whenever the prices for wheat drop in international market. Al-Ahram Weekly (2016) witnessed low productions because it is not grown on large-scale farms. Egypt produces about 4.5 million tons of wheat a year whereas its annual consumption is higher than 12 million tons. In the scenario, the shortfall is met through imports. Though the government buys wheat from the farmers even at the prices much higher than the international market (Shalaby et al. 2011) yet, farmers consider these prices are not high enough to cover their production costs. Being discouraged, they start growing other crops that could generate more profits, forcing the country to rely on imports (Global Arab Network 2009; Shalaby et al. 2011).

5.2.2 Rice

Ranked number 2, rice is another major staple food in Egypt and the most profitable export crop of summer season. The country in 2008 reported the total production 6.9 million tons. The cultivation of rice is increasing and it could be due to a hike on international markets. Its cultivation increases pressure on water resources since rice cultivation is exceptionally water intensive (Global Arab Network 2009; Shalaby et al. 2011). For enhancing the incomes of farmers, Abdelhakam (2005) suggested to practice aquaculture in rice fields as it has proved a successful practice in Nasr Lake and other areas of the country. At occasions, due to water management policy, the government has put ban on rice cultivation but policy proved ineffective. Therefore, the government placed an export ban in order to reduce its price in the domestic market. Rice export enables farmers to realize high prices for their crop and higher profits as compared to other traditional summer crops (Global Arab Network 2009; Shalaby et al. 2011).

5.2.3 Cotton Enjoys an Excellent Reputation

In the past Egyptian cotton enjoyed a very good reputation and the prestigious position worldwide due to its high quality. Although the cotton industry is well developed and quite productive, it still faces many constraints. Areas under cotton crop are decreasing and consequently the production has been declined. Farmers are increasingly losing interest in cotton production because of high input costs and it is also affected due to the international linkages with the textile industry. Realizing the gravity of the problem, the government has launched the farmer friendly policy to enhance cotton cultivation and to support the textile industry, and has started buying yarn produced from the Egyptian cotton at the higher prices (Global Arab Network 2009; Shalaby et al. 2011).

5.2.4 *Sugar Crops*

With its production of 16.8 million tons, Sugarcane is another strategic crop. Egypt produces 1.5 million tons of sugar, and the crop takes up 47.7% of the total farmed area nationwide at 325,700 feddans. Per capita consumption of sugar in Egypt is quite high and an increase in population is further enhancing its demand. Though sugar beet production has been contributing to sugar industry yet recently area under its cultivation is declining resulting lower sugar production. The government supports the sugarcane farmers by buying their crop and in order to reduce dependency on imports and imposing duties on imports. Realizing the importance of the crop, the country has decided to invest in the sector in the years to come (Global Arab Network 2009; Shalaby et al. 2011).

An overview of agricultural sector presented here clearly indicates that most of the strategic crops are under stress and the agricultural sector needs to bring certain changes and make adjustments in the production systems. The government needs to formulate and implement farmers' friendly policies; and the concepts, principles and practices of sustainable agriculture to the farmers at their farms through a suitable strategy (Ministry of Agriculture 2008; Shalaby et al. 2011).

5.3 Challenges and Constraints for a Sustainable Agriculture

Egypt faces a wide range of constraints towards practicing sustainable agriculture; however the shortage of natural resources such as land, water and fertile soil are of prime importance (Seada et al. 2016).

5.3.1 *Heavy Migration of Rural Population*

About 57% of the population lives in rural areas and their livelihood depends upon the agricultural sector (FAOSTAT 2017), as revealed in Fig. 5.3. Though it is important to maintain a fair balance between rural and urban population, the division of rural and urban population from 1990 to 2015 is increasing, as shown in Fig. 5.4.

Due to continued migration from rural areas, the country faces a severe shortage of trained, skilled and qualified labor in the rural areas making the food producers/farmers the net consumers. However, an increased migration from rural areas to urban areas could consequently result many social issues like urban violence (Shalaby et al. 2010). According to El-Laithy (2007) some 29% of the labor force is engaged in agriculture and 43% of the labor force work in the rural areas. About 58% rural population is engaged in agricultural activities. To make rural areas more attractive and to halt migration from the rural areas,

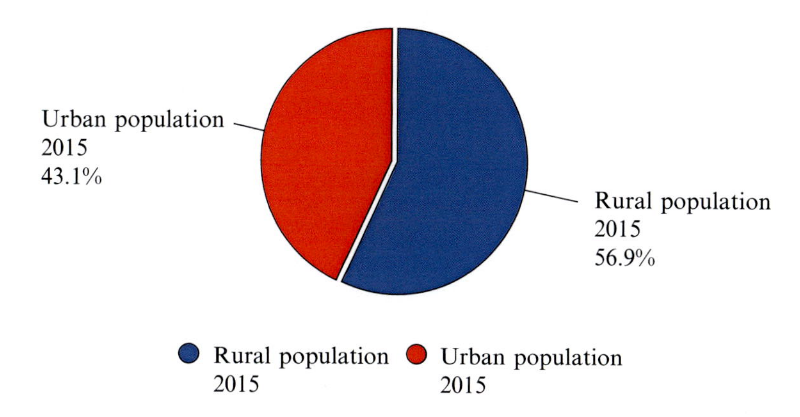

Fig. 5.3 Rural and Urban population in 2015. (Source: http://www.fao.org/faostat/en/#country/59). (Source: FAOSTAT (Jan 03, 2017))

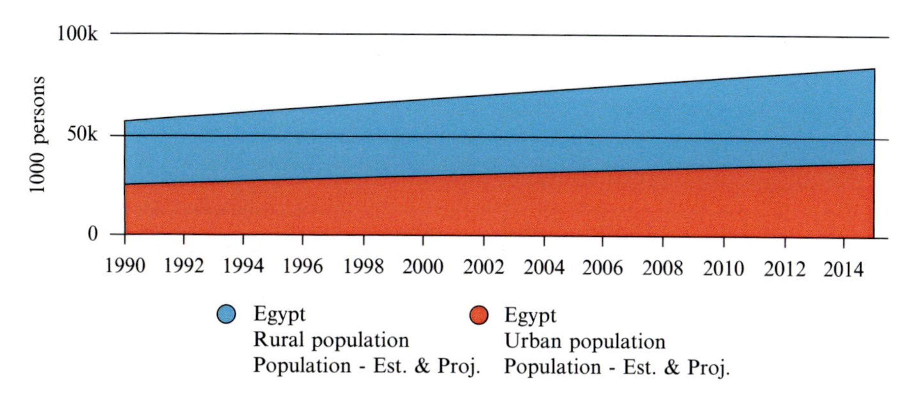

Fig. 5.4 Rural and Urban population 1990–2015. (Source: http://www.fao.org/faostat/en/#country/59). (Source: FAOSTAT (Jan 03, 2017))

there is a need to focus on producing more valuable crops that could generate a livable wage in the suburbs and reduce and prevent urbanization (Shalaby et al. 2010).

5.3.2 *Land Degradation*

Out of total land area of 995,560 Km^2 (384,388 mi^2), only 3% of the total land area, roughly about 3.61 million ha (8.6 million feddans) is brought under cultivation. Lands covering an area of about an area of 2.73 million ha (6.5 million feddan) in the Nile Valley and Delta (old lands) and 0.88 million ha (2.1 million feddans) on the fringes of these regions in new lands reclaimed from the desert are brought under agricultural crops (FAO 2011; ICARDA 2011). Rainfed agriculture is

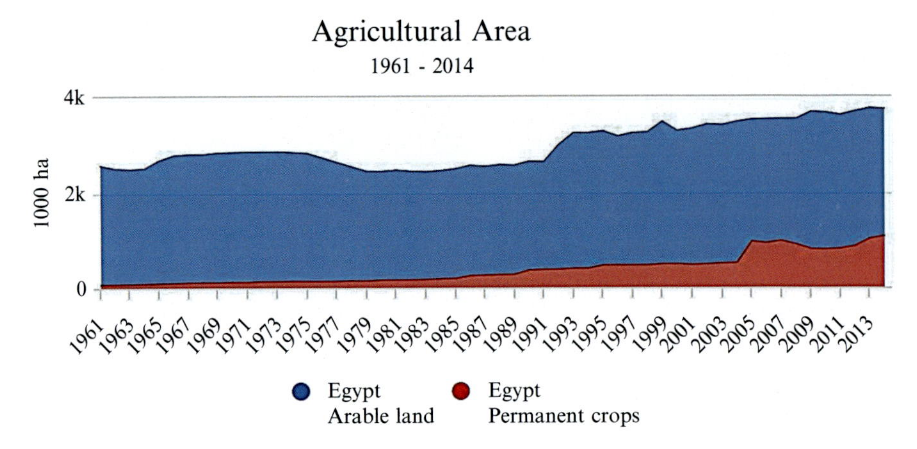

Fig. 5.5 Area under agriculture and arable lands. (Source: FAOSTAT 2017. Available at: http://www.fao.org/faostat/en/#country/59). (Source: FAOSTAT (Jan 03, 2017))

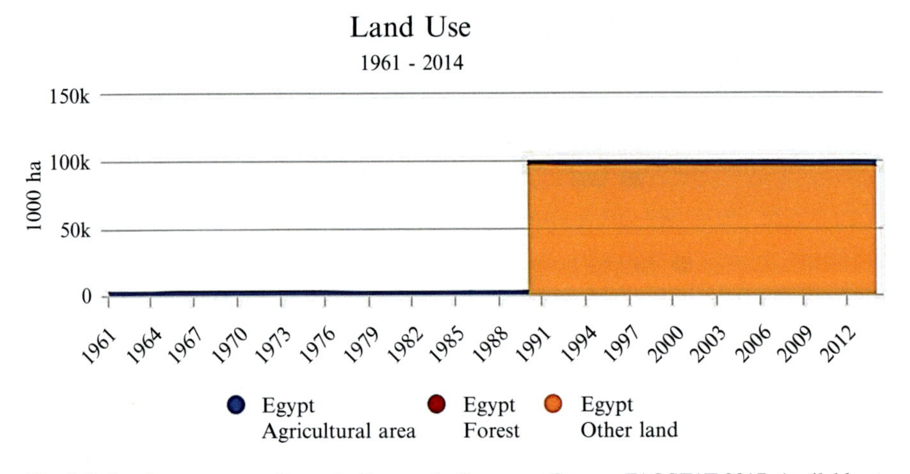

Fig. 5.6 Land use – area under agriculture and other uses. (Source: FAOSTAT 2017. Available at: http://www.fao.org/faostat/en/#country/59). (Source: FAOSTAT (Jan 03, 2017))

practiced on an area of 2.8 mha (6.7 million feddan). Area under agriculture and arable lands are shown in the Figs. 5.5 and 5.6. There is also an extreme shortage of fertile lands and increasing fragmentation of landholdings limit the productive areas. Also, high quality arable land is being lost each year due to desertification, soil deterioration, and increased salinity (Goldfond 2015). An area of about 10,000 ha (22,000 feddan) is being lost due to urbanization annually (MWRI 2005) and the cultivated lands continuously face declining fertility levels (Seada et al. 2016). About 35% agricultural lands are affected by the salinity (ICARDA 2011). Based on these facts, Worldometers (2017) reports that Egypt has the poorest land-per-person ratios in the world with the population density of 96 per Km^2 (248 people per mi^2).

5.3.3 *Small Farms and Fragmented Landholdings*

The agriculture sector in Egypt is dominated by small farms and fragmented landholdings. Farms are predominantly small, some 58% of all farms are less than one feddan and about 81% farms with an area less than 3 feddan cover total of or 38% of the country's entire cultivated area. Only 23% farmers own farms with an area between 1 and 3 feddan. Farmers with small landholdings are unable to maintain an average family of five members. Obviously, farmers with less than one feddan, agricultural land could only generate small incomes, preventing small farms to have access to credit and to good marketing mechanisms further exacerbating this situation (IFAD 2012). Also, it is not possible to practice mechanized farming or farm operations on mass-scale on small and fragmented farms. Small farmers are more worried about feeding their families, therefore they are least interested on internationally recognized standards. They may employ old traditional practices, overuse or misuse agricultural chemicals, adopt outdated technologies; and use tools with low efficiency for land preparation, irrigation, and harvesting. Such practices results increased production costs, reduced yields, low soil fertility and limited marketing opportunities. Being susbsistence farmers, they may have little interest in having the facilities like cold storage infrastructure, transportation systems and market information.

5.3.4 *Water Resources Under Stress*

Egypt – the home of the River Nile is an arid country with no significant rainfall except in a narrow strip along the north coast. The Nile River is the main and almost exclusive source of surface water for Egypt, providing 77% of agriculture's annual water supply (MALR 2009). Scarcity of water is a key constraint on agricultural growth in the country. At present, the average consumption of water for agriculture is about 58 billion m^3/year (ICARDA 2011). Irrigated agriculture of the country primarily depends on Nile River water – although groundwater resources may have minor contributions (ICARDA 2011). Irrigation in the new lands also depends on groundwater and wells (Shalaby et al. 2011). Despite the presence of the Nile River, the country faces an acute shortage of irrigation water. The country is unable to meet its water demands due to the increasing population and the related industrial and agricultural activities (IFAD 2007). Agriculture uses the highest volume of water (86.4%) followed by the domestic consumers (7.8%) as reported by FAO 2011 (Fig. 5.7). Only rice crop consumes more than 10 billion cubic meters of water annually, or more than one-sixth of Egypt's share of Nile water (El-Sayed 2017).

On the other hand, water efficiency is low due to high water losses. Water conveyance efficiency is estimated at 70%, and the mean efficiency of field irrigation systems is estimated at only 50% as reported by the Ministry of Agriculture and Land Reclamation (MALR 2009). Water distribution and management systems

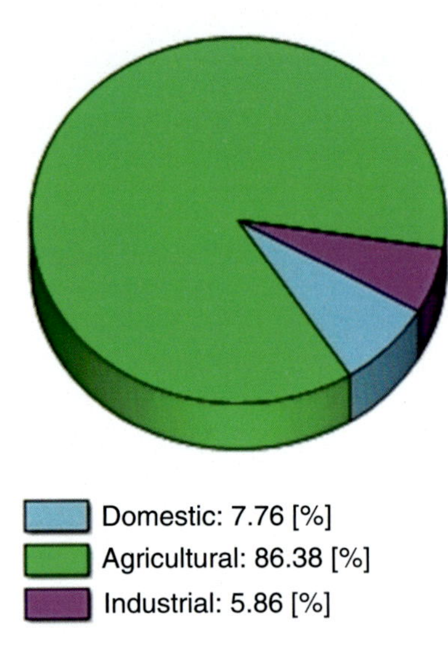

Fig. 5.7 Water use (percentage) by various sectors. (Source: AquaSTAT, FAO of the UN, Accessed on September 29, 2011. http://www.fao.org/nr/water/aquastat/main/index.stm)

have been partially and ineffectively decentralized. The situation is not likely to improve as climate change and population growth combine to raise the risks of inadequate water supplies and conflict over the available supplies – and further pressure is expected from the exploration of Nile River resources by other members of the Nile Basin Commission.

5.3.5 *Climate Change and Its Adverse Effects on Agriculture*

Most of agriculture in Egypt is being practiced in the Nile Delta. Due to its low elevation, Nile delta region is very sensitive and vulnerable to climate change and related Sea Level Rise (SLR). Egypt is listed among the top five countries to be affected with a 1 m sea level rise due to global warming. Egypt ranks fifth in the world that may have an impact of the climate on its urban areas. The production of livestock and the productive potential of many agricultural zones could be reduced due to climate change. A rise in sea levels would have a negative effect on coastal areas, tourism in the Nile Delta region. The natural resources such as coastal zone, water resources, water quality, agricultural land, livestock and fisheries could be vulnerable. In addition to have its impact on GDP, the country could also face environmental crises such as shore erosion, salt-water intrusion, and soil salinity (Batisha 2012). The marginal agricultural areas would be negatively affected and resultantly with increased desertification. High temperatures would increase evaporation and water consumption resulting further stress on the limited water resources of the country. Presently the country faces an annual shortfall of

7 billion cubic meters of water; however the situation could be worse if the rising levels of the Mediterranean Sea cause flood the northern lakes. This in turn would choke off fresh water sources and destroy its fish hatcheries (Batisha 2012; Shaltout et al. 2015).

Scientists believe that the expected rise in temperature would lead to up to 20% decreases in productivity for wheat, barley and maize by 2050 (MALR 2009). Unusual high temperatures and the resultant heat cause dehydration and many deaths, adversely affecting the yields of agricultural crops as well. Scientists believe that a Sea Level Rise (SLR) of 0.5 m in Alexandria might affect 43 port cities in the region, displace more than two million citizens, and cause the financial losses of at least USD 35 billion. It is expected that sea levels in the Nile Delta could rise between 50 and 200 cm by the year 2100. Egypt is also one of the highly vulnerable countries that could experience a significant impact as the Sea Level Rise (SLR) due to climate change. The scientists warn that a 1 m SLR might impact six million people, making Egypt roughly to lose 12–15% of agricultural land in the Nile Delta region. Parts of Alexandria, Behaira, Port Said, Damietta and the Suez governorate are highly vulnerable areas. If the country experiences SLR of 0.5 m and no improvement measures are employed or a business-as-usual scenario prevails, the agriculture sector would be severely and negatively affected (Batisha 2012; Shaltout et al. 2015).

Several initiatives were undertaken in this context, including the Climate Change Risk Management in the country. However, in the situation, Batisha (2012) suggests to formulate policies that could help minimizing the risks through actionable measures. The author considers adaptation to climate change as an emerging prime issue from the perspectives of water resources development, food production, and rural population stabilization. He believes that the country needs to adapt a vulnerability index to identify the most vulnerable regions. He also proposes an adaptation strategy that is based on multi-criterion analysis in order to make the implementable decisions (Table 5.1).

5.4 Strategies to Realize Sustainable Agriculture and Ensure Food Security

Keeping abreast of modern and advanced techniques that support the economic efficiency of agricultural production would result in sustainable agriculture. Following sections discuss measures that if adopted and made the part of farming business, tangible results can be achieved.

Table 5.1 Threats and challenges faced by agricultural sector in Egypt

Agriculture
Due to fragmentation of lands and small landholdings; and adoption of the old traditional cultivation methods yields of many agricultural are low;
High value crops have not received due attention; farmers are unable to maintain a fair balance between the cultivation of high value crops and strategic crops (cotton-wheat-rice-maize);
Farmers don't use new and more efficient irrigation technologies to improve the water-use-efficiency and crop yields;
The government needs to make adjustments in agricultural and rural development policies;
Exporters do not buy about 80% of the crops, produced in the country for not meeting the standards set by the export markets, forcing the farmers to sale them in the local markets;
Natural resources environmental problems
Deterioration of natural resources like lands, water and environment result lower production;
The natural resource constraints include a fragile land base, declining soil fertility, limited water resources, and frequent climatic shocks;
The natural resources base is deteriorating due to burgeoning population pressure, inappropriate agricultural practices, overstocking, deforestation and consequent upon the soil erosion, destruction of habitats for wild fauna;
Biodiversity is vanishing since environment is under severe stress;
Deterioration of land resources due to natural and anthropogenic activities;
Environmental and natural resources degradation; contamination of irrigation water with pollutants and they in turn effect the land qualities adversely;
Land issues
Since an over-whelming majority are the subsistence farmers and about 80% of farmers' own less than or equal to 5 feddan, therefore are not interested in innovative agriculture;
Civil construction and sprawling is swallowing good fertile agricultural lands;
Inadequate access to productive resources, particularly agricultural land;
Land degradation, particularly in the rain-fed and irrigated areas;
Land issues like salinization on irrigated areas; water logging and wind and water erosion hamper crop yields;
Bringing new lands under cultivation through reclamation is difficult and expansive;
Land fragmentation limits the mechanized farming;
The water shortage and drought
Agriculture is the greatest consumer of water in the country, significant loss of water during the irrigation of crops;
In Egypt, there are some areas that receive enough water to practice modern agriculture and some areas suffer from drought and water shortage;
Farmers need to shift to crops that require less water such as wheat and cotton;
Lack of farming systems that could cope with drought and water shortage;
Lack of enough farming technologies to adjust the cropping systems/patterns of various zones according to the availability and supply of water;
Poverty
Some 10.7 million poor live in Egypt. Poverty is predominantly a rural phenomenon as 70% poor live in rural areas;
The poverty rate in rural area is higher than that in urban areas;

(continued)

Table 5.1 (continued)

Poverty varies significantly among rural and urban areas and from region to region. Rural poor people typically include: tenant farmers, small scale farmers, landless laborers, unemployed youth and women;
Need to address issues, evolve farming practices and technologies that could increase agricultural production to combat poverty;
Labor force
Productive labor-force is vital for agriculture and it is not available because the sector can no longer gainfully employ more youth;
Due to increased migration from rural areas, the food producers/farmers themselves have become net consumers.
Lack of trained, skilled and qualified labor;
Increased migration from rural areas to urban satellites, consequently social disruption and increased urban violence are being witnessed;
Employment opportunities
On-farm (agribusiness) and non-farm entrepreneurship/self-employment opportunities are limited and becoming more scarce day-by-day due to lack of intensification and diversification of agriculture sector;
Agricultural activities have declined in the rural areas and are unable to absorb skilled workers, forcing them to move to the urban areas;
Due to non-availability of safety-nets, social dislocations and distortions are widespread;
Women farmers, children and youth
Women in poor families and in rural areas, account for 70% of the total poor in Egypt. They are forced to live inferior lives, facing double discrimination due to poverty and being women;
Gender inequality prevails in the social and economic systems even today being the male dominated society;
Women – the active labor force in the rural areas experience poor nutrition, poor health, high birth-rates and are viewed as unacknowledged labor due to their illiteracy;
Women have limited economic options and less access to social services;
Due to rural poverty, the children leave their homes to seek employment and in cases become the victim to child labor;
Insufficient facilities for the youth to engage themselves in the healthy hobbies.
Inadequate support services
Lack of support system for the resource-poor farmers to get farm inputs like seeds, fertilizers, credit etc.
Farmers are using old primitive cultivation practices and low-level farming technologies;
Small and subsistence landowners have Inadequate access to basic farm services such extension services and technology transfer;
Poor organization and empowerment of extension services at the national level;
Poor policy framework and Institutional constraints
Unequal land distribution and insecurity of land tenure, low public sector investment in physical and social infrastructure in rural areas;
Farmers do not consider agriculture and rural development policies as friendly;
Poor marketing and buying policies and low market prices for some strategic crops;
Lack of organizational and institutional coordination;

(continued)

Table 5.1 (continued)

Institutions and organizations may have conflicts on maintaining gender balance and eradicating poverty etc.;
Few grass-roots and civil society organizations are engaged for the betterment of rural people;
Lack of information on the standards set for exporting agricultural products and commodities and needs of exportable markets;
Need to make adjustments in agricultural, environmental and economic policies; and
Low private sector participation in the developmental plans, policies and projects;

Source: Modified after Shalaby and Baig (2010a); Shalaby et al. (2012)

5.4.1 Improvement Strategies

In order to enhance the shelf-life of the perishable commdities, it is important to pack them appropriately at the pack-houses and ship them in the refrigrated trucks to markets, airports, and warehouses. Such arrangements and steps can bring the higher prices for commodities, crops, produce and products to the farmers and traders.

The countries intending to import require disease free premium quality agricultural commodities from Egypt. Therefore Egypt needs to adopt the screening and grading process to select the commodities with the appropriate sizes and shapes. In addition, efforts should be made to develop new varieties, resistant to diseases and viral attacks. For example, tomatoes should resistant to yellow leaf curl virus, potato varieties should be resistant to brown rot, and fine green bean varieties capable of resulting higher yields. Innovative methods for screening, grading and developing new varieties that address market demand would help increasing exports of horticultural produce (IFAD 2007).

In order to grow more and improve production, it is important for Egypt to enhance its water-use-efficiency. Drip irrigation systems are known to reduce water consumption by up to 50% and, at the same time, increase yields by 30–50%. Therefore these systems must be tested, modified and adopted to cater to the needs of small farmers owning 1/2–1 feddan for horticultural and other crops. In addition to their technical evaluation, they also require financial and economic analysis for their adoption by the smallholders. If they prove technically better and economically viable, then small farmers would upgrade their irrigation systems (IFAD 2007).

5.4.2 Transporting Products to Market

Transporting food commodities and products to the targeted markets remains a challenging task due to the lack of cold storage infrastructure in Upper Egypt. Farmers need to haul their products to the different cities and in temperatures often reaching above 4 °C. To improve the quality of shipments, farmers require small-scale pack houses closer to their farms so that their products are not perished during the travel. Refrigerated trucks should be made available to the qualified vegetable producer organizations to facilitate the transportation of products from farm to pack houses, and from pack houses to the markets, airports, and warehouses. In this way, farmers can maintain the freshness of their produce to receive higher prices for their crops, commoditities and products.

Potential measures and practices that can help realizing sustainable agriculture include:

Using of compost can help reclaiming lands and thus more areas can be brought under cultivation;
Rationalize irrigation water to reclaim and its application on the reclaimed lands;
Organic agricultural products receive higher prices and can help promoting exports;
Low use of chemical fertilizers and pesticides would result cleaner production and environment;
Establishing quality standards for agricultural products would help protecting the environment from pollution;
By employing modern information and communication techniques, Extensionists can reach the bigger segment of the society to make the agricultural sector sound and productive;
Develop marketing facilities, services and agricultural markets to accommodate the small farmers;
Appropriate pre- and post-harvest technologies help improving the quality of agricultural products and commodities;
Establish a special unit for managing agricultural risks by employing modern techniques to monitor, analyze and forecast natural, technical and marketing risks;
Link the farmers, particularly small farmers, with the markets, including the development of marketing systems and channels;
Activate and strengthen the role of the government in exercising supervision on quality standards of both inputs and outputs, banning monopoly and adulteration, and consumer protection;
Strengthen the institutional and organizational mechanisms that support the linkages between local and external marketing;
Take all possible measures to reduce losses throughout the value chain;
Conduct more research horticultural crops due to their greater export potential.
Develop resistant varieties to pests; tolerant to environmental stress and more water use efficient;
Conduct more research in genetics and biotechnology to improve crop varieties with desirable traits;
Launch research field trials though the involvement of Egypt's national research institutions, including the Agricultural Research Center (ARC) and the National Water Research Center (NWRC); and
Make markets and marketing facilities strong as a weak agriculture supply system prevails in the country failing the market demand for high value crops, such as tomatoes and green beans.

Source: Handoussa (2010), MEAS (2011) and Lewis (2012)

5.5 Extension Service in Egypt

Egypt public agricultural extension system as a government service was initiated in 1953. As a ministry-based system, it undertakes its activities at both the ministry (national) and the governorates-districts-villages. At the ministry level, the Central Administration for Agricultural Extension Services (CAAES), one of the seven sectors of the Ministry of Agriculture and Land Reclamation (MALR) is the key agricultural extension organization at the national level, providing technical supervision to extension staff. Shalaby et al. 2010 believe that at present, agricultural extension service is primarily involved in transferring the knowledge, where farmers were viewed as the recipients not participants.

The public sector extension tends to serve the vast majority of small farmers while the private sector, (suppliers of inputs and other services) prefers to take care of the big land owners and corporate farmers. However, if the private and public sector undertake extension activities in partnership, the greater number of farmers could realize more benefits (USAID 2011).

The public extension system has not been sucsessful due to poor wages, lack of continuing education to remain technologically updated, overlapping of functions and competition between research and extension units and among ministerial institutions. The formal extension system has continued to degrade and extension personnel feel marginalized for not having access to the resources. Such limitataions do not enable them contribute much (MEAS 2011).

Universities are an important source of extension advice and guidance. They play an important role in evolving technologies and handing them down to the extension staff for the use of the end users through extension trainings of the extensionists. The colleges of agriculture are engaged in training students to be extension educators (either in the private or public sector). Universities have an essential role in continuing education of extension personnel. With their critical position in linking research to the farmers' needs through the field-based research programs (MEAS 2011).

5.5.1 Benefits of Private Extension

According to MEAS (2011) among all the service providers, the private sector performs better due to the avaialability of the resources, trained professionals and the capacity to implement extention programs in furnishing information to the farmers on agriculture from production to processing and handling. The role of government extension system in Egypt has significantly diminished and mostly it is becoming ineffective, especially for the small farmers who have now moved in agro-export products. However, outreach units of private sector companies have partially filled this gap. Exporters, seed companies, and suppliers have hired extension staff to work with farmers on the introduction of new varieties, cultivation practices, and inputs to address the particular issues. Extension agents hired by the private companies (buyers/traders, wholesalers, exporters or processors) are often extremely skilled, well qualified and competent. Based on their expertise, they successfully predict future yields, identify diseases, pests and adverse conditions; and provide accurate product-specific advice. Since these extension agents are often very active possessing all the qualities needed to be an effective extension agents, therefore they prove very useful for the the buyers/traders, wholesalers, exporters or processors. They accurately estimate the future yields, and identify the pests and diseases while providing an appropriate product-specific advice. With the only limitation, they have selfish sales goals, they do not provide impartial advice to the farmers about the costs and benefits of a particular innovation. Nevertheless, they clearly have a role in the emerging agricultural system, but cannot provide the solutions to all problems faced by Egyptian farming communities.

5.5.2 Problems Associated with Agricultural Extension in Egypt

The National Agricultural Extension Service (NAES) through its working system had been playing a very significant role in enhancing agricultural production in the country. At present, the replacement of old practices with the new modern scientific farming technologies poses the prime challenge for extension. Despite, its good performance on occasions, it is constrained by several issues and problems, limiting its efficiency and effectiveness. The prime issues include problems in Extension Services and organizations, difficulties in bringing behavioral change of the farmers to adopt modern farming technologies capable of enhancing crop yields. The extension workers lack proper education, technical skills and appropriate qualifications to undertake extension activities effectively and efficiently. In order to enhance the efficiency of the extension staff, it is extremly important to equip the extension workers with modern scietific and and technical knowledge and elevate their communication skills to enable them to help farmers elevating their farming capacities to enhance crop yields without harming the natural resources.

National Agricultural Extension Service (NAES) faces shortage of skilled staff, particularly not enough qualified extenionists are available to make impact in the remote areas (Abdelhakam 2005). Extension workers with no transport facilities, low incomes and uncomfortable working conditions are asked to work in the harsh environment. Such poor working conditions cerntainly lower the motivational level of many extension workers. Therefore, they desderve better facilities, suitable rewards, achievable targets and increased budgets especially in the remote rural areas. Shalaby et al. (2010) note that farmers are not provided with sufficient support services regarding the availability of farm inputs like seeds, fertilizers, credit and basic farm services such as the extension services and technology transfer. Farmers are provided with low-level technologies making not much difference to farming. NAES due to its poor organization in the country remains ineffective to make a significant impact on the crop production, farming and the community (Table 5.2).

Table 5.2 Major challenges and prime issues faced by extension service in Egypt

Extension organization experiences a wide communication gap between the top and lower extension staff;
NAES faces difficulties in bringing behavioral change among the farmers to adopt modern farming technologies;
Often extension workers lack essential modern technical knowledge and communication skills to do the extension wok effectively;
Extension workers are hired on low salaries; they are not provided with basic facilities needed to carry out extension work like transport, office equipment;
They lack motivation due to poor working conditions;
They do not feel pride in their work and most often, the change agents deliver extension messages without inculcating missionary spirit.
Extension personnel feel marginalized in the absence of the resources or system, preventing them to contribute and do their job with satisfaction.
Extension agents and staff have computers but no access to Information and Communication Technologies (ICT) and internet connectivity;
Linkages among the farmers, educators, researchers, extension agents, agribusinesses and farmers associations are weak and in cases absent;
Universities produce extension personnel bearing low the ability to meet the technical needs of their potential clients due to poor curriculum and weak courses, hampered by resource constraints, among other issues. These constraints impede both class-room and field based learning.
At the universities, professors primarily focus on class-room teaching and pay low attention on field research realted to extension – that in-turn limits the potential for organizing field based course work.

5.6 Strategies to Improve the Working of National Agricultural Extension Service (NAES)

5.6.1 Adopt Participatory Approach

Participatory approach has proved quite effective in evolving new innovative farming technologies through the participation of the famers and making them popularize among the farming community. NAES can train the local professionals to enable them assisting to the extension workers regarding the development of innovative agriculture through the participatory approaches (Shalaby et al. 2010, 2011). Participatory approach endorsed and advocated by many international organizations like the World Bank, Asian Development Bank, has been well received in both the developed and the underdeveloped countries due to its promising benefits and Egypt is not an exception. In addition, communication skills of extension workers to communicate technical information to the farmers are to be enhaced (Shalaby et al. 2010, 2011). Preferably, by mixing typical instructing techniques with the newly acquired participatory methods, extension workers can help the farmers effectively. A group of able professionals can provide assistance to train the extension workers on the implementation of participatory approach. The critical review of the projects on infrastructure launched with the help of the community by adopting participatory approach have ensured sustainability and created the sense of ownerships as noted by the project designers and managers (Soliman 2007). Extension department needs assistance in how to organize training courses for the extension workers on the implementation of participatory approaches.

5.6.2 *Capacity Building Programs*

Capacity building programs must be launched frequently and on the regular basis to improve their technical and communication for the extension agents, community leaders and progressive farmers. Extension workers are to be educated regularly on the innovations in agriculture. Capacity building programs are also needed for the private sector, all the groups intended to work with farmers like laborers, pack house managers and buyers. In the past, such initiatives have been able to increase both the quality and quantity of vegetable production in Upper Egypt. Universities must link the students with the commercial farms and factories for internships. In order to improve the quality of education, it is also important to establish stronger linkages among the agriculture technical schools, universities, and research institutions and the National Agricultural Extension Service (NAES).

5.6.3 *Capacity Building Through Farmers Fields Schools (FFS)*

Among the extension educational initiatives, Farmer Field Schools (FFS) have succeesfully equipped smallholders to learn farming techniques on the farms by providing them with the oppertunties to actively participate in the entire crop production season. The phrase 'from seed to seed' seems to have relevance here, as learning through FFS enables them to have direct interaction and communication with the agricultural extension service providers. It is, therefore, appropriate to promote the idea of field schools and their adoption, especially for the essential crops (ENID 2014). Present productivity levels in Egypt are below than the productive capacity of the varieties under cultivation, however, there are great possibilities to enhance them by 25–50%, by improving agricultural practices and farm management. This could be only achieved through improving the extension services and introducing FFS.

5.6.4 Role of ICT and the Innovative Extension Education Programs

The delivery of information to the small farmers through telecommunications technology seems very promising and worth considering. In modern Egypt, almost all the farmers use cell phones. Some farmers and their family members have more than one cell phone. Some associations were providing marketing information to the members but more could be done to exploit this reality. On the other hand, the use and availability of computers was highly variable and so is the case on the use of the internet to access information. Extension workers do have the computers, but no internet connectivity in their field offices. Therefore, they should be facilitated by proving wired internet connectivity in their offices (because wired will be both the fastest and the cheapest connectivity) and also they must have access to wireless cellular data service in the rural areas where they work to improve their working. New ICT tools have the potential to overcome the ineffective nature of the formal extension system and give assistance to associations that have the interest to take advantage of this new technology (MEAS 2011). It seems imperative to improve and upgrade the extension workers' technical and communication skills to enhance their efficiencies and make them familiarize with the applications of ICT, Geographic Information Systems (GIS), Computers and Remote sensing etc. Though Internet facility is avaialable to the National Agricultural Extension Service yet its potential benefits have not been fully realized.

5.6.5 Use of Electronic Media

At the moment, Agricultural TV broadcasts two programs: "Secret of the Land" and "Our Green Land". Both the programs of 30 min duration are aired twice a week and are designedto enhance the technical skills of the farmers and provide information on the new innovations. However, such programs sould be aired on daily basis and more in number. The Department of Agricultural Extension organizes the field meetings with specialists and farmers to provide information on various aspects of new innovations. Through effective extension programs by employing an appropriate mix of electronic media, fears and doubts of the famers regarding the new innovations can be addressed (Shalaby et al. 2010, 2011).

5.6.6 Launching of Agricultural Websites to Address Issues

The country has also launched the websites VERCON and RADCON. "The Virtual Extension and Research Communication Network (VERCON)" is successfully disseminating information on various aspects of agriculture to benefit the rural communities. The prime objective of this network is to help farmers solveing their problems, establishing professional links with the extension and research organizations. Another website serving the rural community, Rural and Agricultural Development Communication Network (RADCON) focuses mainly on the overall farming process and rural development. It helps addressing all the issues associated with agriculture and problems faced by the society. The network facilitates the communication process among extension, research, private and public sectors and institutions involved in rural and agricultural development for the uplift of farmers and agri-businesses at rural and village level (Rafea 2010). Extension Service has started employing ICT tools by providing modern scientific and technical agricultural information and disemminating recommendations on all the valuable major crops on CDs (Shalaby et al. 2010).

5.6.7 *Mobile Vans: An Effective Source of Delivering Extension Messages in Remote Areas*

Shalaby et al. (2010) reported that the agricultural etension country is successfully disseminating the innovations to the farmers and rural people by using the mobile vans. The extension service using mobile vehicles equipped with TV, video and microphones to create awareness and educate farmers on farming in the villages. Videos on various crops, farming practices, crop diseases and insect-pest issues, information on innovations are prepared in the plain and simple language and provide ample information. They also disemminate and make avaialable the various recommendations made for each crop.

5.6.8 *Print Media: For Educated Farmers*

National Agricultural Extension Service (NAES) in Egypt uses many and multiple channels for disseminating agricultural information and play an important role in educating the fatmers and tranferring technologies (Baig and Aldosari 2013; Sani et al. 2014; Kassem et al. 2017). In order to exchange large volume of information to realize agricultural improvement and development, mass media channels have proved very valuable tools (Ariyo et al. 2013; Uzezi 2015). In Egypt, Specialized Research Centers and National Agricultural Extension Service (NAES) disseminate technical information on innovations to the litrate farmers in the form of publications and pamphlets for creating awareness as reported by EL-Gamal 2015. Although many agricultural extension pamphlets have been developed and disseminated in the Egypt,the evaluation of their effectiveness has not been conducted in a systematic manner so far. Therefore, Kassem et al. (2017) conducted a study to evaluate the readability and usefulness of agricultural extension pamphlets focusing on small-scale litrate farmers. In addition, National Agricultural Extension Service (NAES) in Egypt also publishes a bi-monthly Agricultural Extension Magazine, providing information on agricultural practices and associated innovations. Another magazine named "The New Land" is also published specially to help farmers on new reclaimed lands with the information on various aspects of farming like: the cultivation of new crops, the recommended farming technologies, irrigation systems (sprinkler and drip irrigation) and other farm operations (Shalaby et al. 2010, 2011).

5.6.9 Reorganization and Reinvigoration of Government Extension Service

The national Agricultural Extension Service can help in realizing higher yields and sustainable farming communities. However, of all the extension services and information providers, farmers rated the government extension system as the least useful. Today, extension activities are offered in one-direction, the transfer of knowledge practice. In this model, farmers are considered recipients not participants. However, in order to be more useful, the Agricultural Extension Services will have to move from supply-led information to information customized to cater the actual needs of the rural communities. For example, to realize sustainable cotton production, the extension workers and farmers need join hands in a way that meets the needs of the famers and the objectives of the extension service. The country needs to reorganize the public extension and outreach programs to be more responsive to farmers' needs. However, to ensure the sustainability of information delivery, public, private and volunteer associations must work together to meet the challeneges faced the agricultural sector (MEAS 2011).

5.6.10 Development of Extension Staff and Their Professional Growth

Delivery of efective delivery of extension advice depends upon the technical and communication skills of the extension workers. Therefore it seems imperative to launch the capacity building program to improve their technical knowledge and enhancing their communication skills.

5.6.11 Extension Staff Needs to Have Job Description

Extension staff usually perform multiple tasks other than extension activities. With the clear job description for every extension professional, it would be easy to hold responsible him for his actions. Sometimes they are involved in the non-professional activities that are not desired by their organization. Such activities carry them away from their professional assignments of addressing the real issues of farmers, resulting in the loss of time, money, resources and effectiveness. Clear description of the tasks required from an extension worker would help him improving his working efficiency (Shalaby et al. 2010, 2011).

5.6.12 Extension Services for Women and Youth

Though women actively participate in all the farming activities starting from sowing to the harvesting yet they face poor nutrition, poor health, high birth-rates and are recognized as the unacknowledged labour. Despite their signficant contribution, NES does not accord due importance to the capacity building programs for women Extension Agents, even though women's empowerment is extremely essential to ensure the sustainability of the improvement measures (Shalaby et al. 2010, 2011)

5.7 Conclusions and Recommendations

Agriculture contributes to the economy and development of Egypt significantly. However, the existing set of issues, challenges, limitations like: smallholdings, land fragmentation, labor intensive cultivation, old farming methods, severe shortage of irrigation water and practice of traditional wasteful water irrigation methods, population growth, continued migration of farmers from the rural areas disrupting the rural fabric etc. place the current small scale in severe stress.

Egypt sustains 10.7 million poor while 70% of them residing in rural areas. The country needs to design and adopt the policies for the development of agriculture must also focus on poverty reduction in rural areas. Whereas the scientists need to address issues and evolve innovative farming practices and technologies to increase agricultural production to combat poverty. To implement sustainable development, a workable strategy must have the following steps to consider:

- Make farming areas more attractive, capable of producing more valuable crops that could generate jobs to reduce and prevent urbanization.
- Enable small farmers to cultivate high-value crops to realize higher economic returns.
- Water – an essential put to practice agriculture is becoming a scarce input in the country, therefore, its quality and quantity needs wise management and consumption. In order to realize the sustainable management of water resources, it is important to control water pollution.
- Strengthen rural organizations to address the poverty issues and to enhance rural income generations – both within and outside agriculture sectors. Non-farm income generation sources need to be explored and made the part of rural culture.

- Government needs to provide and promote agro-processing facilities in the rural areas. Small-scale projects can enhance the economic value of agricultural products. Obviously value-added products fetch higher prices.
- High value crops deserve more attention as they can bring more economic returns to the farmers.
- Technical and financial assistance improve the quality and production levels of the crops, create more export opportunities while at the same time can protect the environment.
- Due to extensive migration of young farmers to the cities and the gulf countries, the farmers have become the consumers of their own agricultural commodities; they are losing their status of producers and suppliers of basic food items. Therefore, there is a need to convert villages as producers and to enable them to meet the food requirements of the cities by launching small-scale projects to make their farming economically viable and productive to attract them back to their villages and farming business.
- For realizing sustainable agriculture, it is imperative to formulate and launch vibrant agricultural, economic, environmental and rural development policies. Above all, there is a need to better coordination, cooperation and accommodate policies framed towards sustainable development.
- Egypt needs to adjust the cropping systems/patterns in various zones depending upon the availability of water. Crops with low water requirements such as wheat and cotton should receive preference.
- Rice farmers to adopt aquaculture in their rice fields in order to strengthen food security programs particularly in Nasr Lake and other areas of the country.
- Capacity building programs are needed for both the male and female extension agents to improve their technical knowledge and enhance their communication skills.
- Extension services should also launch programs for the women farmers as they participate from sowing to harvesting.
- Through the introduction of extension education, the technical skills and organizational capacity of both the poor rural men and women needs to be elevated, upgraded and strengthened to take advantage of rural on- and off-farm economic opportunities.
- Greater focused efforts are needed to initiate small enterprises to create more employment opportunities in rural areas. This could be achieved mainly through the provision of vocational training and availability of better financial services.
- Stronger public-private (farmers' organizations, farmer marketing associations, water users' organizations and community development associations) partnerships are needed to benefit the rural households and bring sustainability in the delivery of information.

5.8 In Closing

A robust agriculture that could help developing sustainable agriculture based on scientific concepts and principles would ensure continued supplies of food commodities and the improved economic livelihoods for rural dwellers. Focused efforts are needed to increase the availability of water, rehabilitate primitive and traditional agricultural facilities, adopt improved farming techniques, and develop new market opportunities for world-class Egyptian products (i.e. honey, cotton, fruits and vegetables) that will increase small-scale farmer income, create employment opportunities and improve food security. Along with the increase in sustainable agriculture, the country requires a vibrant extension system and an efficient national service. Better facilities these goals, setting suitable rewards, achievable targets and increased budgets would help improve the working and strengthen the organization of agricultural extension. Together, the agriculture sector and extension service have a significant and positive role in achieving sustainable agriculture. Hence, we conclude that innovative crop production combined with the corrective measures would make agriculture sustainable leading to ensure food security in the country.

Acknowledgements The authors are extremely grateful and express their gratitude to the Saudi Society of Agricultural Sciences for extending us all the possible help and sincere cooperation toward the completion of this piece of work and research.

References

Abdelhakam, A. M. (2005) Chapter ii: Agricultural extension in Egypt. In *Proceedings of FAO regional Workshop on Options of reform for agricultural extension in the near east*. Amman, Jordan 2–4 October, 2004. Food and Agriculture Organization of the United Nations Regional Office for the near east, Cairo, 2005, pp. 22–26.

Al-Ahram Weekly. (2016). Egypt's land shortage. http://weekly.ahram.org.eg/News/15779.aspx

Ariyo, O. C., Ariyo, M. O., Okelola, O. E., Aasa, O. S., Awotide, O. G., Aaron, A. J., & Oni, O. B. (2013). Assessment of the role of mass media in the dissemination of agricultural technologies among farmers in Kaduna North local government area of Kaduna State, Nigeria. *Journal of Biology, Agriculture, and Healthcare., 3*, 19–28.

Baig, M. B., & Aldosari, F. (2013). Agricultural extension in Asia: Constraints and options for improvement. *The Journal of Animal & Plant Sciences, 23*(2), 619–632.

Batisha, A. F. (2012). Adaptation of sea level rise in Nile Delta due to climate change. *Journal of Earth Sciences Climate Change, 3*, 114. https://doi.org/10.4172/2157-7617.1000114.

CAPMAS. (2016). http://www.capmas.gov.eg/ (the website of the Central Agency for Public Mobilization and Statistics).

El-Din, O. K. (2007). Current situation. Chapter 3. In. Kruseman, Gideon and Vullings Wies (Eds.) Rural development policy in Egypt towards 2025. targeted conditional income support: A Suitable option. Alterra-Rapport, 1526. pp. 29–32.

El-Gamal, H. M. A. (2015). Study of some characteristics of agricultural extension pamphlets and its relation with readability level, MSc Thesis. Mansoura University, Egypt.

El-Laithy, H. (2007). Chapter 9: Employment, income and marketing. In G. Kruseman & W. Vullings (Eds.), *Rural development policy in Egypt towards 2025. Targeted conditional income support: A suitable option* (pp. 103–121). Alterra-Rapport, 1526.

El-Sayed, M. (2017). *Egyptian invention cuts rice irrigation water by half.* Available at: http://www.scidev.net/global/design/news/egyptian-invention-rice-irrigation-water.html

ENID. (2014). *Reforming the agricultural extension services in Egypt: The case of the farmers field schools in Qena*. Egypt Network for Integrated Development. Policy brief 29. Available at: http://www.enid.org.eg/Uploads/PDF/PB29_agric_extension.pdf

FAO. (2011). FAO achievements in Egypt. FAO representation in Egypt July 2011. Available at: http://www.fao.org/3/a-ba0006e.pdf

FAO. (2017). Country briefs – Egypt reference May 28, 2017 available at: http://www.fao.org/giews/countrybrief/country.jsp?code=EGY&lang=en

FAOSTAT. (2017). Egypt. Country indicators. Available at: http://www.fao.org/faostat/en/#country/59

Global Arab Network. (2009). Modernization of agriculture in Egypt. Bulletin of the Arab-British Chamber of Commerce. Egypt Economic Report, Bank Audi, June 2009.

Goldfond, J. (2015). Food insecurity and climate change in Egypt. Available at: http://www.worldpress.org/article.cfm/Food-Insecurity-and-Climate-Change-in-Egypt. May 4, 2015.

Handoussa, H. (2010). Situational analysis: Key development challenges facing Egypt. Report of the Situational Analysis Taskforce, Cairo, Egypt. pp. 1–110.

ICARDA. (2011). *Water and agriculture in Egypt*. Technical paper based on the Egypt-Australia-ICARDA Workshop on On-farm Water-use Efficiency, July 2011, Cairo-Egypt.

IFAD. (2007). *Egypt: Smallholder contract farming for high-value and organic agricultural exports Near East and North Africa Division*. Programme Management Department.

IFAD. (2012). *Arab Republic of Egypt – Country strategic opportunities programme*.

IFPRI-WFP. (2013). *Tackling Egypt's rising food insecurity in a time of transition*. A Joint IFPRI-WEP-CAPMAS Country Policy Note. May 2013. Available at: http://ebrary.ifpri.org/utils/getfile/collection/p15738coll2/id/127559/filename/127770.pdf

Kassem, H. S., Abdel-magieed, M. A., El-Gamal, H. M., & Aldosari, F. (2017). Effect of readability on farmers' knowledge: An assessment of some agricultural Extension pamphlets. *Life Science Journal, 14*(4), 16–22. ISSN: 1097-8135 (Print) / ISSN: 2372-613X (Online). http://www.lifesciencesite.com. 3. https://doi.org/10.7537/marslsj140417.03

Lewis, Lowell N. (2012). Egypt's future depends on agriculture and wisdom. Available at: http://aic.ucdavis.edu/calmed/August%2008,%20final%202d%20Egypt.pdf

MALR. (2009). *Sustainable agricultural development strategy towards 2030*. Cairo: Ministry of Agriculture and Land Reclamation.

MEAS. (2011). *USAID/Egypt office of productive sector development*. Scoping mission: Assessment of agricultural advisory services in upper Egypt sustaining active and efficient associations.

Ministry of Agriculture. (2008). *Economic and social development plan for the Year 2008/2009. Report of the ministry of agriculture*. Cairo: Arab Republic of Egypt.

Ministry of Agriculture. (2011). Agricultural production in Egypt. Available at: http://www.agregypt.gov.eg/5-.

MWRI. (2005). Ministry of Water Resources and Irrigation (MWRI). 2005. National Water Resources Plan 2017: Water for the Future, Cairo: MWRI.

Rafea, A. (2010). *Experience with building a rural and agricultural development communication network*. Scientific and Technical Information and Rural Development IAALD XIIIth World Congress, Montpellier, 26–29 April 2010.

Sani, L., Boadi, B. Y., Oladokun, O., & Kalusopa, T. (2014). The generation and dissemination of agricultural information to farmers in Nigeria: A review. *Journal of Agriculture and Veterinary Science., 7*, 102–111.

Seada, T., Mohamed, R., Fletscher, T., Abouleish, H., Abouleish-Boes, M. (2016). *Comparative study of organic and conventional food production systems in Egypt. Version 1.0*. January 2016. Prepared by the Carbon Footprint Center (CFC), Heliopolis University for Sustainable Development (HU) and the Academy of Scientific Research and Technology (ASRT). El Horreya Heliopolis, Cairo, Egypt.

Shalaby, M. Y., & Baig, M. B. (2010a). Rural development in Egypt: threats and challenges. Banat University of agricultural sciences and veterinary medicine, timisoara, Romania. Lucrări Ştiinţifice. *Management Agricology. Serial I, XII*(2), 169–182 ISSN:14531410. Available at http://www.usab-tm.ro/pdf/2010/management_2010.pdf

Shalaby, M. Y., Baig, M. B., & Al-Shayaa, M. S. (2010). Agricultural extension in Egypt: Issues and options for improvement. *Arab Gulf Journal of Scientific Research, Bahrain, 28*(4), 205–213.

Shalaby, M. Y., & Baig, M. B. (2010b). Rural development in Egypt: threats and challenges. Banat University of agricultural sciences and veterinary medicine, timisoara, Romania. Lucrări Ştiinţifice. *Management Agricology. Serial I, XII*(2), 169–182 ISSN:14531410. Available at http://www.usab-tm.ro/pdf/2010/management_2010.pdf

Shalaby, M. Y., Al-Zahrani, K. H., Baig, M. B., & Straquadine, G. S. (2012). Realizing sustainable agriculture through rural extension and environmental friendly farming technologies in Egypt: Basic ingredients. *Bulgarian Journal of Agricultural Science, 18*, 836–846.

Shaltout, M., Tonbol, K., & Omstedt, A. (2015). Sea-level change and projected future flooding along the Egyptian Mediterranean coast. *Oceanologia, 57*(4), 293–307. https://doi.org/10.1016/j.oceano.2015.06.004.

Soliman, A. (2007). *Agricultural and rural development notes. Building infrastructure and social capital in rural Egypt* (pp. 1–4). Washington, DC: The World Bank.

USAID. (2011). Scoping mission: Assessment of agricultural advisory services in Upper Egypt. Sustaining active and efficient associations. August 2011. This Discussion Paper was produced as part of the United States Agency for International Development (USAID) project "Modernizing Extension and Advisory Services" (MEAS). www.meas-extension.org Available at: https://www.agrilinks.org/sites/default/files/resource/files/MEAS%20Country%20Report%20EGYPT%20-%20May%202011.pdf

USAID. (2017). Agriculture and food security. Available at: https://www.usaid.gov/egypt/agriculture-and-food-security

Uzezi, O. P. (2015). Challenges of information dissemination to rural communities: A case of Niger-Delta communities, Nigeria. *Journal of Emerging Trends in Computing and InformationSciences, 6*, 350–354.

Vella, J. (2012). *The future of food and water security in New Egypt*. Dalkeith, Australia: Published by Future Directions International Pty Ltd.

Worldometers. 2017. *Egypt population*. Available at: http://www.worldometers.info/world-population/egypt-population/

Dr. Mirza Barjees Baig is working as a Professor of Agricultural Extension and Rural Development at the King Saud University, Riyadh, Saudi Arabia. He has received his education in both social and natural sciences from USA. He completed his Ph.D. in Extension Education for Natural Resource Management from the University of Idaho, Moscow, Idaho, USA. He earned his MS degree in International Agricultural Extension in 1992 from the Utah State University, Logan, Utah, USA and was placed on the "Roll of Honour". During his doctoral program, he was honored with "1995 Outstanding Graduate Student Award". Dr. Baig has published extensively in the national and international journals. He has also presented extension education and natural resource management extensively at various international conferences. Particularly issues like degradation of natural resources, deteriorating environment and their relationship with society/community are his areas of interest. He has attempted to develop strategies for conserving natural resources, promoting environment and developing sustainable communities through rural development programs. Dr. Baig started his scientific career in 1983 as a researcher at the Pakistan Agricultural Research Council, Islamabad, Pakistan. He has been associated with the University of Guelph, Ontario, Canada as the Special Graduate Faculty in the School of Environmental Design and Rural Planning from 2000-2005. He served as a Foreign Professor at the Allama Iqbal Open University (AIOU) through Higher Education Commission of Pakistan, from 2005–2009. Dr. Baig the member of IUCN – Commission on Environmental, Economic and Social Policy (CEESP). He is also the member of the Assessment Committee of the Intergovernmental Education Organization, United Nations, EDU Administrative Office, Brussels, Belgium. He serves on the editorial Boards of many International Journals Dr. Baig is the member of many national and international professional organizations.

Prof. Dr. Gary S. Straquadine serves as the Interim Chancellor, Vice Chancellor for the Academic Programs, and Vice Provost at the Utah State University – Eastern, Price USA. He did Ph.D. from the Ohio State University, USA. Presently he also leads the applied sciences division of the USU-Eastern campus. He is responsible for faculty development and evaluation, program enhancement, and accreditation. In addition to his heavy administrative assignments, he manages to find time to teach some undergraduate and graduate courses and supervise graduate student research. Being an extension educator, he has passion for the economic development of the community through education and has also successfully developed significant relations with agricultural leadership in the private and public sector. He has also served as the Chair, Agricultural Comm, Educ, and Leadership, at The Ohio State University, USA. Before accepting the present position as the Vice Provost, he served on many positions as the Department Head; Associate Dean; Dean and Executive Director, USU-Tooele Regional Camp and the Vice Provost (Academic). His professional interests include extension education, sustainable agriculture, food security, statistics in education, community development, motivation of youth and outreach educational programs. He has also helped several under developing countries improving their agriculture and educational programs.

Prof. Ajmal Mahmood Qureshi serves as the Senior Associate, at the Asia Center, Faculty of Arts and Sciences, Harvard University, Cambridge, MA, USA. Previously, he has served at the United Nations as the FAO Resident Representative (Ambassador) for over seven years in China, multiple accreditations to North Korea and Mongolia, regarded as one of the most challenging assignments in the UN System. He has represented the Organization to the host government, diplomatic missions, bilateral and multilateral donors, international organizations, NGOs and other stakeholders. His services have been recognized at the highest administrative and political levels. He was awarded the title of "Senior Advisor and Professor at Chinese Academy of Agricultural Sciences". He successfully implemented a large array of technical assistance projects throughout the country including Tibet. In addition, through a large number of regional projects, Chinese expertise was used in training and capacity building of countries in the Asia Pacific Region. He was the pioneer to launch of Food Security Project in Sichuan Province that was named as an ideal model and was being replicated in over hundred countries. He also assisted in the preparation of China's Agenda 21, holding of the Roundtable on Agenda 21 and was the keynote speaker on the "Environmental Issues". In the restructuring of the FAO, Prof. Qureshi contributed to the preparation of the Organization's strategic plan in the context of China's priorities.

As Chairman of the Donors Group in China, Prof. Qureshi helped mobilize millions of dollars for the agriculture, forestry and fisheries sectors of China. For over eight years, at the Ash Centre of Democratic Governance and Innovation at Harvard Kennedy School, he has worked on the issues of governance and democracy. Prof. Qureshi has served in North Korea, Mongolia, Uganda and Represented the United Nations to the host governments, diplomatic missions, bilateral and international organizations and all other stakeholders. He was the Member of the High Level Steering Committee advising the President of Uganda on the implementation of the Plan for Modernization of Agriculture. Through his offices, he had close interactions with the multilateral donors; inter alia, USAID, World Bank, UN System, Asian Development Bank, SIDA, AUSAID and NORAD. As Head of International Cooperation, he led the country's delegation to the Governing Boards of Rome based UN Organizations: FAO, WFP and IFAD. Prof. Qureshi also organized and assisted in holding numerous international events like: FAO Ministerial level Regional Conference for Asia and Pacific in Islamabad, attended by 28 Ministers. He has the honor to serve as the Consul General (Head of Mission) of Pakistan in Istanbul, Turkey.

He received his education in Pakistan, France and the USA with major focus on political science, international relations and development economics. At the Harvard Kennedy School, he dealt with issues of governance, democracy and innovation; and under the leadership training programs, inter-acted and collaborated with senior officials from China, Viet Nam and Indonesia. He has published extensively in Harvard journals, bearing on China and DPRK's Food Security. He has also been named as the "Most Distinguished Alumni" at the Boston University. He is also the Member of the Board of Directors of the United Nations Association of Greater Boston (UNAGB).

Dr. Asaf Hajiyev is a Professor at the Baku State University and the Executive Member (Academician) of the Azerbaijan National Academy of Sciences. He holds the Dr. Sci. degree from Bauman Moscow State Technical University, and has done Ph.D. and post-doctoral research at Lomonosov Moscow State University. He has the vast research and teaching experience. He serves as the Chair of the Department of Probability and Statistics, Chair, of the Department of Controlled Queues, Institute of Control Systems at the Azerbaijan National Academy of Sciences; Chair of the Department of Theory of Probability and Mathematical Statistics, Baku State University; and the Senior Scientific Researcher, Department of Probability Statistics, Royal Institute Technology in Stockholm. Being a renowned researcher, he has served at the several universities around the world including China, Germany, Italy, Portugal, Sweden, Turkey, and USA. He serves on the editorial boards of many prestigious national and international academic journals. He has been an organizing member and the Keynote speaker at the numerous international conferences. He has been honored with many prestigious awards like: Azerbaijan Lenin Komsomol Prize Winner on Science and Engineering; Grand Prize at the International Conference "Management Science and Engineering Management" Macao; and Grand Prize at the İnternational Conference "Management Science and Engineering Management" at Islamabad, Pakistan.

He is the Honorary Academician of Academy of Sciences of Moldova, Foreign Member of the Mongolian National Academy of Sciences, Member of TWAS (The World Academy of Sciences), Honorary Professor of Chengdu University (China) and the elected member of the International Statistical Institute.

He has also the honor of holding the Office of the Vice-President of the Parliamentary Assembly of the Black Sea Economic Cooperation Organization and since 2015; he serves as the Secretary General of the same organization (PABSEC). He shares his talents and expertise on the boards of many international academic organizations and institutions. He has more than 135 peer reviewed scientific publications to his credit, published in the highly reputed journals.

Dr. Aymn F. Abou Hadid serves as the Professor at the Arid Land Agricultural Studies and Research Institute (ALARI), Faculty of Agriculture, Ain Shams University (ASU), Cairo, Egypt. Presently he is involved in professional activities related to Agriculture and rural Development, Agro Climatology Environmental Stresses, Remote Sensing, Water/Plant/Environment Relations, Crop Water Requirements, Irrigation of Horticultural Crops, Protected Cultivation, Soilless Culture, Vegetable Production, Breeding, Seed Production. He is the former President of Agricultural Research Center, Ministry of Agriculture and Land Reclamation (from 2007 to 2011). He was also the Head of Egyptian Environmental Affair Agency at the Ministry of environment. He has also served as the Minister of Agriculture for 2 terms (from Jan 2011 to July 2011, and from July 2013 to June 2014). He received his education at UCW, Aberystwyth- UK and Al Azhar University-Cairo Egypt. He has been associated with many International Organizations. He has been honored with many medals and numerous prestigious awards to acknowledge his scientific services and scholarly contributions. His researches have been published in the form of book chapters and research papers in the high ranking scientific journals.

Chapter 6
Dynamics of Food Security in India: Declining Per Capita Availability Despite Increasing Production

Pooja Pal, Himangana Gupta, Raj Kumar Gupta, and Tilak Raj

6.1 Introduction

Concern for food security can be traced back, in addition to the world food crisis of 1972–1974, to the Universal Declaration of Human Rights in 1948 (United Nations 1949), which recognized the right to food as a core element of an adequate level of living. Food security as a concept emerged at the World Food Conference of the Food and Agriculture Organisation (FAO) in 1974 (United Nations 1975). It is centered on two sub-concepts: food availability which refers to the supply of food available at local, national or international levels; and food entitlement which refers to the capability of individuals and households to obtain food. It suggests that people do not usually starve because of an insufficient supply of food but because they have insufficient resources, including money, to acquire it (Sen 1981).

The World Bank, FAO, and the US Agency for International Development (USAID) define food security as "access by all people at all times to sufficient food to meet dietary needs for a productive and healthy life" (USAID 1992). According to the World Food Summit (1996) "Food security exists when all people, at all times, have physical and economic access to sufficient safe and nutritious food that meets their dietary needs and food preferences for an active and healthy life".

P. Pal (✉) · T. Raj
University Business School, Panjab University, Chandigarh, India
e-mail: traj@pu.ac.in

H. Gupta
National Communication Cell, Ministry of Environment, Forest and Climate Change, Government of India, New Delhi, Delhi, India

R. K. Gupta
Journalist and Independent Analyst in Environment, Food and Social Policy, New Delhi, India

M. Behnassi et al. (eds.), *Climate Change, Food Security and Natural Resource Management*, https://doi.org/10.1007/978-3-319-97091-2_6

Soaring food prices can dramatically impact poor households, exacerbating food insecurity and creating social tensions. According to the latest World Bank data, safety nets are insufficient or non-existent in many developing countries. At least 60% of people in developing countries – and nearly 80% in the world's poorest countries – lack effective safety net coverage. Stamoulis and Zezza (2003) list safety nets as: (i) targeted direct feeding programs including school meals, feeding of expectant and nursing mothers as well as children under five through primary health centers, soup kitchens and special canteens; (ii) food-for-work programs for providing support to households while developing useful infrastructure such as small-scale irrigation, rural roads, buildings for rural health centers and schools; and (iii) income-transfer programs, including food stamps, subsidized rations and other targeted measures for poor households.

India has emerged as the country facing some of the greatest challenges pertaining to food security based on a range of relevant indicators, including food availability, prevalence of undernourishment, and poor anthropometric indicators of child malnutrition (Sharma and Gulati 2012). Therefore, expanding and strengthening the system of delivery of subsidized food should be a matter of high priority for state policy. However, consumer food subsidies and the Public Distribution System (PDS) came under attack in the policy pronouncements of the government of India in the post-1991 regime of structural adjustment (Swaminathan 1996).

In this chapter, we study the state of declining per capita foodgrain availability and access in India despite record agricultural production and rising food subsidy. We review the present government policies and resolutions for food management and the reasons as to why the present policies have not been able to address the food availability problem in the country. The data were taken from the government records in the Economic Survey of India and the Annual Reports of the Food Corporation of India (FCI). We find that in the last decade, the availability was less than the average for eight out of 10 years. This happened in the years of record agricultural production. When people had less to eat, the food subsidy bill was rising consistently and the government foodgrain stocks swelled to unmanageable level. In 2011 and 2012, the stocks with the FCI were double the buffer norm meant to smooth out the fluctuations in agricultural production. Interestingly, the correlation coefficient of per capita net availability with subsidy was low at 0.19 but the correlation of subsidy with agricultural production was high at 0.91.

6.2 Progress in Global Food Security

The four dimensions of food security – availability, access, utilization, and stability – are better understood when presented through a suite of indicators which makes it a complex condition (FAO et al. 2013). During the World Summit on Food Security in 2009, it was agreed "to undertake all necessary actions required at national, regional, and global levels and by all states and governments to halt immediately the increase in – and to significantly reduce – the number of people suffering from hunger, malnutrition and food insecurity" (FAO 2009). A shortage of quality food and poor feeding practices have contributed to making underweight prevalence among children the highest in the world in Southern Asia. It will be difficult to meet the hunger-reduction target in many regions of the developing world due to economic crises and rising food prices. The disconnect between poverty reduction and the persistence of hunger has brought renewed attention to the mechanisms governing access to food in the developing world (United Nations 2011). The proportion of undernourished people—those individuals not being able to obtain enough food regularly to conduct an active and healthy life—decreased from 23.3% in 1990–1992 to 12.9% in 2015. About 795 million people are estimated to be suffering from chronic hunger who are not getting enough food to conduct an active life. This figure is 216 million less than in 1990–1992, a reduction of 21.4%, despite the world population having increased by 1.9 billion (FAO et al. 2015).

The Millennium Development Goal 1 (MDG) was to 'halve, between 1990 and 2015, the proportion of people whose income is less than $1.25 a day, achieve full and productive employment and decent work for all, including women and young people and halve the proportion of people who suffer from hunger. The hunger target condition has been almost met at the global level. As many as 72 of the 129 countries monitored for progress have halved the hunger rate (FAO et al. 2015). The global poverty rate of $1.25 a day fell in 2010 to less than half the 1990 rate (United Nations 2014). Several organizations have worked together to achieve lower poverty rates globally. The World Food Programme (WFP), which was born in 1961, supports national, local and regional food security and nutrition plans. One of the goals of the WFP is to assist governments and communities to establish or rebuild livelihoods, connect to markets and manage food systems. WFP also seeks to reduce undernutrition levels and mortality due to undernutrition where urgent action is required to save lives and avoid irreparable harm to health, including through provision and distribution of specialized nutritious foods (WFP 2013).

6.3 Materials and Methods

The historical data for net foodgrain availability and PDS offtake were obtained from the 2016 Economic Survey of India. The data on carrying cost, food subsidies were taken from the Annual reports of the FCI. The Economic Survey of India gives the per capita net foodgrain availability in grams per day. For the data to make more sense, it was converted to kg per year by multiplying the figures with 365 and dividing it by 1000. The subsidy figures were found to be at the current prices. Therefore, these figures were brought to the constant 2004–2005 prices using the index numbers of Wholesale Price Index (WPI) from the Economic Survey of India.

$$Adjusted\,Value = Current\ Price * \frac{WPI\ of\ 2004-2005}{WPI\ of\ Current\,Year}$$

For simplicity, and also because different sources treated the years differently, all data for a financial year were shown under the ending year. This means that the data from the 2002 to 2003 financial year has been treated as the data for the year 2003. Maximum, minimum, average and standard deviation were calculated on data obtained in this way. Correlation coefficients were calculated for the pairs of net availability vs. carrying cost, net availability vs. total food subsidy, and food subsidy vs. PDS offtake.

6.4 Food Availability, Climate Change, and Biodiversity

Food security is inherently interlinked with other current global challenges related to economy and climate change. The uncertainty surrounding food production and distribution systems due to the adverse effects of climate change would result in added pressure on food security, nutrition, and livelihoods of entire communities (Kattumuri 2011). Climate impacts on the poor's food security are particularly harsh because the majority of them depend on agriculture as a source of food and income (Braun 2008). Climate change has the potential to affect all food security dimensions, food production and distribution channels, human health, livelihood assets, as well as change purchasing power and market flows. Extreme climatic events can damage or destroy transport and distribution infrastructure and adversely affect other non-agricultural parts of the food system (FAO 2008).

Agricultural yields are likely to decrease with even a slight climate change where crops are near their maximum temperature tolerance and where dry-land and non-irrigated agriculture predominates. Even without the effects of climate change on crop yields, but with the present levels of population growth and economic growth, world cereal production is estimated at 3286 million tons (mt) in 2060 compared with 1795 mt in 1990 (Parry et al. 2005).

Climate change carries with it increasing uncertainty in the form of seasonality of food scarcity and hunger with millions in South Asia facing recurring food insecurity each year. There is more uncertainty in the seasonal events such as rains, draughts and floods, and their duration and intensity. In the rural areas, the fine edge between survival and destitution hinges on the advent of the monsoons. Failure of a monsoon can lead to starvation at the household level, and deplete buffer stocks at the country level. Despite increasing urbanisation, millions of people remain dependent on agriculture for food as well as wage income. Seasonality impacts not only food production, but also the availability of wage income from farm or off-farm jobs (Ramachandran 2011).

Biodiversity is the key to future food security as the majority of today's modern crop and livestock varieties are derived from their wild relatives (Sunderland 2011). About 7000 species of plants have been cultivated or collected for consumption in human history. Presently, only about 30 crops provide 95% of human food energy needs, with 5 cereal crops (rice, wheat, maize, millet and sorghum) providing 60% of the energy intake of the world's population (Secretariat of CBD 2013).

6.5 Food Security Issues in India

The last major famine in India was the Great Bengal Famine which occurred in 1943 when the country was under the British rule and claimed between 1.5 and 4 million lives. Later studies on the phenomenon showed that it was not as much the shortage of food but access to it that caused the problem. Because of the crop failure in 1942, a majority of population, comprising subsistence farmers, had neither food for own consumption nor money to buy food in the market. According to the theory proposed by Indian Nobel Laureate Amartya Sen (1981) "famines often take place in situations of moderate to good food availability, without any significant decline of food supply per head". His study focused on "people's ability to command food through legal means available in the society (including the use of production possibilities, trade opportunities, entitlements vis-a-vis the state, etc.)".

There has been no major famine in India since 1943 (Devereux 2006) and Sen attributes this trend of decline or disappearance of famines after independence to a democratic system of governance and a free press – not to increased food production (Iqbal and You 2001). However, there are sporadic reports of starvation deaths from some pockets of the country, especially Kalahandi region of Odisha. On a visit to India in 2003 amid reports of starvation deaths, Sen remarked that "we must distinguish between the role of democracy in preventing famine and the comparative ineffectiveness of democracy in preventing regular undernourishment" (New York Times 2003).

Though India has left the story of famines behind, per capita net availability of foodgrains has been declining since its peak in 1991 (Table 6.1). On the other hand, the per capita demand for foodgrains is growing due to changing consumption patterns. The combination of rising demand and declining availability is a food crisis, which is manifested in rising price of cereals and pulses in the market (Kumar et al. 2012). The declining foodgrain availability is happening despite the rising food subsidy bill. The figures, however, do not catch the seriousness of the problem. In their book *Hunger and Public Action*, Drèze and Sen (1989) note that "nearly four million people die prematurely in India every year from malnutrition and related problems. That's more than the number who perished during the entire Bengal famine".

Despite the Green Revolution of the mid-1960s and White Revolution in later years – which make India the first in the world in milk production, second in rice, wheat, sugarcane, groundnut, vegetables, fruits, livestock, fisheries and poultry – food security is still a sensitive issue in India. The Green Revolution of mid-1960s, saved the nation from the 'drought of the century' in 1987. Surplus stocks procured and stored in earlier years were used instead of resorting to large-scale imports. Efforts towards ensuring food security to millions of poor households intensified in the 1990s. Under the global hunger index (GHI), India comes at the 97th rank with the prevalence of undernourishment, child wasting, child stunting, and child mortality at 28.5% of total population hungry (IFPRI 2012) with over 43% of children under five underweight (IIPS and Macro International 2007).

Table 6.1 Per capita net availability of foodgrains, production, PDS offtake, carrying cost of stocks, and food subsidy

	Per Caput		Production	PDS Offtake	Carrying cost	Subsidy
Year	Gm/day	kg/year	(Million tons)		(INR million at 2004–2005 prices)	
1991	510.10	186.19				
1992	468.80	171.11				
1993	464.10	169.40				
1994	471.20	171.99				
1995	495.40	180.82				
1996	475.20	173.45				
1997	503.10	183.63				
1998	447.00	163.16				
1999	465.70	169.98				
2000	454.40	165.86	171.8			
2001	416.20	151.91	162.5			
2002	494.10	180.35	174.5			
2003	437.60	159.72	143.2	23.20	51647.7	119113.6
2004	462.70	168.89	173.5	28.30	27209.2	226809.4
2005	422.40	154.18	162.1	31.00	8123.8	207543.3
2006	445.30	162.53	170.8	31.80	4103.0	204136.9
2007	442.80	161.62	177.7	32.80	5469.5	215465.0
2008	436.00	159.14	197.2	34.70	6357.8	257731.6
2009	444.00	162.06	192.4	41.30	28507.9	276091.0
2010	437.10	159.54	178.0	43.70	45066.9	327776.1
2011	453.60	165.56	198.2	47.90	44217.6	393541.0
2012	450.30	164.36	211.9	44.90	42531.1	440083.7
2013	401.40	146.51	208.9	44.50	50247.7	480687.2
2014	491.40	179.36	215.1	43.50	56137.8	503897.2

Source: Ministry of Finance 2016 and FCI (Annual Reports)
Average = 167.14. Standard deviation = 10.10. Maximum = 186.19 Minimum = 146.51 (in kg/year)
Correlation coefficient of per capita availability and carrying cost = 0.20
Correlation coefficient of per capita availability and subsidy = 0.19
Correlation coefficient of total subsidy and PDS offtake = 0.88
Correlation coefficient of total subsidy and foodgrain production = 0.91

6.6 Government Policies to Manage the Food Security Challenge

India has moved a long way from the critical food shortages of the 1950s that triggered massive food aid imports under the US law P.L. 480, also known as "food for peace". In order to enhance the access to food, several policies on price and market regulation were introduced in 1965. Two new institutions were created: the Agricultural Prices Commission, a government agency to determine the level of support prices for various crops and several other administrative controls; and the FCI, with the responsibility for purchase, storage, movement and distribution of food through the PDS, as well as acting as the main handling agent for imported grain and its distribution. The Targeted Public Distribution System (TPDS) and the Mid-Day Meal Scheme (approximately 120 million children are signed up) are two large government food distribution schemes in India.

TPDS was introduced in 1997 to benefit the poor and to keep the budgetary food subsidies under control to the desired extent following failure of the earlier PDS system (GoI 2005). Under the scheme, special cards were issued to families below poverty line (BPL) and foodgrains were distributed at a lower price for these families compared to those above the poverty line. The entire population was divided into three categories – BPL (Below Poverty Line), APL (Above Poverty Line) and AAY (Antyodaya Anna Yojana). The BPL families were given 35 kg of foodgrains per month at subsidized price. AAY were provided a monthly provision of 35 kg of foodgrains at specially subsidized rates of INR 2 per kg for wheat and INR 3 for rice (Dev and Sharma 2010).

India now has many government programs such as TPDS including AAY, nutrition programs like mid-day meals, Integrated Child Development Services (ICDS), etc. to improve food and nutrition security. In addition, workfare programs, like National Rural Employment Guarantee Scheme (NREGS), and self-employment programs, including Employment Assurance Programme, have been also adopted with the objective to guarantee 100 days of work for poor households in rural India. Under the 2009 Food Entitlements Act, all Antodaya cardholders shall be entitled to a monthly quota of at least 50 kg of foodgrains per family at a price not exceeding INR 1/kg under the PDS. Coarse grains shall be made available through the PDS at subsidized rates, wherever people prefer these, such that the total of all foodgrains, including coarse grains provided is at least 50 kg per month per family. Under the Act, every household and those covered by the AAY Scheme, shall be entitled to and provided a Food Entitlements Card (FEC). All FEC holders shall be entitled to a monthly quota of at least 50 kg of foodgrains per family at a price not exceeding INR 3/kg for rice and INR 2/kg for wheat, under the PDS. About 25 million poorest of the poor people are covered under the scheme.

In 2011, the Government had passed a National Food Security Bill in the Parliament which had raised more questions than it had proposed to answer. Government's own Commission for Agricultural Costs and Prices (CACP) prepared a discussion paper (Gulati et al. 2012) which raised questions over several provisions of the Bill. One of the contentious issues was the *Force Majeure* clause (Clause 52) which said that "the Central Government, or the State Governments, shall not be liable for any claim by persons belonging to the priority households or general households or other groups entitled under this Act for loss/damage/compensation, arising out of failure of supply of foodgrains or meals when such failure of supply is due to conditions such as war, flood, drought, fire, cyclone, earthquake or any act of God" (GoI 2011). But, the CACP report pointed out that it is precisely in these times and conditions that a failure of market forces, volatility in prices and resultant distress occur. It is at times like this that the poor and vulnerable would depend on the Government to ensure their food security (Gulati et al. 2012).

The other issue with the bill was the eventual plan to move to cash transfer system. If the right to food security moves to the cash transfer system where those entitled to food subsidy will have to buy rice and wheat directly from the market, at the market price, and the subsidy will be directly paid into their bank accounts, then what will happen to the elaborate PDS put in place by the Government? If half a million PDS outlets are closed, there is going to be a large-scale unemployment. And what will happen to the food procurement system by the FCI that provides price support to farmers? Will farmers be thrown to the mercy of market forces?

However, the National Food Security Act was passed in 2013 under which every person belonging to priority households (identified under section 10 of the Act), shall be entitled to receive 5 kg of foodgrains per person per month at subsidized prices. Regarding the *Force Majeure* clause, the Act mentions that the Central Government, or the State Government, shall be liable for a claim by any person entitled under this Act, except in the case of war, flood, drought, fire, cyclone or earthquake affecting the regular supply of foodgrains or meals to such person under this Act. Provided that in such a case, the Central Government may declare whether or not any such situation affecting the regular supply of foodgrains or meals to such person has arisen or exists (GoI 2013).

6.7 Problems with Food Management and Present Food Scenario

Today, the problem of food management system faced by the country is not the foodgrain shortage but consists of finding ways and means of managing the accumulated surplus (Virmani and Rajeev 2002). Economic survey of India (Ministry of Finance 2013) has called for an urgent attention to efficient food stocks management, timely off-loading of stocks, and a stable and predictable trade policy. A recent analysis showed that, on an average, the costs of maintaining buffer stocks of rice and wheat are higher than procurement costs in domestic or international markets (Dorosh 2008). So, the need to off-load them has become more of an immediate concern than poverty relief, and off-loading stocks has a fiscal cost. Heavy input subsidies and technological change, coupled with farm price support policies, have led to heavy accumulation of foodgrain stocks with the government, and the internal carry-over costs have increased, while at the same time the hard core poor continued to suffer from food insecurity (Rao 1994). A study by the International Food Policy Research Institute (IFPRI) finds that the government spent \$3.40 to transfer \$1.00 to the poor (Coady 2004). In spite of the present government schemes of distribution, the number of undernourished people in India remains to be 194 million, 15.2% of the total undernourished in the world with a GHI of 28.5 (Table 6.2).

Table 6.2 Present food scenario in India and the World

	World	India
Food produced (million tons 2016)[a, b]	2577	252
Undernourished (millions 2015)[c]	795	194
Proportion of undernourished people (%)[c]	10.9	15.2
Global Hunger Index[d]	–	28.5
Total cereals exports (millions tons 2015)[e]	39,622	12.42
Total cereals imports (millions tons 2015)[e]	–	0.72

Sources of data:
[a]FAO Cereal Supply and Demand Brief
[b]Annual Report (2015–2016), Ministry of Agriculture
[c]The State of Food Insecurity in the World 2015
[d]Global Hunger Index 2016
[e]Agricultural Market Information System

6.8 Results and Analysis

The long-term trend of net foodgrain availability from 1991 to 2014 shows fluctuations from year to year (Table 6.1). The availability ranges from 186.19 to 146.51 kg per person per year, the average being 167.14 kg per person with a standard deviation of 10.10. The decline was over 34 kg per person in a year. In the first decade and a half of the twenty-first century, the foodgrain availability was below the long-term average in 11 out of 14 years. During this period, the GDP grew at an average historic high rate of over 7% (Table 6.3).

The value declined consistently from its peak of 186.19 kg per person in 1991 to the bottom of 146.51 in 2013 (Table 6.1). This difference becomes starker when one considers that an average family of five had 198.4 kg of foodgrain less to eat in 2013 than in 1991. This happened in the years of record agricultural production. When people had less to eat, the food subsidy bill was rising consistently and the Government foodgrain stocks swelled to unmanageable level. Interestingly, the correlation coefficient of per capita net availability with subsidy was low of 0.19 but the correlation of subsidy with agricultural production was high at 0.91.

The food subsidy ballooned nearly four times (at inflation-adjusted constant prices) from INR 119.113 billion ($1.75 billion at current exchange rate) in 2003 to INR 503.897 billion ($7.4 billion) in 2014. That means that when the people got less to eat, the Government was running up huge subsidy bill (Fig. 6.1). This is because FCI was stocking up on grains and running huge carrying costs. In 2011 and 2012, the stocks with the FCI were double the buffer norm meant to smooth out the fluctuations in agricultural production. The carrying cost of foodgrain stocks fluctuated wildly during the decade. It dropped to a low of inflation-adjusted INR 4.103 billion in 2006 and shot 13-fold to reach INR 56.138 billion in 2014 (Table 6.1). At the same time, the correlation coefficient of per capita foodgrain availability with carrying cost was 0.20. The correlation of availability with total subsidy was 0.19. However, the correlation coefficient of food subsidy vs. PDS offtake is as high as 0.88 (Fig. 6.2).

Table 6.3 GDP growth at 2004–2005 prices

Year	GDP Growth
2001	3.6
2002	5.0
2003	3.9
2004	7.9
2005	7.9
2006	9.3
2007	9.2
2008	10.2
2009	3.7
2010	8.5
2011	9.8
2012	6.9
2013	5.3
2014	6.6
2015	7.3
2016	7.5

Source: Data derived from Economic Survey of India 2015–2016

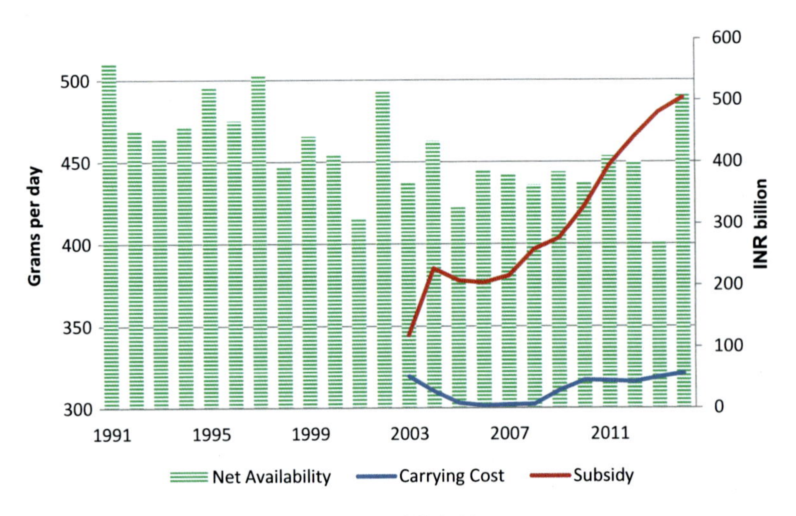

Fig. 6.1 Foodgrain availability, carrying cost and Subsidy

This stands to reason since food subsidy is obtained by adding the consumer subsidy to the carrying cost. The consumer subsidy is the difference between the economic cost and the price at which the Government issues the foodgrain to the public or central issue price (CIP). The economic cost consists of procurement price, storage, transportation, carrying cost and administrative costs of the

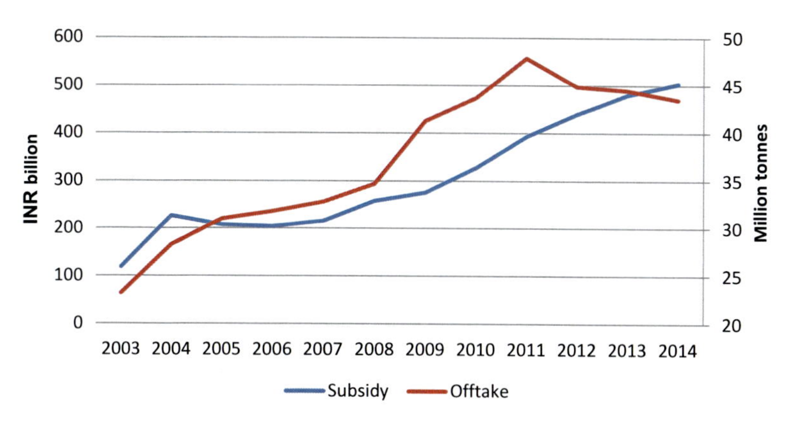

Fig. 6.2 Food subsidy and food offtake from PDS. (Source: Economic Survey (2015–2016) and Annual Reports of FCI)

FCI. When the carrying cost increases, it also adds to the economic cost of foodgrains. Therefore, the subsidy increases when the offtake increases.

$$Consumer\,Subsidy = Economic\,Cost - Central\,issue\,Price\left(CIP\right)$$

Carrying cost is influenced by the size of the stocks, expenses on storage and handling, interest on capital, freight, and storage loss. Interest accounts for the major share of the carrying cost and the storage charges come next (Sharma 2012). The difference between the economic cost and CIP is the consumer subsidy which is reimbursed by the Central Government to the FCI. The economic cost of foodgrains has been increasing during the last 10 years whereas the issue prices have remained unchanged since July 2000 for BPL families, December 2000 for AAY households, and since July 2002 for APL families. In real terms, the current issue prices are around 55% of the original CIPs implying that rice and wheat are available at almost half the price through the PDS (Gulati et al. 2012).

The main reasons for the increase in food subsidy include rise in minimum support prices, accumulation of stocks, rising economic costs of foodgrains, high offtake and constant CIP. Table 6.1 shows that the subsidy bill, apart from a dip in the years 2005 and 2006, increased constantly while the foodgrains available in the country decreased or remained low in the years of bumper crops. In the year 2008, there was a record production (Table 6.4). Out of this, 14.4 million tons were exported and 17 million tons were added to the government stocks. In 2009, another 11.5 million tons went to the Government stocks. This reduced the availability for domestic consumption, while the subsidy increased. It is not surprising that the total food subsidy has a high correlation of 0.91 with agricultural production.

Against the buffer norm of 31.9 million tons of rice and wheat (on 1st July of each year), total Central Pool stocks were more than double at 80.5 million tons in 2012 while the FCI has a total storage capacity of only 62 million tons. The rest of

Table 6.4 Production, net imports, change in Government stocks, and net availability of cereals

Net availability of cereals (million tons)				
1	2	3	4	5
2000	171.8	−1.4	13.9	156.6
2001	162.5	−4.5	12.3	145.6
2002	174.5	−8.5	−9.9	175.9
2003	143.2	−7.1	−23.2	159.3
2004	173.5	−7.7	−3.3	169.1
2005	162.1	−7.2	−2.4	157.3
2006	170.8	−3.8	−1.8	168.8
2007	177.7	−7.0	1.7	169.0
2008	197.2	−14.4	17.0	165.9
2009	192.4	−7.2	11.5	173.7
2010	178.0	−4.7	−0.5	173.8
2011	198.2	−9.6	8.3	180.1
2012	211.9	−19.8	11.2	181.0
2013	208.9	−71.9	−23.6	160.6
2014	215.1	−19.4	−6.0	201.6

Source: Annual reports of Food Corporation of India
[1]Year, [2]Production, [3]Net Imports, [4]Change in Government Stocks, [5]Net availability (2 + 3 + 4)

the stocks is either lying in the open or kept in hired warehouses. Overstocking has fiscal implications, apart from distorting the free functioning of foodgrain market (Gulati et al. 2012).

6.8.1 *Reasons for Excessive Food Stocks*

The main reasons for excessive food stocks in the FCI godowns is the shift in the consumption pattern of the population in favor of superior food items like milk, vegetables, fruits, and animal foods. Thus, the growth of aggregate demand for cereals in the country is slowing down because of deceleration in the pace of population growth and a shift in consumer preference towards non-cereals (Virmani and Rajeev 2002). The other main factor that has contributed to excess stocks is high procurement price. The fact that in recent years there has been a tendency among successive governments to fix Minimum Support Prices (MSP) for paddy and wheat in excess of the levels prescribed by the CACP. While this increases farmers' incentive to produce more, it has raised the market prices and has reduced the demand for cereals (Virmani and Rajeev 2002).

The buffer stock policy, but, has served more purposes than it has caused problems. The main objective of buffer stock policy was to stabilize the availability and prices of foodgrains, and, thereby, to achieve national food security (FCI). The policy is used by the government to support/stabilize the market price for both the producer and the consumer (Sutopo et al. 2010). Buffer stocks have guaranteed the availability of foodgrains in drought years so that famines can be avoided. It helps at the time of shortages, avoiding external pressure and prevents sudden flight of foreign exchange to finance imports (Gulati et al. 1996).

However, maintaining buffer stocks is not the only, or necessarily the best, method to stabilize foodgrain prices and availability. As too low buffer stocks will jeopardize food security, carrying too high stocks is costly and inflationary. Therefore, exporting foodgrains in the years of surplus production and importing them in the years of short crops may regulate stocks, but the decision whether to store or trade should be based on the comparison of the cost of storage with the gains from exporting now and importing later (Gulati et al. 1996).

6.8.2 *Reasons for Decreasing Food Availability: Past, Present and Future*

Though the subsidy is the price a country of predominantly poor people pays for food security, the phenomenon peculiar to India is that the subsidy goes up along with the reduction in foodgrain availability to the people. India uses an extensive variety of instruments which include in-kind transfers and public works program to tackle food security albeit with disappointing outcomes (Sharma and Gulati 2012). According to Tyagi (1990), government policies were not considered as particularly successful in protecting the interests of the vulnerable sections of society, either producers or consumers. As with the increasing role of the government in the marketing of foodgrains, not only did the subsidy bill rose, but it adversely affected the efficiency of the marketing system.

Though, PDS has helped to manage droughts and improve the overall availability of foodgrains, the availability of subsidized foodgrains varied across states. In some states, foodgrains distributed through PDS form a substantial portion of the per capita foodgrain consumption. Unfortunately, the distribution of foodgrains had not been targeted to states with high poverty levels (Gulati et al. 1996). Also, as far as the coverage and performance of the Indian PDS was concerned, leakages into the free market was a serious problem, as was the weak targeting in the program (Ahluwalia 1993) due to which TPDS was introduced. In spite of rising subsidy, food availability remained an issue of concern because in partial subsidy schemes, the welfare of agents (both producers and consumers) depend not only on the price at which the government buys and sells food but also on the market price of food. Therefore, the impact of a food subsidy scheme depends on how the market price of food varies with the subsidy (Ramaswami and Balakrishnan 2002). Also, governments in poor countries are often under pressure, from international creditors, to trim subsidies. Such measures encounter political resistance especially when organized interests capture subsidies. Food subsidies, thus, are badly targeted and many of the poor receive insignificant amounts of subsidy and depend on the market to access supplies. In spite of this, a reduction in the food subsidy is not in their interest as the reduction in subsidy increases the market price of food (Ramaswami and Balakrishnan 2002).

6.9 Conclusion

The long-term trend of net foodgrain availability shows fluctuations from year to year. The availability ranges from 186.19 to 146.51 kg per person per year, the average being 167.14 kg per person with a standard deviation of 10.10. It declined consistently from its peak of 186.19 kg per person in 1991 to the bottom of 146.51 kg per person in 2013. The decline was nearly 40 kg per person in a year. This difference becomes starker when one considers that an average family of five had 198 kg of foodgrain less to eat in 2013 than in 1991. In the last decade and a half, the availability was less than the average for 11 out of 14 years. This happened in the years of record agricultural production. When people had less to eat, the food subsidy bill was rising consistently and the government foodgrain stocks swelled to unmanageable level. In 2011 and 2012, the stocks with the FCI were double the buffer norm meant to smooth out the fluctuations in agricultural production. The result was that for managing the food security, the government has to spend a lot of money in terms of carrying cost. By conserving the stock, government reduces the actual supply of food that further leads to less consumption in the country. For resolving the issue, government needs to reduce the total stock with FCI which will actually help in increasing the supply of food in the country and also reduce the food subsidy by less cost of buffer stock.

Acknowledgements The authors are thankful to the University Grants Commission (UGC), New Delhi for providing funding support for this research in the form of Senior Research Fellowship.

References

Ahluwalia, D. (1993). Public distribution of food in India: Coverage, targeting and leakages. *Food Policy, 18*, 33–54.

von Braun, J. (2008). *The impact of rising food prices and climate change on the ultra poor*. Washington, DC: International Food Policy Research Institute.

Coady, D. P. (2004). *Designing and evaluation social safety nets: Theory, evidence and policy conclusion*. Washington, DC: International Food Policy Research Institute.

Dev, S. M., & Sharma, A. N. (2010). *Food security in India: Performance, challenges and policies*. New Delhi: Oxfam India.

Devereux, S. (2006). *The new famines*. London: Routledge.

Dorosh, P. A. (2008). Food price stabilisation and food security: International experience. *Bulletin of Indonesian Economic Studies, 44*, 93–114. https://doi.org/10.1080/00074910802001603.

Drèze, J., & Sen, A. (1989). *Hunger and public action*. Oxford: Oxford University Press ISBN: 978-0-19-828365-2.

FAO. (2009). *Declaration of the World summit on food security*. Rome: Food and Agriculture Organization.

FAO. (2008). *Climate change and food security: A framework document*. 107.

FAO, IFAD, WFP. (2013). *The state of food insecurity in the world, 2013: The multiple dimensions of food security*. Rome: FAO.

FAO, IFAD, WFP. (2015). The state of food insecurity in the world. Meeting the 2015 international hunger targets: Taking stock of uneven progress. FAO, Rome.

FCI (Annual Reports) http://fci.gov.in/finances.php?view=5.
GoI. (2005). *Performance evaluation of Targeted Public Distribution System (TPDS)*. New Delhi: Programme Evaluation Organisation, Planning Commission, Government of India.
GoI. (2011). The National Food Security Bill, 2011.
GoI. (2013). The National Food Security Act, 2013.
Gulati, A., Gujral, J., & Nandakumar, T. (2012). *National food security bill: Challenges and options*. New Delhi: Commission for Agricultural Costs and Prices.
Gulati, A., Sharma, P., & Kähkönen, S. (1996). The Food Corporation of India: Successes and failures in Indian foodgrain marketing. Center for Institutional Reform and the Informal Sector.
IFPRI. (2012). *Global hunger index 2012*. Bonn: International Food Policy Research Institute.
IIPS and Macro International. (2007). *National family health survey, 2005–06* (Vol. II). Mumbai: International Institute for population Sciences.
Iqbal, F., & You, J.-I. (2001). *Democracy, market economics, and development: An Asian perspective*. Washington, DC: World Bank.
Kattumuri, R. (2011). Food security and the Targeted Public Distribution System in India. Asia Research Centre.
Kumar, M. D., Sivamohan, M. V. K., & Narayanamoorthy, A. (2012). The food security challenge of the food-land-water nexus in India. *Food Security, 4*, 539–556. https://doi.org/10.1007/s12571-012-0204-1.
Ministry of Finance. (2013). *Economic survey 2012–2013*. New Delhi: Ministry of Finance, Government of India.
Ministry of Finance. (2016). *Economic Survey 2015–2016*. New Delhi: Ministry of Finance, Government of India.
New York Times. (2003). Does democracy avert famine? In: N. Y. Times. http://www.nytimes.com/2003/03/01/arts/does-democracy-avert-famine.html?pagewanted=all&src=pm. Accessed 16 May 2013.
Parry, M., Rosenzweig, C., & Livermore, M. (2005). Climate change, global food supply and risk of hunger. *Philosophical Transactions of the Royal Society B Biological Science, 360*, 2125–2138. https://doi.org/10.1098/rstb.2005.1751.
Ramachandran, N. (2011). Climate change, seasonality and hunger: The South Asian experience. In M. Behnassi, S. Draggan, & S. Yaya (Eds.), *Global Food Insecurity: Rethinking Agricultural and Rural Development. Paradigm policy* (pp. 201–215). New York: Springer.
Ramaswami, B., & Balakrishnan, P. (2002). Food prices and the efficiency of public intervention: The case of the public distribution system in India. *Food Policy, 27*, 419–436. https://doi.org/10.1016/S0306-9192(02)00047-7.
Rao, C. H. H. (1994). *Agricultural growth, rural poverty and environmental degradation in India*. Oxford: Oxford University Press.
Secretariat of CBD. (2013). *Biodiversity for food security and nutrition*. UNEP-CBD-FAO.
Sen, A. (1981). Ingredients of famine analysis: Availability and entitlements. *Quarterly Journal of Economics, 96*, 433–464.
Sharma, P., & Gulati, A. (2012). *Approaches to food security in Brazil, China, India, Malaysia, Mexico and Nigeria: Lessons for developing countries*.
Sharma, V. P. (2012). *Food subsidy in India: Trends, causes and policy reform options*. Ahmedabad: Indian Institute of Management.
Stamoulis, K. G., & Zezza, A. (2003). *A conceptual framework for national agricultural, rural development, and food security strategies and policies*. Food and Agriculture Organization of the United Nations. Agricultural and Development Economics Division.
Sunderland, T. C. H. (2011). Food security: Why is biodiversity important? *International Forestry Review, 13*(3), 265.
Sutopo, W., Bahagia, S. N., Cakravastia, A., & Samadhi, T. A. (2010). *A buffer stocks model for stabilizing price of staple food with considering the expectation of non speculative wholesaler*. In Proceedings of the World Congress on Engineering 2010 Vol III WCE 2010 June 30–July 2 2010 London, UK.

Swaminathan, M. (1996). Structural adjustment, food security and system of public distribution of food. *Economic and Political Weekly, 21*, 1665–1672.
Tyagi, D. S. (1990). *Managing India's food economy – Problems and alternatives*. New Delhi: Sage Publications.
United Nations. (1949). *United nations universal declaration of human rights 1948*. United Nations.
United Nations. (1975). *Report of the world food conference, Rome, 5–16 November 1974*. New York: United Nations.
United Nations. (2011). *The millennium development goals report 2011*. New York.
United Nations. (2014). *United Nations millennium development goals*. http://www.un.org/millenniumgoals/poverty.shtml. Accessed 17 Aug 2014.
USAID. (1992). *Food security*. http://www.usaid.gov/guatemala/food-security. Accessed 14 May 2013.
Virmani, A., & Rajeev, P. V. (2002). *Excess food stocks, PDS and procurement policy*. New Delhi: Planning Commission.
WFP. (2013). *WFP Strategic plan (2014–2017)*. Rome: World Food Programme.
World Food Summit. (1996). WHO|Food Security. In: WHO. http://www.who.int/trade/glossary/story028/en/. Accessed 14 May 2013.

Dr. Pooja Pal is Assistant Professor in the Post Graduate Government College for Girls, Sector 11, Chandigarh, India. She did her Master's in Commerce from the University Business School at Panjab University and is doctorate in management studies. She conducted research on the effects the climate change policies on the profitability and competitiveness of energy-intensive sectors of Indian industry. The title of her doctoral thesis is 'Cost and Competitiveness Implications of climate policy for Energy Intensive Sectors of Indian Economy.'

Dr. Himangana Gupta is an expert in climate change and biodiversity policy and diplomacy. In her present job, she is coordinating with scientists and climate experts to compile the Biennial Update Reports and Third National Communication to the UNFCCC. She is a doctorate in environment science with specialisation in climate change and biodiversity policy and was a University Gold medallist in M Sc. She has written research papers in reputed international and national journals on current state of climate negotiations, forestry, industrial efficiency, rural livelihoods and women in climate change mitigation and adaptation.

Dr. Raj Kumar Gupta is an anthropologist by education and journalist by profession. He did his Master of Science in Human Biology in 1977 and worked with three national dailies including National Herald, Indian Express in New Delhi and Ahmedabad. He was the chief of economic bureau with The Observer of Business and Politics in New Delhi. He wrote extensively on agriculture, food security, international trade, environment, and social policies and their interlinkages.

Dr. Tilak Raj is Assistant Professor in the University Business School of Panjab University, Chandigarh, India. He specializes in economics and general management. He has written several research papers on food security, agriculture, and small and marginal farmers in the hill state of Himachal Pradesh in northern India.

Chapter 7
The European Union as a Player in the Global Food Security

Szilárd Podruzsik and Olaf Pollmann

7.1 Hunger and Poverty

The current living conditions on our planet with an entire population of approximately 7.5 billion people (2017) cannot be guaranteed as about one ninth of the world population (795 million) cannot be provided with enough food. Ninety-eight percent of the world's undernourished people live in developing countries e.g. Asia (526 million), Sub-Saharan Africa (214 million), Latin America and the Caribbean (37 million). Countries such as Bangladesh, Benin, Burkina Faso, Ethiopia, India, Ghana, Malawi, Mexico, Mozambique, Peru, Senegal and Uganda suffer most of all from hunger.

In terms of gender, 60% of people suffering from hunger and the shortage of food are women and 50% of pregnant women in developing countries lack proper maternal care, resulting in approximately 300 thousand maternal deaths annually from childbirth. One sixth of infants in developing countries are born with a low birth weight and almost every 10 s a child dies from hunger-related diseases. Nearly half of all deaths of children under 5 are attributable to under-nutrition. With these environmental conditions especially women, children and elderly people are trapped in the circle of poverty and malnutrition (Fig. 7.1).

In addition to malnutrition, approximately 37 million people, 50% of whom are women, live with HIV/AIDS.

About 896 million people in developing countries, i.e. nearly 10% of the entire world population, live on USD 1.90 (at purchasing power in 2012) a day or less (World Bank Group 2016).

S. Podruzsik (✉)
Department of Agricultural Economics and Rural Development, Corvinus University of Budapest, Budapest, Hungary
e-mail: szilard.podruzsik@uni-corvinus.hu

O. Pollmann
SCENSO – Scientific Environmental Solutions, Sankt Augustin, Germany
e-mail: o.pollmann@scenso.de

M. Behnassi et al. (eds.), *Climate Change, Food Security and Natural Resource Management*, https://doi.org/10.1007/978-3-319-97091-2_7

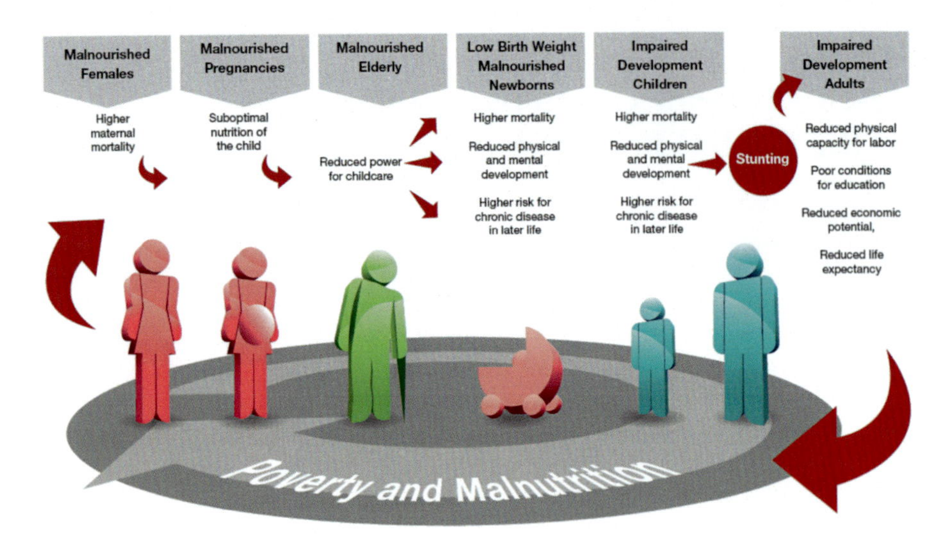

Fig. 7.1 Trapped in the cycle of hunger, generation after generation. (Source: Biesalski 2013)

The agricultural sector plays an important role in fighting poverty, as it is the backbone of economy in developing countries. About 70% of the world's poorest people live in rural areas and depend on agriculture and related activities for their livelihood. The burgeoning population pressure in developing countries has stressed the resources required for agriculture, which has also impacted on the quality of soil. Soil quality has decreased over time because of changing environmental conditions, such as climate change, worsening access to water, the increasing volume of cash crops farming. As a result of these facts about 50% of hungry people live in farming families. About 663 million people lack access to clean water and about 2.4 billion people did not have adequate sanitation in 2015. This situation even contradicts fundamental human rights.

Overall, hunger and poverty are related to various aspects of disadvantages, like the lack of adequate provision of necessary goods, water and food – a situation which needs to be remedied immediately.

As Ann Wigmore, a Lithuanian holistic health practitioner, nutritionist and health educator mentioned: *"The food you eat can be either the safest and most powerful form of medicine or the slowest form of poison."* Science has to bring equity in the provision of basic standards to the population worldwide and has to find solutions of how to support the world equally with all necessities to a satisfying and healthy life (Biesalski 2013).

7.2 Sustainable Development Goals (SDGs)

Based on the Millennium Development Goals (MDGs) adopted by 189 member states of the United Nations (UN) and expired in 2015, the member states of the United Nations accepted and signed an action plan, Sustainable Development Goals

Fig. 7.2 Millennium and sustainable development goals. (Source: European Commission)

(Fig. 7.2) to transform the present state of the world. MDGs served as guidelines for governments to strengthen the political will to work out their national polices and contribute to raising public awareness and allocate national resources for the purpose of poverty eradication. Although MDGs produced remarkable results such as stronger cooperation with the private sector and a radical decrease in poverty, new and complex challenges required innovative approaches during a broad consultation process to tackle them effectively. There are also additional tasks to enhance development, such as to lessen the differences between rural and urban areas, the differences in the availability of resources in developing and developed countries or climate change.

The tasks set out in the new plan 2030 Agenda for Sustainable Development are to contribute and strengthen the responsibilities of the actors towards sustainable development for the next 15 years. The plan came into force in January 2016 and puts an outstanding emphasis on global poverty as the greatest challenge on the planet. All aims announced in the plan focus on peoples' living standards directly and indirectly.

The most ambitious part of the plan is to reduce poverty and end hunger by 2030. As one of the elements of food security, the reduction of malnutrition is also a priority in the plan. Focusing on the allocation of financial and natural resources in developing countries, the essential tools need to improve nutrition and promote sustainable agriculture. The partnership and participation of international organisations in the accomplishment of the objectives are crucial for final results.

Relating to food security, nutrition and sustainable agriculture, Goal No 2 in the plan formulates the achievement of food security and the necessity to end hunger, to improve nutrition and to promote sustainable agriculture. These sub-goals target the agricultural sector and different groups of people that are the most vulnerable in terms of food insecurity. The goal encompasses the protection of infants under five, women, and older generations by increasing the availability of sufficient food supply, improving agricultural productivity, and supporting small agricultural producers, as well as by preventing environmental degradation and strengthening resilience to the changing environment. It promotes investment into technology, rural infrastructure and research done through international cooperation.

The criticism of the plan focuses on its number of goals and their sub-goals. Arising costs are also a huge burden on countries to build social safety nets to eliminate poverty and invest into infrastructure development in agriculture, water management or transport (UN 2015).

7.3 Food Security Policy in the EU

The growing demand for food safety and safe food poses crucial challenges for the European Union. Beside the increasing population and global food demand, economic growth and urbanisation are also key drivers affecting food security due to dietary changes. These challenges require a strategic response to the question: how to satisfy and secure the changing demand in a more coordinated and coherent food system. Since the competition for land is getting stronger, and the availability of land is limited, key solutions lie in the transformation of the agricultural system, in the prevention and maintenance of rural environment, and in retaining the balance between food consumption and production at different levels of the food supply chain. It is most desirable to maintain a demand-driven food system and an attitude of responsible consumption (Fig. 7.3).

From the aspect of the environmental challenge, water scarcity or soil degradation play an important role in food security, which in turn determine food availability. As climate change has a serious impact on productivity determining food prices, poor people, both producers and customers, are the most vulnerable (Maggio et al. 2015).

In its initial response the European Union set up new objectives for the Common Agricultural Policy to manage the challenges relating to rural society, the aging

Fig. 7.3 Global food security vision 2030. (Source: Maggio et al. 2015:11)

population, land management and migration. In the focus of these objectives were food and environment security and rural areas.

The EU adopted and adjusted its strategy to upcoming challenges. In its policy papers, such as the Food Security Regulation, Advancing the Food Security Agenda to Achieve MDGs or Thematic Strategic Paper 2007–2010 formulated the goals aiming at food security and expressed its commitment to achieving them.

These strategic documents emphasize the national, regional and more importantly the global dimension of food security to decrease poverty and hunger and to protect the most vulnerable population including children under 5 years of age and women.

In relation to development goals, food and nutrition security policy covers four areas in the EU. The aim of this policy is to fulfil development goals and present a multi-dimensional approach incorporating nutrition, health care and water availability. They take the key challenges relating to food availability into consideration. These are inadequate policies, land degradation, water scarcity and low productivity. The adjusted policies intend to generate positive effects on productivity, food availability for poor people in developing countries.

Access to food is determined by physical, economic and social elements. The EU's strategy highlights the importance of food quality and the appropriate utilisation of food as these have considerable effects on poor peoples' diets and malnutrition. Price and production, food stocks and early warning systems reflect the importance of food security.

The outstanding importance of the elimination of hunger is reflected in investment into food production facilities and projects that address nutrition and safety measures. The EU focuses on creating awareness on the consequences of malnutrition. The strategic priorities relate to a stronger mobilisation and political commitment, interventions on nutrition, and investment into applied research and information systems that are relevant at both national and international levels.

The solution of nutrition problems also requires treatment on a global scale and a stronger cooperation among countries through various initiatives and measures. The EU's contribution to the development is significant. A good example is the European Commission's initiatives in Sahel (Supporting Horn of African Resilience) to reduce hunger and poverty in one of the poorest regions of the world. The strategy for the resilience initiative mainly focuses on Mauritania, Mali and Niger, and regions that have close relations with these countries, all of which have strong effects on EU citizens. Sahel supports the resilience of the Horn of Africa in response to the crises caused by drought. The EU participates in global alliance with donor countries and other African States. The humanitarian and development actions support the coordination and assistance in the targeted countries (EC 2014).

Access to food is also an important element of the EU food security policy. The main barrier in terms of food affordability is the lack of income generating activities to enable people to have access to food. Employment opportunities and income generating activities would make food more affordable for poor people in rural and urban areas. The global food crises highlighted the importance of social transfers in reducing vulnerability, hunger and poverty. These social transfers, such as cash, food-for-work or vouchers ensure the access to food in short term. Protective and productive social transfers are the potential solutions to food insecurity.

A new communication policy comprising ten steps supports the design of national strategies for resilience, disaster management and warning systems. Beside the realisation of its policy, the EU is also committed to the monitoring and measurement of the impact of its actions.

7.4 The Contribution of the EU to the Improvement of Food Security

The European community takes part in and is committed to a worldwide development on the basis of solidarity. The Treaty of Rome launched the European Development Fund (EDF) in 1957. The Fund started to operate in 1959. The aim of this European instrument is to provide aid to African, Caribbean and Pacific countries (ACP), overseas countries and territories (OCTs). Although the EDF is financed by the member states, its budget is separated from the EU's budget. The 11th EDF as an intergovernmental agreement was signed in 2013. It covers economic development, social and human development, regional cooperation and integration and supports a more flexible and prompt reaction to unexpected events in the period of 2014–2020.

The EU is an active player in actions reacting to the challenges of hunger and malnutrition. The aim of the Food Security Thematic Programme (FSTP) is to support poor people and reduce the malnutrition of the most vulnerable population.

Regarding food security, national activities involve bilateral discussions at not only regional but also at continental and international levels in cooperation with the private sector. The cooperation with G8 resulted in a New Alliance with an aim to lift 50 million people out of poverty in 10 years. The EU is in the forefront of this initiative. The aim of this cooperation with partner governments on the basis of the EU policy framework is to eliminate undernutrition. At the same time, the movement called Scaling Up Nutrition (SUN) reacts to the problem of increasing undernutrition in partnership with governments, civil society, private companies, research institutions, the UN organisations and the World Bank (EC 2017).

The EU actively collaborates with the Food and Agriculture Organisation (FAO), and the International Fund For Agricultural Development (IFAD), too. The ever-stronger cooperation between the EU and FAO also aims to reduce hunger and poverty. Knowledge transfer between the two organisations and rural areas plays an important part in the process.

The EU emphasized that its activity on food security should be in line with international governance to ensure its efficiency. New challenges require the coordinated response of the EU, that is why the measures are adjusted to the development goals and they ensure the allocation and efficient use of available resources. To ensure food security, the EU invests into technology development, supports farmers, aids national and regional food security strategies and assists the harmonisation of the international systems of governance.

On the basis of the agreement between the EU and the FAO the minimum number of target countries participating in the FAO-EU partnership is set to 35 with the aim to improve and develop food security, nutrition, sustainable agriculture and resilience. The need of countries and demand-driven measures are supported with a budget of EUR 73.5 million allocated by the two organisations. In line with the SDGs, the technological and political cooperation and coordination target the reduction of poverty and hunger with two strings of the 5-year programme: Food and Nutrition Security Impact, Resilience, Sustainability and Transformation (FIRST) which provides policy assistance and capacity assistance to support local governments and administrations; and Information for Nutrition Food Security and Resilience for Decision Making (INFORMED) that relates to resilience measures and provides up-to-date information to policy as well as decision makers to withstand the food crisis.

The EU budget dedicated to the Development and Cooperation Instruments contributes to the financial basis of the initiatives in the framework of the Global Public Goods and Challenge (GPGC) program. The EU Trust Fund for Africa was launched with a budget of EUR 66.5 million to address climatic phenomena, e.g. El Nino is to eliminate the results of floods and droughts. The EU focuses on countries with the most severe food insecurity, such as: Ethiopia, Somalia, South Sudan and the Trust Fund supports the fast and collective responses to emergencies in the world. All these specific tools are to counter the negative impacts of climate change (FAO 2015). The priorities of the Trust Fund relate to economic programmes, resilience, migration management, stability and governance. The budget of the Trust Fund is secured by the European Commission and it is supplemented with new resources such as the Development Cooperation Instrument (DCI). The aims of the actions under the priorities are to improve employment by creating new jobs in local communities, providing services to the most vulnerable groups, assisting migration management by the contribution to the national and regional strategies which last but not least play and important role in supporting overall governance in target areas (EC 2015).

7.5 The Institutional Framework of Development Policy

Decision making is a complex process globally and especially in the European Union. The initiatives of the European Commission are pre-evaluated. After the assessment of the impacts on the economy, society and the environment, the possible policy is formulated in consultation with a wide range of interested groups to ensure the reduction of bureaucracy and the approval by the interested groups.

The Council of the European Union composed of national ministers of member states has the right to review the EC's initiative and make suggestions for improvements. The European Parliament, the co-decision making organization assist on matters related to legislation and the budgetary process and scrutinises the implementation of the EU policy. In case of disagreement with the initiative, the European Parliament has its own right to refuse it.

The strategic guidance arrives from the heads of state and government who constitute the European Council, without formal legislative actions.

The European External Action Service (EEAS) collaborates with the EU, as well as with international institutions such as the United Nations. The EEAS is responsible for the coherence of the EU actions in line with the development objectives, in cooperation with international collaborators.

The European Investment Bank being the financial institution for the EU remains the largest financial player in the common market, entrusted with the mandate to provide financial instruments for sound and sustainable investment projects in developing countries.

The EU policies and finances are audited by the European Court of Auditors. Its role is to evaluate the expenditures made and revenues generated from the European budget in different areas including development cooperation (Fig. 7.4).

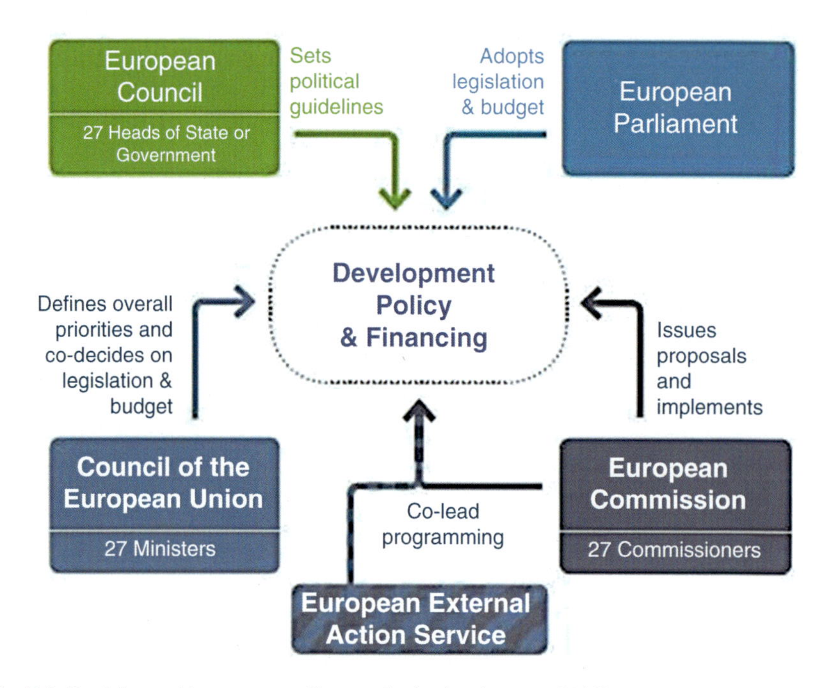

Fig. 7.4 Decision making process. (Source: Seeks development 2012)

7.6 Responsibilities of Affected Regions and Countries

During the past centuries, most of the countries supported focused mainly on the donor-taker-mentality. This relationship needs to be changed. Nowadays, it is more than important to support less-developed and developing countries in terms of donation, consultation or even manpower. However, equally important is that the countries receiving funds take over the responsibility and take their own initiatives. These countries know their most urgent demands best and are better aware of their immediate obstacles and circumstances.

The Federal Ministry for Economic Cooperation and Development (BMZ) in the Federal Republic of Germany published the document "Marshall plan with Africa" (2001/2017) with the aim of partnering Africa and Europe for development and peace. With this plan, Europe and especially Germany will tackle the problems of peace, poverty and climate change in order to support Africa to become autonomous and self responsible.

Africa was reckoned as the continent of resources. Africa has 15% of global oil reserves, 40% of global gold reserves and 80% of global platinum reserves (Fig. 7.5). These facts clearly show a main focus on export within economically strong

Fig. 7.5 Natural resources in Africa. (Source: CIA Factbook 2016)

countries and Europe. To support and make Africa sustainable, it would have been more substantial to invest in knowledge, develop capacity building programs and create employment opportunities. For the continent of 54 countries with an average age of 18, and the likelihood of the population having doubled by 2050, it would be more appropriate to invest into people and the economy to prevent migration and escape.

Investing into the education of young people would ensure the future of a country. With the support of Europe to gain brain on the continent, issues like climate change, food security, global economic growth will become manageable or even successful highlights of a well-functioning capacity development. If the support was only the transfer of funding with the possibility given to beneficiaries to make investment decisions themselves, the resulting development would be different from what international donors expect – because of the different understanding of successful and effective support and demand. The problems and possible solutions need to be discussed before financial support will be provided by Europe and other donors.

7.7 The Collaboration of Economic Unions

The strategy of the EU to be the global player in food security and main fighter against poverty through funding has not yielded the expected success so far. To sustain international funding, the involvement of economic unions and/or other funding agencies seems to be essential for success. Especially in Africa, economic unions are strong partners on the way of economic growth and eradicating poverty. Trade cooperation is implemented in the framework of either the Arab Maghreb Union (UMA) as the Common Market for Eastern and Southern Africa (COMESA) or the Community of Sahel-Saharan States (CEN-SAD) as the East African Community (EAC). In addition to intergovernmental organisations such as the Southern African Development Community (SADC), some regional cooperation exists to raise the living standard of people in the region, such as the Economic Community of Central African States (ECCAS) and the Economic Community of West African States (ECOWAS).

Several of the Regional Economic Communities (RECs) overlap in terms of their membership: for example, East Africa, Kenya and Uganda are members of both the EAC and COMESA, whereas Tanzania, also a member of the EAC, left COMESA and joined SADC in 2001. This multiple and confusing membership creates duplication and sometimes competition in activities, while placing an additional burden on the already overburdened staff responsible for foreign affairs to attend all the various summits and other meetings.

Even by recognising the overlap within these economic communities, the overall benefit for supporting poor countries outweigh the disadvantage. With a strong community network and economic support, large-scale educational insufficiencies can be overcome. Within this network capacity development, knowledge and scientific research can be supported and uplifted onto an internationally recognised level.

7.8 Conclusion

Among global challenges, food security and nutrition appear as priority areas that require solutions to save vulnerable population on the Earth. It would not be possible to come up with feasible solutions in the absence of strong and concrete decisions and without the support and cooperation of developed countries. The elimination of poverty and combating hunger are not just development issues, they are also humanitarian issues.

From the outset, among other development initiatives and development aims, food security has been the topmost priority of the EU. The European Union takes the responsibility to provide support to vulnerable societies through its institutional and financial framework. The EU cooperates with international organisations such as the United Nations and takes part in alliances to work out programmes that contribute to increasing the availability of food and its affordability in efficient ways. Alongside different development goals, the allocation of financial resources depends on the needs of people and is considered as an investment to improve living conditions and living standards in developing countries. The European Union is among the most important donors worldwide.

However, the nations receiving support from the EU must also fulfil their responsibilities to initiate and continue the collaboration with other developing countries and make efforts to achieve food security through implementing relevant policies and identifying specific gaps where finances could be particularly required. This would also help streamline available financial resources.

References

Biesalski, H. K. (2013). *Hidden hunger*. Berlin: Springer.

CIA Factbook. (2016). Report: From fragility to resilience: Managing natural resources in fragile situations in Africa.

European Commission. (2014). *EU approach to resilience: Learning from food crises (Factsheet)*. http://ec.europa.eu/echo/files/aid/countries/factsheets/thematic/resilience_africa_en.pdf

European Commission. (2015). *The EU emergency trust fund for Africa*. https://ec.europa.eu/europeaid/regions/africa/eu-emergency-trust-fund-africa_en

European Commission. (2017). *Food and nutrition security – International relations*. https://ec.europa.eu/europeaid/sectors/food-and-agriculture/food-and-nutrition-security/international-relations_en

FAO. (2015). *European Union and FAO launch new programmes to boost food and nutrition security, sustainable agriculture and resilience*. http://www.fao.org/news/story/en/item/298350/icode/

Maggio, A., Van Criekinge, T., & Malingreau J. P. (2015). *Global food security 2030 assessing trends with a view to guiding future EU policies, JRC science and policy reports*. European Commission DG Joint Research Centre.

SEEK development. (2012). *Global development policy making in the European Union*. www.seekdevelopment.org/seek_donor_profile_eu_april_2012.pdf

The Federal Ministry for Economic Cooperation and Development (BMZ). (2017). *Africa and Europe – A new partnership for development, peace and a better future*.

United Nation. (2015). *Transforming our world: The 2030 agenda for sustainable development*. http://www.un.org/ga/search/view_doc.asp?symbol=A/RES/70/1&Lang=E

World Bank Group. (2016). *Global monitoring report 2015/2016: Development goals in an era of demographic change*. Washington, DC: World Bank. https://doi.org/10.1596/978-1-4648-0669-8. License: Creative Commons Attribution CC BY 3.0 IGO.

Dr. Szilárd Podruzsik holds a PhD in economics from the University of Economic Sciences and Public Administration in Budapest. He works for the Corvinus University of Budapest as a senior lecturer. His research fields cover the areas of agriculture and food industry. Currently, his research focuses on the food consumer welfare, food logistics and its process optimisation. In his research, he applies different models to help stimulate, estimate and evaluate the relevant sectors.

Olaf Pollmann is a civil engineering and natural scientist. He is holding a Doctorate (Dr.-Ing./PhD) in the field of environmental-informatics from the Technical University of Darmstadt, Germany and a second Doctorate (Dr. rer. nat./PhD) in the field of sustainable resource management from the North-West University, South Africa. Olaf Pollmann is a visiting scientist and extraordinary senior lecturer at the North-West University and CEO of the company SCENSO – Scientific Environmental Solutions in Germany. He is also deputy head of the section "African Service Centers" in West (WASCAL) and Southern Africa (SASSCAL) on behalf of the Federal Ministry of Education and Research (BMBF).

Part II
Climate, Water, Soil and Agriculture: Managing the Linkages

Chapter 8
Climate Change Impacts on Water Supply System of the Middle Draa Valley in South Morocco

Ahmed Karmaoui, Guido Minucci, Mohammed Messouli, Mohammed Yacoubi Khebiza, Issam Ifaadassan, and Abdelaziz Babqiqi

8.1 Introduction

Climate change presents a risk for water resource in the developing countries in Africa where agriculture is the main economic activity (Diao et al. 2010). Global climate models suggest that temperatures are expected to increase by 2–6 °C by the end of this century (IPCC 2014). In fact, basins under water stress are located in North Africa in the Mediterranean region (Bates et al. 2008) and the Draa basin in the south of the High Atlas Mountains (Morocco). In 2014, the demand for water exceeded the available supply by more than 25% (Karmaoui et al. 2015a). Fresh water sustains inland water ecosystems (rivers, lakes, and wetlands), providing cultural, regulatory, and supporting services that contribute directly and indirectly to human well-being through recreation, scenic values, and maintenance of fisheries (Aylward et al. 2005). In fact, it is the basis of other ecosystem services, maintains ecological balance and aids socio-economic development.

The paper explores the incidence of socio-economic impacts and climate change on water demand. First, the analysis is carried out using Statistical downscaling models (SDSM software), to draw future projections of two meteorological

A. Karmaoui (✉)
Department of Environmental Sciences (LHEA-URAC 33), Faculty of Sciences Semlalia, Marrakech & Southern Center for Culture and Sciences, Zagora, Morocco

G. Minucci
Department of Planning and Urban Studies, Politecnico di Milano, Milan, Italy

M. Messouli · M. Y. Khebiza · I. Ifaadassan
Department of Environmental Sciences (LHEA-URAC 33), Faculty of Sciences Semlalia, Cadi Ayyad University, Marrakech, Morocco
e-mail: issam.ifa@laposte.ne

A. Babqiqi
Regional Observatory of the Environment and Sustainable Development, Moroccan State Secretariat for Environment, Marrakech, Morocco

M. Behnassi et al. (eds.), *Climate Change, Food Security and Natural Resource Management*, https://doi.org/10.1007/978-3-319-97091-2_8

quantities (precipitation and temperature); and the second, Water Evaluation and Planning System (WEAP) is used for the management of water resources. The outputs of these analyses could be helpful to support decision-making in matters related to eventual climate change and anthropogenic impacts on water resources, for future urban, agricultural, and environmental uses. The outputs can support decision making on:

- How can future projections be used into water resources planning at local scale?
- How can decision-making tools be used to quantify the eventual impacts of climate change on water resource in the Middle Draa Valley (MDV)?

The main objective of this paper is to examine how climatic and anthropogenic factors impact water supply; focusing on projections from 2010 to 2099.

8.2 A Brief Introduction to the Study Area

The MDV is an oasean region located in the middle part of Draa Basin. It was declared a Biosphere Reserve by UNESCO in 2000. The region is characterized by low population density (17,5 inhabitants / km^2) and a heterogeneous spatial distribution. Census data for 2014 in the province of Zagora (MDV) reported a population of 307,306 inhabitants, where 256,558 were located in the rural area. Most communes of the Middle Draa Valley recorded a growth rate, but in some communes the population change has recorded negative values as M'Hamid commune in the downstream of the valley. This is due to migration to urban centers and abroad (Karmaoui et al. 2015b).

Surface water resources in the Middle Draa Valley consist of the Draa Wadi fed by Mansour Eddahbi Dam. In the MDV, the oases occupy about 26,000 ha in six palm groves, dominated by the date palm promoting micro-hot and humid climate conducive to diversified agricultural production (Karmaoui et al. 2014a). This region is highly vulnerable to drought events, which are frequent and severe and have devastating impacts on population and economy.

8.3 Materials and Methods

The methodology used in this paper is based on the use of two tools:

- The SDSM model to develop climate scenarios that will be used for the WEAP software;
- The WEAP model to assess the water vulnerability used in Upper Draa Valley (Karmaoui et al. 2014b).

Data (water demand) was collected from the ONEE (Office national d'Electricité et d'Eau potable), the ORMVAO (Office régionale de mise en valeur agricole

Table 8.1 Demand sites of MDV

Demand sites		Sites of supply	
Urban sites	Agricultural sites	Groundwater	Water surface
Ouarzazate	Mezguita	Mezguita	Draa Wadi
Zagora	Tinzouline	Tinzouline	Mansour Eddahbi
Agdez	Ternata	Ternata	Dam
	Fezouata	Fezouata	
	Ktaoua and M'hamid	Ktaoua and M'hamid	

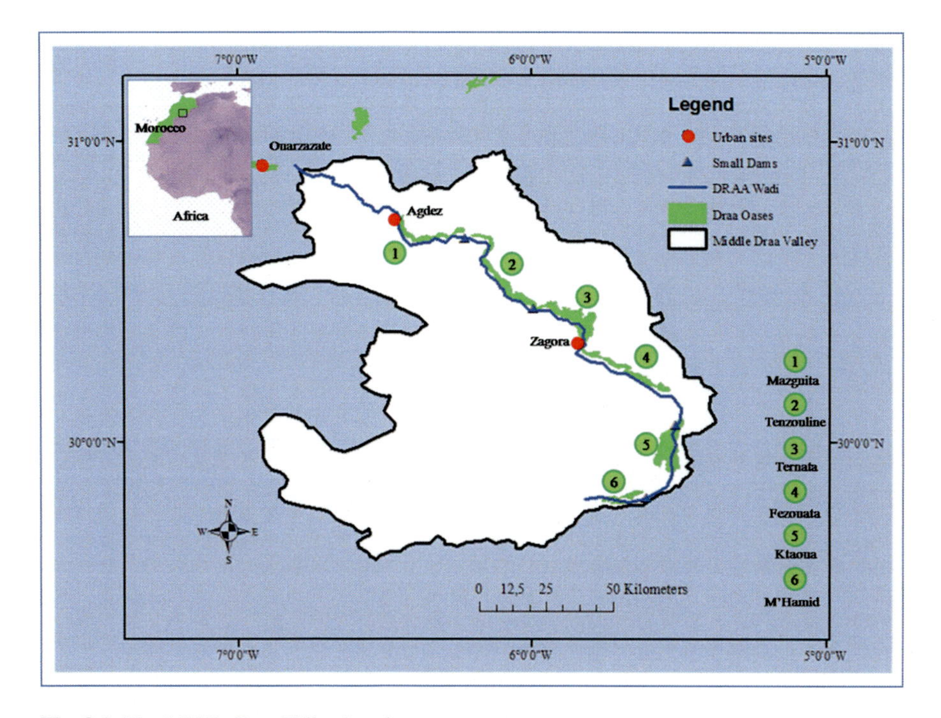

Fig. 8.1 The Middle Draa Valley location

d'Ouarzazate) and ABHO (Agence du basin hydraulique d'Ouarzazate) for climatic data and the dam outflow and inflow.

The adopted modeling process aims at exploring the vulnerability scenarios using the meteorological quantities of precipitation and temperatures in the two climate change scenarios A2 and B2. Table 8.1, Figs. 8.1 and 8.2 show the selected demand and supply sites for this study. Three urban centers derive their water supply from the Draa valley groundwater. Four agricultural sites derive water both from the six-groundwater sites and from the two water surface sites (Draa Wadi and Mansour Eddahbi Dam).

WEAP can be used for urban and agricultural systems. Figure 8.1 shows the study area and the approximate location of the nine demand sites (all demand and

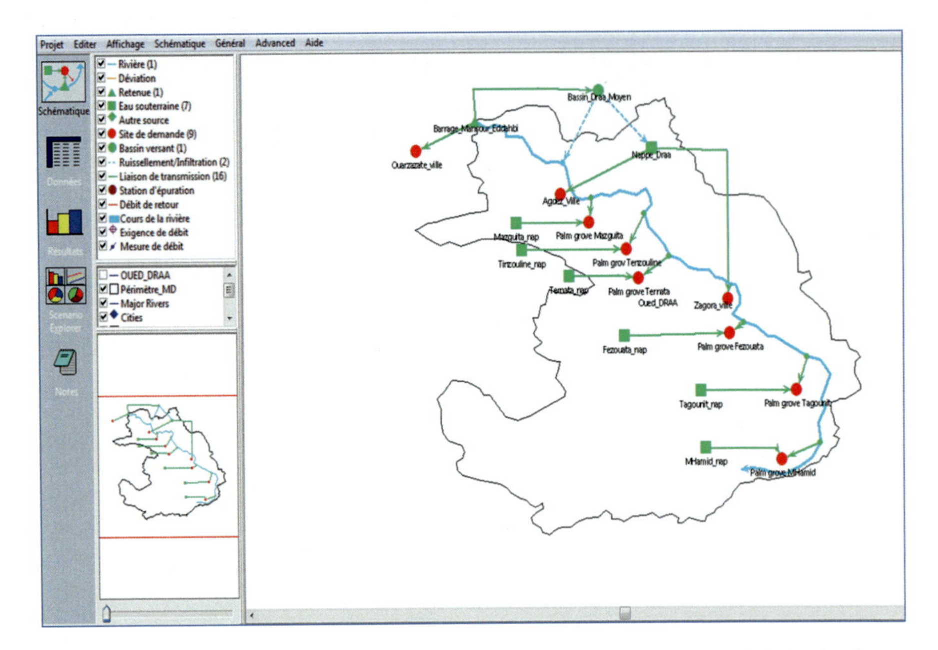

Fig. 8.2 Diagram of the WEAP model including the Middle Draa valley and all demand and supply sites, south east of Morocco

supply sites in Middle Draa Valley) simulated using WEAP. Ouarzazate urban centre (near the Mansour Eddahbi Dam) is also shown in Fig. 8.2 although it is not a part of the Middle Draa Valley, because it shares the reserves of this dam (the main source of water of the Middle Draa Valley). The model schematic (Fig. 8.2) shows WEAP node-network and the GIS layer of the Middle Draa Valley. Water demand is aggregated into three urban demand sites (Ouarzazate, Zagora and Agdez) and six agricultural demand sites (Mezguita, Tinzouline, Ternata, Fezouata, Ktaoua and M'Hamid), see also Fig. 8.1. These demand sites are supplied water from both groundwater and surface water.

8.3.1 Elaboration of Climate Change Scenarios A2 and B2 at Local Scale Using SDSM

The Model of statistical downscaling (SDSM) is a tool conceived to assess the impact of local climate change. These vulnerability scenarios were generated by using the meteorological quantities of rainfall and mean temperatures in the two climate change scenarios A2 and B2. To project the climate change scenario, we have used the SDSM. This software allows predicting the mean temperature and precipitation for the selected period of 2010–2080 based on climatic data of the period 1961–2000. For calibration and validation of data, we calculated the Root Mean Square Error (RMSE) between the monthly mean temperature observed and modeled between January 1981 and December 2000 for the parameter of average temperature, and compared between monthly total rainfall (mm) observed and that calculated by SDSM during the same period, based on data prepared by Babqiqi (2014).

8.3.2 Elaboration of Socio-economic Scenarios for the WEAP Model 'Water Evaluation and Planning System'

WEAP is a software that integrates physical hydrology with priority-driven water resources allocation, and is specifically built to support policy and planning (Mehta et al. 2013). The object-oriented approach and all equations are detailed in Yates et al. (2005). Using WEAP, scenarios can be built and then compared to assess their impacts; all scenarios start from a common year, for which the model Current Accounts data are established (Sieber et al. 2005).

Two principal scenarios are used, which are: firstly, the reference scenario is with population growth at a rate of three in urban area (RGPH 2004). The Reference scenario is the scenario in which the current situation (2010) is extended to the future (2011–2080). The current situation for the reference scenario was set at 2010 since a complete set of data is only available for this year, whereas for the years 2011–2015, data are not available or are only partially available. However, it is worth noting that no major changes are imposed in this scenario and that slight changes occurred only in the cropping pattern during these years (2011–2015). Besides the reference scenario, one other scenario is analyzed. This represents the high rate of population growth that we estimated at 4%. This value is the maximum value recorded in the part that encompasses the valley. In fact, urbanization has a phenomenal dimension with an average (annual) growth rate of 5.75% in 44 years (regional) compared to 3.06% for national urban (the Moroccan country) during the same period (RBOSM 2008).

8.4 Results

8.4.1 Elaboration of Climate Change Scenarios in the Draa Valley

For the case of Middle Draa, the SDSM program reproduces the average temperature with a degree of precision that remains sufficient to assess the impact of climate change on water demand (Figs. 8.2 and 8.3). The RMSE is in the order of 0.31 for the mean temperature. Similarly, the accuracy of the modeled seasonal mean

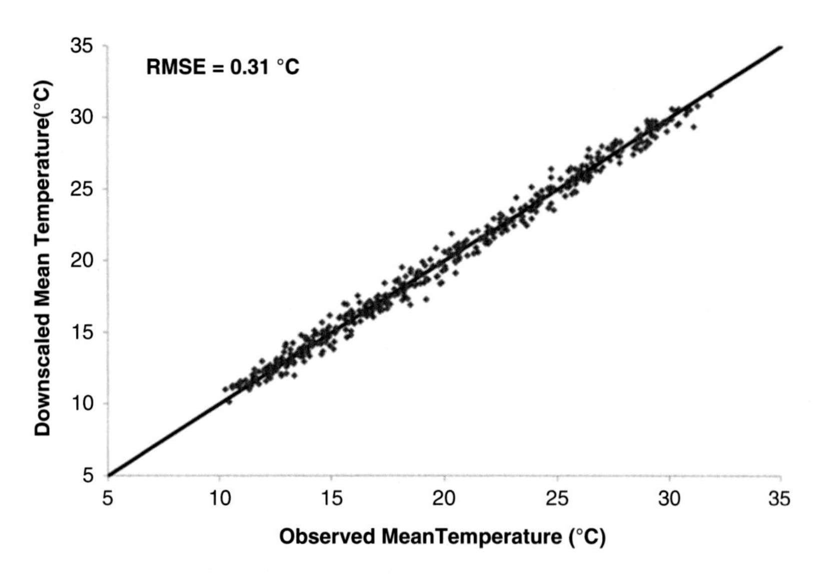

Fig. 8.3 Scatter plot with the RMSE between monthly mean temperatures observed and modeled between the period January 1981 and December 2000

temperatures is 0.37, 0.25, 0.19 and 0.18 °C respectively for spring, summer, autumn and winter.

RMSE values were taken into account to use climate change scenarios developed by the SDSM on WEAP, adjusting the values predicted by SDSM, and by shifting slightly the SDSM near RMSE values. According to Fig. 8.3, there is a strong correlation between past and predicted values by SDSM; this correlation is perfect with a correlation coefficient of 1.

Figure 8.4 shows a comparison between the average monthly precipitation observed and reproduced from the technical downscaling SDSM over the 1981–2000 periods. There is a general agreement between these two types of data. The maximum difference is 8.3 mm and occurs during the autumn season (September, October, and November) when the SDSM underestimates the amount of rainfall actually observed.

Also, Fig. 8.3 shows that while downscaled values have the same trend as the observed values, the actual values are different. The observed values are higher than the downscaled values during most of the months of the year (from November to May). In the months of June to September, the downscaled values are higher than the observed values. Overall, SDSM reproduces the annual rainfall with an RMS error of 4.6 mm. So as in the case of temperature, the accuracy of the SDSM technique is acceptable for most studies assessing the impact of climate change on water resources.

During the period 2001–2080, and under the A2 climate change scenario (Figs. 8.5 and 8.6), in winter for example, the SDSM model predict an increase in mean temperature from 1 to 3 °C, and a decrease of rainfall from 0% to 19%. In autumn, during the same period, we see an increase from 0.5 to 4 °C and a decrease

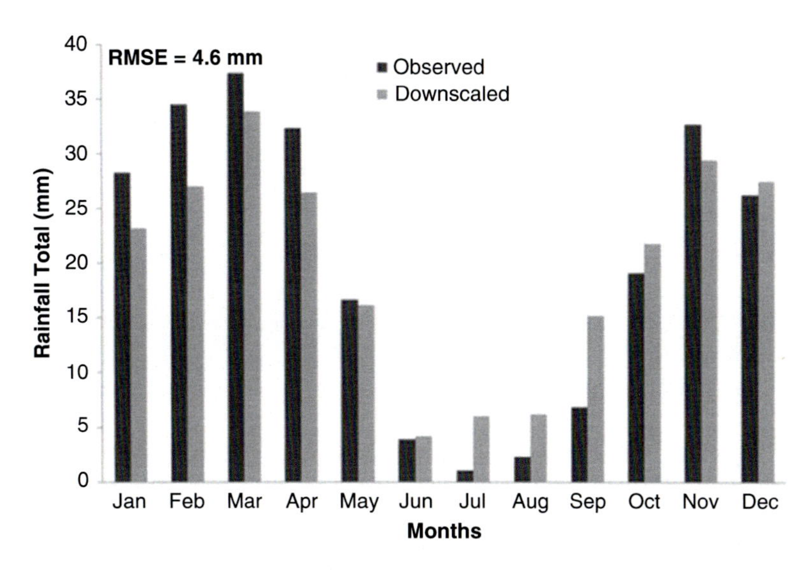

Fig. 8.4 Total rainfall: Comparison between total monthly rainfall (mm) observed and calculated from SDSM under the period of 1981–2000

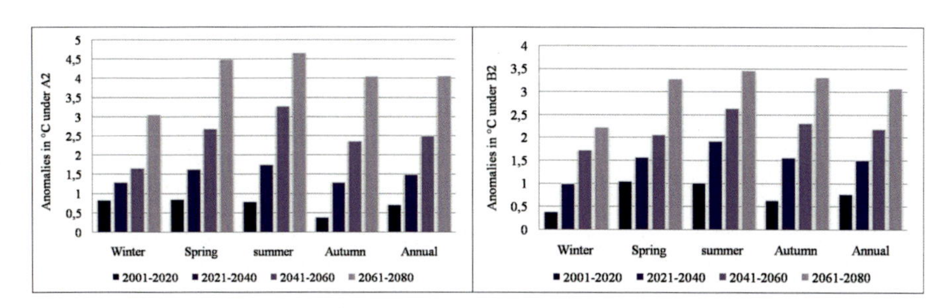

Fig. 8.5 Seasonal and annual anomalies (°C) of mean temperature for the three future horizons 2020, 2050 and 2080 and for both A2 and B2 scenarios at MDV

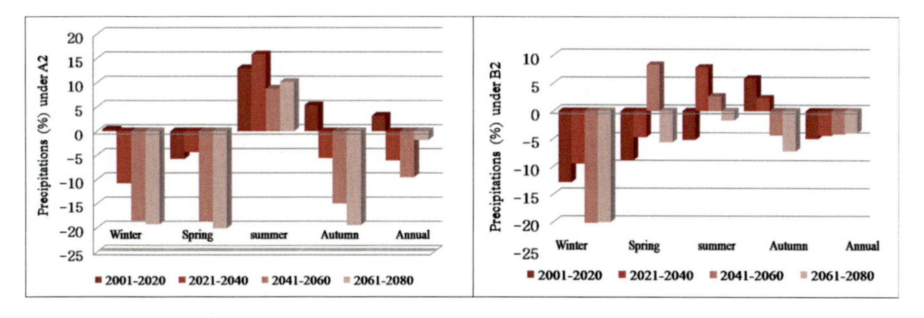

Fig. 8.6 Percentage change in the level of cumulative seasonal (winter, spring and autumn) and annual for the three future horizons 2020, 2050 and 2080 and for both A2 and B2 scenarios at MDV

from 5% to 20% in rainfall. Under B2, and in winter season, the mean temperature will increase from 0.5 to 2.3 °C, the precipitation will decrease from 12% to 20%. In autumn, the mean temperature may increase from 0.6 to 3.2 °C, and the precipitation may decrease from 5% to 7%.

The calculation of future climate anomalies (2011–2040, 2041–2070 and 2071–2099) relative to the current climate (1961–2000) for mean temperature and precipitation shows increased temperatures and decreased precipitation for the mentioned periods. In fact, the results predict an increase in mean temperatures. Temperatures could rise from 1.4 to 3 °C between 2011 and 2070 in scenario A2 and from 1.6 to 2.7 °C in B2. The decrease in precipitation is estimated to be up to approximately 5.1–15% in A2 and 3.9–8.9% in B2.

The changes in precipitation and temperature resulting from climate change are expected to reduce the agricultural and urban water supply and impact water demand significantly. The trends observed under the two scenarios were used in the WEAP software to predict the water supply and demand in the Middle Draa Valley.

8.4.2 Water Model in the Middle Draa Valley

The data on water resources in MDV consisting of three urban cities (Agdez, Zagora and Ouarzazate) and six separated palm groves was compiled and subsequently incorporated into WEAP.

Regarding the water demand for the three urban centers of Ouarzazate, Zagora and Agdez, the model predicts (Fig. 8.7) an increase in water demand in Agdez city

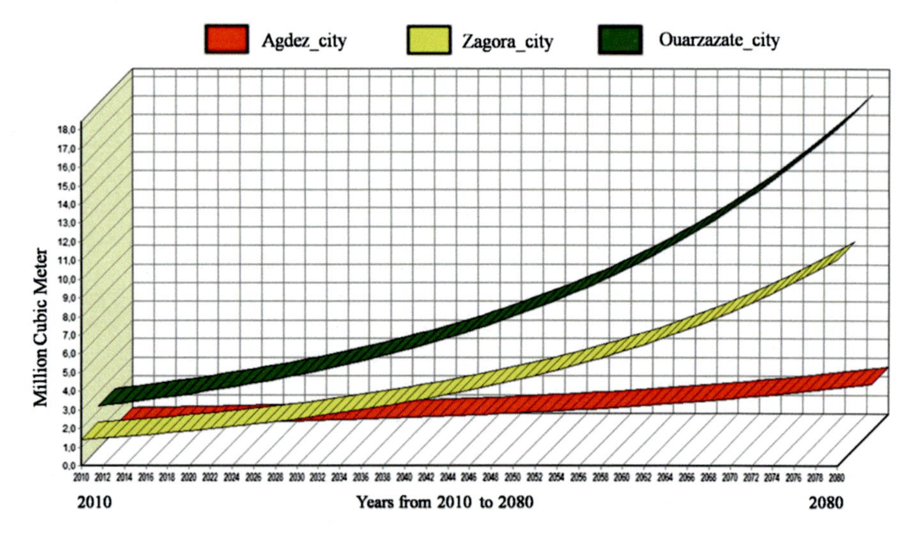

Fig. 8.7 Water demand of the tree cities (Ouarzazate, Zagora and Agdez), under reference scenario

from 0.3 Million Cubic Meters (MM^3) in 2010 to 0.6 MM^3 in 2030 (two times the water demand in 2030 and in Zagora city from 1.4 MM^3 (2010) to 2.5 MM^3 (2030) and in Ouarzazate city from 2.3 MM^3 in 2010 to 4.1 MM^3 in 2030.

For water demand for the three urban centers together (Fig. 8.8) under the three selected scenarios (Reference scenario and two climate change scenarios A2 and B2), the WEAP model predicts an increase in demand from 0.3 MM^3 in 2010 to 2 MM^3 in 2030 to 27 MM^3 in 2060. The B2 scenario indicates that total water demand will increase from 0.3 MM^3 in base year 2010 to 1.2 MM^3 in 2030 and 8.3 MM^3 in 2060. However, under reference scenarios, the demand will increase from 0.3 MM^3 in 2010 to 0.7 MM^3 in 2030 and 2.5 MM^3 in 2060.

Figure 8.9 shows the water demand of the six palm groves (Mazguita, Tinzouline, Ternata, Fezouata, Tagounite and M'Hamid) classified by water availability and

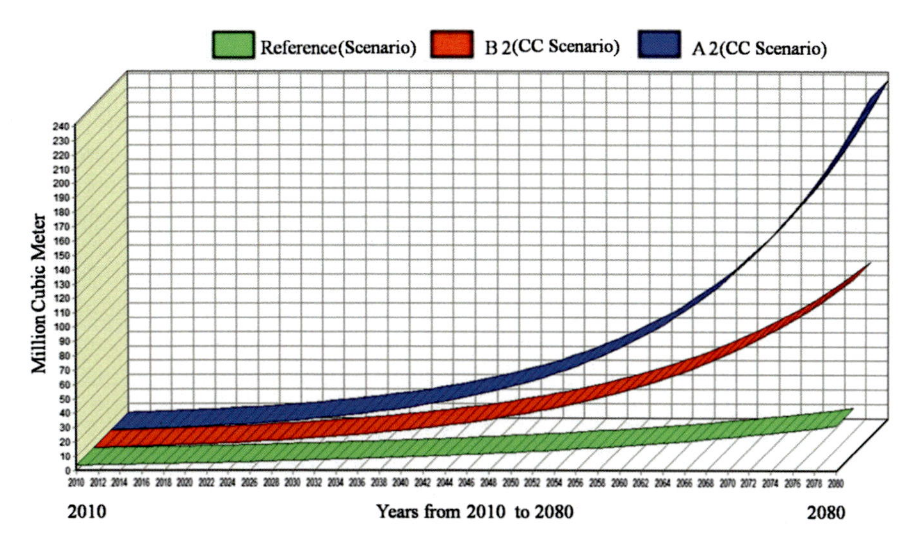

Fig. 8.8 Water demand of all urban sites (the tree cities: Ouarzazate, Zagora and Agdez), under three scenarios (CC: Climate Change A2 and B2 and reference scenario) in Million Cubic Meter (MM^3)

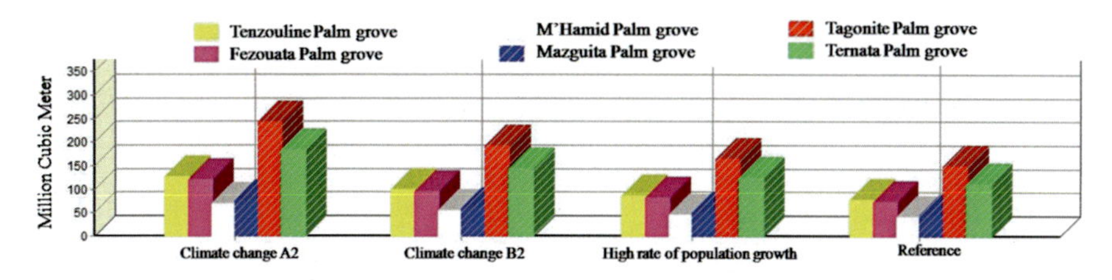

Fig. 8.9 Water demand (not including loss and reuse) for the six palm groves under 4 scenarios (Climate change A2 and B2, High rate of population growth and the reference scenario) for the year 2020

agricultural land. Under the reference scenario, as a result of reduced losses, the water volume delivered to consumers would increase in Tinzouline palm grove for example from 52.4 MM^3 in 2010 to 77.9 MM^3 in 2020 (Fig. 8.9). The increase is about 30% in the six palm groves by 2020. Demand coverage to all these palm groves would increase under Climate change A2 scenario following B2 and the high rate population growth compared to reference model.

The region is hilly and is crossed by nearby Wadis flowing north-west into the Iriki Lake (the extreme downstream). Although much of Draa is classified as a water-scarce region. Water is supplied to the urban area and surrounding rural areas by groundwater sites. The urban population increases by about 3% (RGPH 2004), which potentially swells water usage. Water demand would steadily rise as a result of population increasing at 4% annually from 2010.

The waterworks system infrastructure and operational capacity is the primary challenge in this region. The utility has several plans under way to increase water production. These include: increased water abstraction from groundwater sites; sourcing secondary abstraction sites; extension of the waterworks system network; groundwater abstraction to supply a few local institutions; and hydropower generation to reduce electricity costs. The climate change can affect the land class inflows and outflows (runoff area, precipitation, irrigation, increase of soil moisture, groundwater flow, evapotranspiration and soil moisture decrease) (Fig. 8.9). This impact is a threat to water supply, and food (land productivity) security in this vulnerable region. The changes in land class inflows and outflows starts to be visible from 2050 in the Reference scenario (Fig. 8.10a). However, this unbalance is visible from the year 2015 under climate change B2 scenario for example (Fig. 8.10b); the situation will be serious under climate change B2 scenario. We can also find that the sum of the outflow and the inflow for the reference scenario appears higher than inflow and outflow under climate change scenario.

8.5 Discussion

The A2 and B2 scenarios were used because they are closest to the trajectory of the evolution of Moroccan society and changes associated with climate indicators (Gommes et al. 2008). The models developed in this paper highlight key findings for each demand site (in agricultural and urban sectors). The agricultural sector is the biggest consumer of water resources (Karmaoui et al. 2015a; Heidecke and Thomas 2010). For the distribution of water resources in the different users (sectors), and from the total amount of exploitable water resources; 96.66% is used for agriculture, 2.70% for domestic, 0.28% for tourism, and 0.36% for economic activities (SADAM 2003).

Four scenarios were designed to investigate the effectiveness of policy options in the area (Climate change A2 and B2, reference scenario and the high rate of population growth scenario).

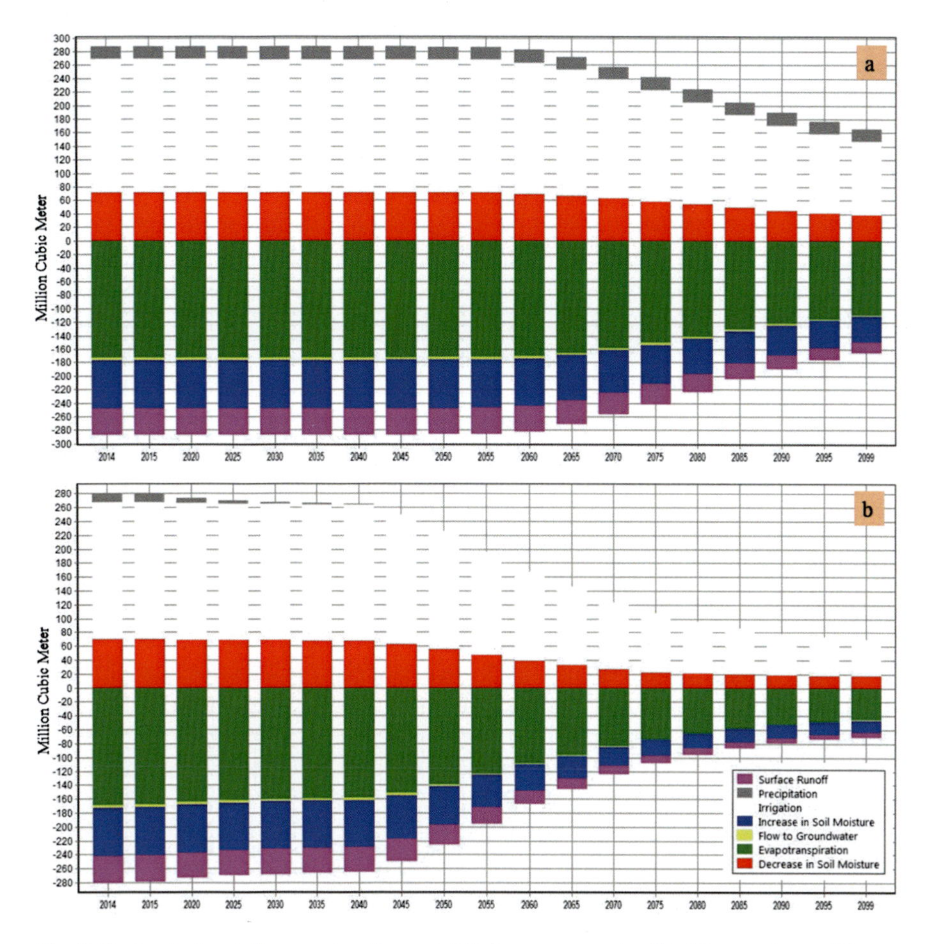

Fig. 8.10 Land class inflows and outflows. (**a**) Reference scenario; (**b**) Climate change B2 scenario from 2014 to 2099. The outflows are represented as negative values and the inflows as positive values

Based on the climatic model developed in this paper, an increase in temperature by 4.6 °C in summer (A2), 3.5 °C (B2) and 3 °C (A2), 2.2 (B2) in 2080. The rainfall in winter, will decrease by 19% under the A2 scenario and by 21% under B2 scenario by the end of 2080. The water resources are becoming scarce due to decrease in precipitation and temperature increase (Fig. 8.10). The coverage rate for water irrigation in Middle Draa valley was 91% in 2004; this coverage became 74% in 2014 (Karmaoui et al. 2015a).

Under climate scenarios, water availability continues to be insufficient because of the repeated droughts. According to Karmaoui et al. (2015a), the Middle Draa Valley will suffer an increase in dry years in the 2010–2099 period. The results of the two models show that under the A2 climate change scenario, there will be more dry years during this period than under B2 scenarios. In addition to drought, this region experiences a high rate of evaporation, which impacts the soil moisture and then the soil productivity and the decrease of vegetation cover. The anthropogenic pressure is the second aspect of degradation in the area. This starts by a traditional irrigation (submersion irrigation) that aggravates the water availability, an increased population growth affecting the water resource (water surface and groundwater). In short- and long-term, climate change and population growth place additional impacts on water resources. Since the construction of Mansour Eddahbi Dam in 1972, the number of motor pumps has increased steadily. It is estimated that in 1977, the six oases had about 2000 pumps, and in 1985 this number had doubled; in 2005, the number of motor pumps had increased to nearly 7000 (CMV 2005) for the whole MDV and more than 10,000 in 2011 (Chelleri et al. 2014). Extraction of water exceeds the natural recharge of aquifers, this poses an important challenge.

Water is scarce in arid and semi-arid regions as in the case of the Middle Draa Valley. Drought and population pressure in this area are impacting this resource. Groundwater is constantly declining for the last 30 years (Fig. 8.11), for the six palm groves of the MDV.

The rapid and continuous decline in groundwater level of M'Hamid is due to the growth of the use of water and the degradation of favorable conditions for groundwater recharge. The deterioration of vegetation cover limits the possibilities of infiltration and consequently, groundwater recharge is reduced (Zainabi 2003).

In this valley, the river system is fed by releases from the Mansour Eddahbi Dam. In parallel with the impact of drought, this dam is subjected to the phenomenon of siltation. Indeed, the capacity of the dam was reduced by approximately 25% in 1998 (Diekkruger et al. 2010). This trend will undoubtedly impact the production of hydraulic energy of this dam.

Another aspect of water shortage is the orientation of farmers to more profitable crops but it would be harmful in the long run, especially the cultivation of watermel-

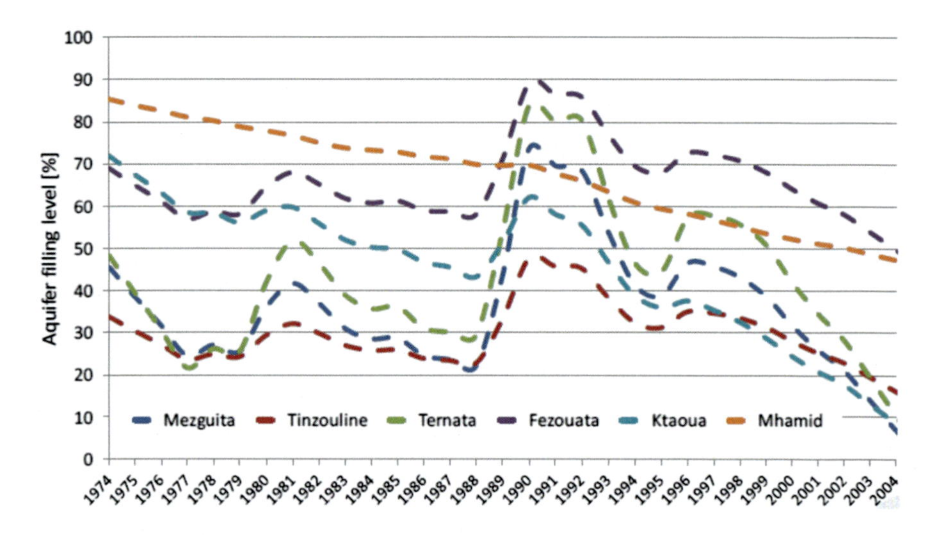

Fig. 8.11 Level of groundwater resources in the MDV (Source: IMPETUS project)

Fig. 8.12 Evolution of watermelon area in the same zone (Feija): raw satellite imagery from U.S. Geological Survey (USGS): www.landsatlook.usgs.gov/. Cloud: 20%. Sensors: TM, ETM+, OLI, Transparency Visible. Spatial resolution: 2 km. (Source: Karmaoui et al. 2014a)

ons, which dramatizes the demand for water (Karmaoui et al. 2014a). In fact, from Fig. 8.12, the watermelon area is constantly increasing. This evolution can be clearly seen in Fig. 8.5 processed by Karmaoui et al. (2014a)

This type of exploitation and all above-mentioned aspects of resource and aquatic ecosystems degradation have increased food insecurity due to which people are migrating to big urban centers and poverty is increasing. This will certainly lead to conflicts and overexploitation of ecosystems and subsequently widespread desertification. Recognizing this situation, we must consider several options including: remove mud from dams; monitor the water quantity and quality; re-use of wastewater; desalination; rainwater aggregation; economic irrigation; and water management.

The importance of such options is being recognized through initiatives such as Moroccan green plan or "Plan Maroc vert", the national program of water irrigation economy (called PAPNEEI in Morocco), and the provisioning drinkable water Program of rural population (called PAGER in Morocco).

Under the expected climate scenario, water availability continues to be non-sufficient at current water capacity. Ouarzazate represents the most challenging of the three urban sites. The need for major infrastructure to collect water and better water management at farm scale level are most urgent in Draa. Mehta et al. (2013) reported that governments continue to rely on donor support for major infrastructure investments in the water and sanitation sector. In addition, the international donors have sponsored small-scale projects re-establishing both traditional techniques such as rainwater harvesting or financing the development of modern drip-irrigation technologies (Jobbins et al. 2015).

MDV is facing water constraints as to how to use the available water to meet urban and agricultural demands. Policies should integrate the supply and demands to address water stress issues. This paper anticipates hydrologic change in order to choice the management decisions to answer the water scarcity problem in this region. All four scenarios show an increasing trend in water requirements with time. The A2 climate change scenario exhibits the most pronounced increase. These increasing supply requirements are due to increasing summer temperatures for each of the two climate change scenarios. Most notably, these models provide useful indications of the timing of major investment in infrastructure improvements and expansion, and the size of the expansion required.

8.6 Conclusion

This paper presents an impact study of climate change scenarios on the water supply system of the Middle Draa Valley, combining downscaled climate scenarios with a numerical model of water supply system developed through the WEAP software.

Based on the climate change scenarios, the Middle Draa Valley will face prolonged drought from 2014 to 2080. For all scenarios, the increase in crop water demands meant that irrigation districts would not be able to meet their irrigation demands. Climate change will affect the storage levels in both Mansour Eddahbi Dam in upstream and the groundwater reservoir in downstream. The results predict increases in mean temperatures by 1.4–3 °C in scenario A2 and by 1.6–2.7 °C in scenario B2 from 2011 to 2070; and a decrease in precipitation by approximately 5.1–15% in A2 scenario and by 3.9–8.9% in B2 scenario. These conditions will result in increased water demand throughout the region. Irrigation efficiency and shifts in cropping patterns can reduce the demand in the agricultural sector for other sectors like tourism.

For the use of these decision support tools, we must take into account the degree of certainty of the tested scenarios (climate and socio-economic changes) and the complexity of water issues.

Acknowledgements Sincere thanks go out to the reviewers who helped in improving this work. We would like to thank SEI for the WEAP software license granted to the first author. We would like to acknowledge Dr. Ben Salem A. and Dr. Rochdane S. for assisting in use of WEAP software.

References

Aylward, B., Bandyopadhyay, J., Belausteguigotia, J. C., Borkey, P., Cassar, A. Z., Meadors, L., & Voutchkov, N. (2005). Freshwater ecosystem services. *Ecosystems and human well-being: Policy responses, 3*, 213–256.

Babqiqi, A. (2014). Changements Climatiques au Maroc : Etude du cas de la Région de Marrakech Tensift Al Haouz et implications sur l'agriculture à l'horizon 2030. Thèse de doctorat. Département de biologie, Faculté de Semlalia. Université Cadi Ayyad, Marrakech.

Bates BC, Kundzewicz ZW, et Palutikof J. (2008). Le changement climatique et l'eau, document technique publié par le Groupe d'experts intergouvernemental sur l'évolution du climat, Secrétariat du GIEC, Genève, 236 p.

Chelleri, L., Minucci, G., Ruiz, A., & Karmaoui, A. (2014). Responses to drought and desertification in the Moroccan Drâa Valley Region: Resilience at the expense of sustainability? *International Journal of Climate Change: Impacts & Responses, 5*(2), 17–33.

CMV. (2005). *Distribution de l'eau dans le Draa Moyen*. Morocco: Office régional de mise en valeur Agricole de Ouarzazate.

Diao, X., Hazell, P., & Thurlow, J. (2010). The role of agriculture in African development. *World Development, 38*(10), 1375–1383.

Diekkruger, B., Busche, H., Klose, A., Klose, S., Rademacher, C., & Schulz, O. (2010). Impact of global change on hydrology and soil degradation-scenario analysis for the semi-arid Drâa catchment (South Morocco).In *The global dimensions of change in River Basin*. (pp. 5–13).

GCPH (2004) General census of the population and housing. https://rgph2014.hcp.ma/ Accessed 09 May 2015.
Gommes, R., Kanamaru, H., El Hairech, T., Rosillon, D., Babqiqi, A., et al. (2008). *World bank – Morocco study on the impact of climate change on the agricultural sector, FAO component: Impacts on crop yields* (p. 105). Maroc: FAO and Agricultural Ministry of Morocco.
Heidecke, C., & Thomas, H. (2010). The impact of water pricing in an arid river basin in Morocco considering the conjunctive use of ground-and surface water, water quality aspects and climate change. In *The global dimensions of change in River Basins*.
IPCC. (2014). Climate change 2014: Impacts, adaptation, and vulnerability. In V. R. Barros, C. B. Field, D. J. Dokken, M. D. Mastrandrea, K. J. Mach, T. E. Bilir, M. Chatterjee, K. L. Ebi, Y. O. Estrada, R. C. Genova, B. Girma, E. S. Kissel, A. N. Levy, S. MacCracken, P. R. Mastrandrea, & L. L. White (Eds.), *Part B: Regional aspects. Contribution of working group II to the fifth assessment report of the intergovernmental panel on climate change* (p. 688). Cambridge: Cambridge University Press.
Jobbins, G., Kalpakian, J., Chriyaa, A., Legrouri, A., & El Mzouri, E.H. (2015). To what end? Drip irrigation and the water–energy–food nexus in Morocco. *International Journal of Water Resources Development*, 1–14. Ahead-of-print.
Karmaoui, A., Messouli, M., Yacoubi Khebiza, M., & Ifaadassan, I. (2014a). Environmental vulnerability to climate change and anthropogenic impacts in dryland, (pilot study: Middle Draa valley, South Morocco). *Journal of Earth Science and Climatic Change, S11*, 002. https://doi.org/10.4172/2157-7617.S11-002.
Karmaoui, A., Messouli, M., Ifaadassan, I., & Khebiza, M. Y. (2014b). A multidisciplinary approach to assess the environmental vulnerability at local scale in context of climate change (pilot study in Upper Draa Valley, South Morocco). *Global Journal of Technology and Optimization, 6*, 167. https://doi.org/10.4172/2229-8711.1000167.
Karmaoui, A., Ifaadassan, I., Babqiqi, A., Messouli, M., & Yacoubi Khebiza, M. (2015a). Analysis of the water supply-demand relationship in the Draa valley basin, Morocco, under climate change and socio-economic scenarios. *Journal of Scientific Research & Reports*. https://doi.org/10.9734/JSRR/2016/21536.
Karmaoui, A., Ifaadassan, I., Messouli, M., & Khebiza, M. Y. (2015b). Sustainability of the Moroccan oasean system (Case study: Middle Draa Valley). *Global Journal of Technology and Optimization, 6*, 170. https://doi.org/10.4172/2229-8711.1000170.
Mehta, V. K., Aslam, O., Dale, L., Miller, N., & Purkey, D. R. (2013). Scenario-based water resources planning for utilities in the Lake Victoria region. *Physics and Chemistry of the Earth, Parts A/B/C, 61*, 22–31.
RBOSM. (2008). Réserve de Biosphère des Oasis du Sud Marocain. Plan cadre de gestion de la Réserve de Biosphère des Oasis du Sud Marocain (RBOSM).
SADAM. (2003). Strategie d'amenagement et de developpement des oasis au MAROC. Analyse, Diagnostic, Typologie des Oasis. Première phase, 2003. Direction de l'Aménagement du Territoire DIRASSAT. MATEE-DAT/DIRASSET.
Sieber, J., Swartz, C., & Huber-Lee, A. H. (2005). *Water evaluation and planning system (WEAP): User guide*. Boston: Stockholm Environment Institute.
Yates, D., Sieber, J., Purkey, D., & Huber-Lee, A. (2005). WEAP21—a demand-, priority-,and preference-driven water planning model. *Water International, 30*, 487–500.
Zainabi, À. (2003). La Vallée du Dra- Développement Alternatif et Action Communautaire. World Development Report.

Dr. Ahmed Karmaoui is an Associate Researcher at the Faculty of Sciences Semlalia, Cadi Ayyad University of Marrakech. His main research areas are water resources, environmental vulnerability, ecosystem services and climate change impacts and response.

Dr. Guido Minucci is a Postdoctoral Fellow in urban studies at the Politecnico di Milano. With a background in urban and regional planning and disaster studies, his research addresses damage assessment and risk assessment. Guido get involved in several projects dealing with natural disasters, urban vulnerability and risk management (EU-project IDEA, KNOW-4-DRR, ENSURE and M.I.A.R.I.A). He has co-founded the international networks UR-Net (Urban Resilience research Network) and the university civil protection group LARES-Lombardia.

Prof. Mohammed Messouli, with 30 years of professional experience in the fields of environment, climate change and sustainable development, he has a very good knowledge of the policy aspects of impact, vulnerability and adaptation to climate change. He contributed to the IPCC in the drafting of the fifth assessment report and a large number of meeting internationally. He was a member of the Scientific Committee of CoP22.

Prof. Mohammed Yacoubi Khebiza, PhD, is a Professor at Department of Environmental Sciences and the Director of the laboratory (LHEA-URAC 33), Faculty of Sciences Semlalia, Cadi Ayyad University. He has been working as the Team Investigator of various profession consultancy jobs and research activities on environmental vulnerability and climate change adaptations, and is a member of the Moroccan association of biotechnology and protection of natural resources.

Mr. Issam Ifaadassan is a professor in Science Didactics and a PhD candidate in Bioclimatology and vulnerability of agrosystems to climate and anthropogenic changes at the Faculty of Sciences Semlalia, Cadi Ayyad University. Issam is a member of the Moroccan association of biotechnology and protection of natural resources and a member of the research group on the impact; vulnerability and adaptation to climate change.

Dr. Abdelaziz Babqiqi is the Director of the Regional Observatory for Environment and Sustainable Development in Marrakech, Moroccan State Secretariat for Environment, Morocco.

Chapter 9
Determination of Date Palm Water Requirements in Saudi Arabia

Abdulrasoul Al-Omran, Fahad Alshammari, Samir Eid, and Mahmoud Nadeem

9.1 Introduction

Date palm, *Phoenix dactylifera* L., is one of the oldest fruit trees in the world. The number of date palms is about 120 million worldwide, of which 70 million palms can be found in the Arab world (Zaid 2002). The place of origin of the date palm is uncertain. Some researchers claim that the date palm first originated in Babel, Iraq, while others believe that it originated in Dareen or Hofuf, Saudi Arabia (Fig. 9.1). The date palm is a perennial, the females of which normally begin to bear dates within an average of five years from the time of planting of the offshoot. The date palm reaches an age of about 150 years.

In Saudi Arabia oases, date palm trees stand tall with their branches outstretched towards heaven and their roots anchored deep into the earth. These dense green groves have been a treasured part of the Saudi landscape for generations, both for their beauty and their utility. Since ancient times, the date palm has been a source of food for the inhabitants of the Arabian Peninsula, and its branches have granted shade from the strong desert sun (Figs. 9.2 and 9.3).

The Government of the Kingdom of Saudi Arabia represented by the Ministry of Water, Environment and Agriculture has exerted incessant efforts to develop the agricultural sector. Continuous support and care was allocated to the date palm production sector in particular, due to the important role of this blessed tree in the realization of food security and its historical relation to the Saudi population. Several farmers

A. Al-Omran (✉) · M. Nadeem
Soil Science Department, King Saud University, Riyadh, Saudi Arabia
e-mail: rasoul@ksu.edu.sa; menadeem@ksu.edu.sa

F. Alshammari · S. Eid
Ministry of Environment, Water and Agriculture, Riyadh, Saudi Arabia

M. Behnassi et al. (eds.), *Climate Change, Food Security and Natural Resource Management*, https://doi.org/10.1007/978-3-319-97091-2_9

Fig. 9.1 Location of date palm fields in eight different regions of Saudi Arabia. (Source: Al-Shemeri 2016)

Fig. 9.2 Date palm production in Saudi Arabia

Fig. 9.3 Drip irrigation for Date palm production in Saudi Arabia

with the support of the government, have started cultivating high quality varieties of date palms. Concern regarding the problems of marketing and processing of dates has increased as investment in these fields was encouraged. The Saudi farmers' concern with agriculture has increased, and date palm orchards were established on appropriate modern scientific basis avoiding traditional methods of cultivation. Numerous modern projects for date palm plantation and production were established in many parts of the Kingdom. Table 9.1 shows the estimated number of date Palm trees of different regions in the Kingdom of Saudi Arabia for the years from 1999 up to 2015, while Figs. 9.4 and 9.5 show the estimated area and dates palm production (General Authority for Statistics 2015).

The Kingdom of Saudi Arabia is considered as one of the pioneer countries in date palm cultivation and dates production. The current date production (for the year 2015) is estimated by over than million tons with an increase of about 45% in the last twenty years. The cultivated areas of date palm have also increased and reached about 140 thousand hectares in the year 2004 with an increase of 57% during the same period. The number of date palms in the Kingdom are estimated to be 28 million and about 400 different date varieties are found in different agricultural areas of the Kingdom. Each area in the Kingdom is characterized by certain date palm varieties.

Table 9.1 Estimated number of date palm trees by region for the years 1999 to 2015

Region	1999	2002	2003	2004	2005	2015
Riyadh	4160565	4493410	4702830	4941944	4972529	7030731
Makkah	1781466	1923983	1890100	1773830	2027431	1237568
Madina	2299666	2483639	2701372	2810870	2843902	4619640
Qasseim	3120558	3370203	3188705	3790032	3922561	6979753
Eastern Province	2579856	2786244	2907115	2919608	2544652	3731759
Assir	2289709	2472886	1698691	1900168	1829933	1027431
Tabuk	627595	677803	895940	899863	870023	834358
Hail	1315040	1420243	1988091	1780201	1696804	1773442
Northan Frontier	1751	1891	1942	1826	2047	23089
Jizan	7746	8366	11172	8738	7360	8581
Najiran	437512	472513	387136	443709	608488	385623
Albaha	102956	111192	88952	124894	145177	70612
Jouf	580768	627229	862065	892174	1155076	848217
Total	19305188	20849602	21324111	22287857	22625983	28572819

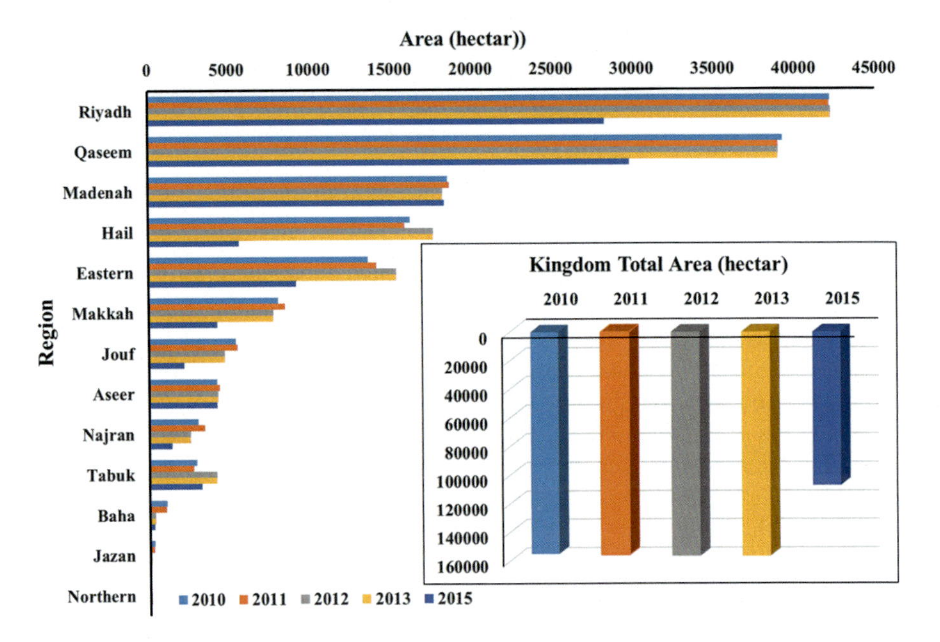

Fig. 9.4 The estimated areas of dates palm production in Saudi Arabia. (Source: General Authority for Statistics 2015)

9.2 Location and Climate

The Kingdom of Saudi Arabia is situated in southwest Asia between latitude 16–32 North and longitude 35–65 East. The total area of the Kingdom is about 2.25 million square kilometers which represents about 80% of the area of the Arabian Peninsula. The Kingdom's area extends from the Red Sea in the West to the Arabian

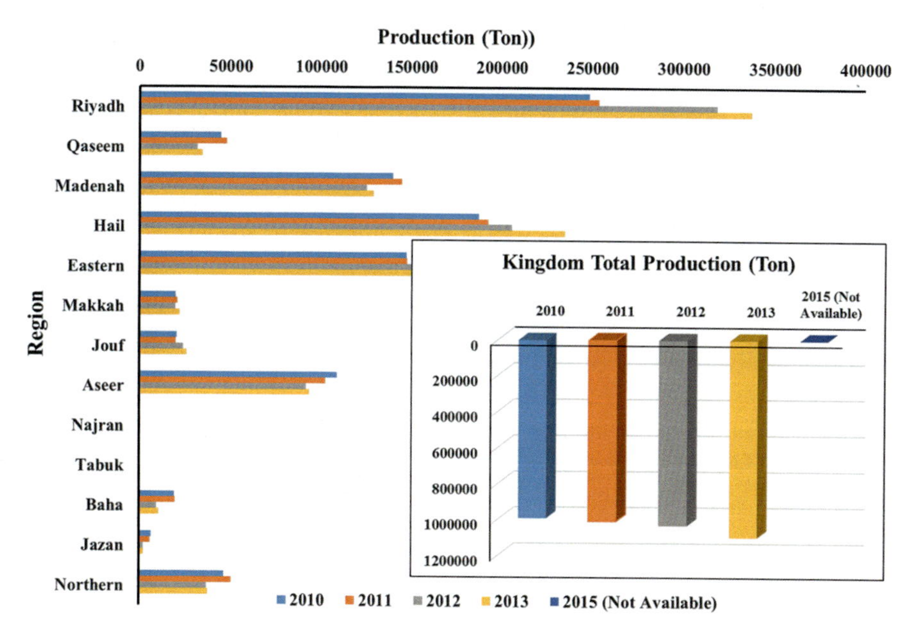

Fig. 9.5 The estimated of dates production in Saudi Arabia. (Source: General Authority for Statistics 2015)

Gulf in the East. The vast area of the Kingdom together with its geographic location led to diversity in its terrain and geological formation which had brought about relative advantages to certain parts of the Kingdom. Most of the regions in the Kingdom fall within the dry tropical zone for west continents. The climate of the Kingdom is characterized by hot, dry and long summer where temperatures in certain areas sometimes reach more than 45° C during the months of June, July and August, generally with an average temperature of around 35° C during the summer. On the other hand, the winter season falls during the months of December, January and February when temperatures in certain areas often drop to less than zero degree. Levels of humidity rise during summer season and range between 35 and 85%, whereas during the winter season, it ranges between 35 and 70%. The levels of humidity are generally considered high during the summer months in the western coastal areas compared to the eastern coast area. In the middle inland areas, levels of humidity are relatively low ranging between 15 and 35% during summer months, and rising a little during winter season, ranging between 20 and 70% (Al-Omran et al. 2002).

Despite the Kingdom, spreads over more than two million square kilometers, most its areas are considered the driest in the world. The average annual rainfall is about 100 mm, most of which occurs between December and March. Sometimes, the average annual rainfall in some areas of the southern region of the Kingdom reaches 400 mm. The Kingdom is exposed to north, northeasterly or north-westerly dry winds during most of the year, while the western winds coming from the Mediterranean during winter and the seasonal southwestern winds during summer cause the rainfall in different areas of the Kingdom. Rates of evaporation in the Kingdom are high due to high air temperatures, scarcity of rainfall and drought.

9.2.1 Water Resources

Water used for irrigation purposes is counted as one of the most important factors affecting the agricultural sector. The most important sources of water are underground water, rainwater, and treated wastewater. The Ministry of Agriculture being aware of the importance of water conservation and the rationalization of water consumption in date palm plantation.. The Ministry further encouraged the farmers and agricultural companies to adopt appropriate irrigation methods and follow the concept 'More Crop per Less Drop' which means more crop production by using less quantities of water instead of the concept of 'More Crop per Drop' which means more crop production by using the same quantity of water. In this context, the Ministry is striving hard to create awareness for the use modern irrigation systems such as Drip irrigation or the use of Pivot Sprinklers and avoidance of traditional methods such as flood irrigation. The ministry provides agricultural licenses only in the precondition of use of modern irrigation systems.

9.2.2 Soil Salinity and Date Palm

Date palm growth is influenced by soil salinity, which results in loss of productivity. The soluble salts present in soil are mainly: Na, Ca, Mg, Cl, and SO_4. Richards (1954) defines saline soils as soils that have an electrical conductivity (ECe) higher than 4 dsm^{-1} at 25°C, with a sodium absorption ratio (SAR) of less than 15 and pH generally less than 8.5. In the Saudi Arabia, most of the soils are suitable for date palm plantation since these soils have lower values of ECe, usually less than 4 and are well aerated (Heakal and AlAwajy 1989). However, Saudi Arabia's date palm can also be grown in saline soil where ECe reaches up to 16–20 dsm^{-1} (Bashour et al. 1983) in some areas of central region, and between 2–8 dsm^{-1} in Al-Hasa oasis (Al-Barrak 1990), and more than 20 dsm^{-1} in coastal soils of Al-Hasa (Al-Barrak 1997). The electrical conductivity of main aquifers in Saudi Arabia ranged between 2 and 5 dsm^{-1} (Al-Omran et al. 2005). According to Ayers and Westcot (1985), date palm can tolerate an ECe of 4.0 dsm^{-1} with an ECw of 2.7 dsm^{-1} without losing any yield (Maas and Hoffman 1977).

9.2.3 Date Palm Spacing

The spacing between date palms differs worldwide. The recommendation for date palm spacing for the farmers in Saudi Arabia is 10 m × 10 m i.e. 100 offshoot/ha (Ministry of Agriculture 2000). This spacing allows a sufficient sunlight even after the plants grow tall in 7–10 years, and allow sufficient working space within the field. In some regions, the spacing is smaller than recommended which is 7 m × 7 m, about 204 trees /ha. In some other old oasis such as Al-Hasa, it is common to grow some other crop with the date palms. Vegetables crops are grown while the date palms are still young, and after the date palms grow tall enough to allow cultivation under them, the planting of fruit trees such as citrus among them is advisable. In other areas such as on the edges of the basin, the date palm is grown with alfalfa and citrus, figs and pomegranate as associated crops. Some private farmers also use different spacing such as, 8 × 8 m. However, narrower spacing is not advisable.

9.2.4 Irrigation Methods

Irrigation is the timely application of water to a crop when it is really in need of any water applied when not necessary, is a waste of a precious commodity. Irrigation must take place where the roots of the plant can easily reach it. It is of no use to the plant if water is applied where the roots cannot reach it. For a date palm tree, if the soil is divided into four layers of equal depth from top to bottom, 40% of all roots can be found in the top layer, 30% in the second layer, 20% in the third layer and the remaining 10% in the last layer. The same percentages apply in concentric rings around the plant. The same percentage of water will also be extracted from the soil in the different layers due to the presence of the roots in these respective layers.

Different irrigation techniques are available, but not all of them are suitable for date palm irrigation. The flood irrigation method is the oldest method known, and is also the method most widely used at the old date palm farms in all regions of Saudi Arabia. Recently the new irrigation systems like drip irrigation (surface drip irrigation and subsurface drip irrigation) and bubbler have been introduced and are in use on the modern farms.

9.2.5 Irrigation Water Quality

Many researchers have reported on the evaluation of irrigation water quality in different regions of Saudi Arabia, including Riyadh region, Al-Hasa oasis, Al-Qassium, Al-Kharj, and some selected regions (Hamza et al. 1975; Mee 1983; Al-Omran et al. 2005; Al-Jaloud and Hussain 1992; MAW 1985; Jahangir et al. 1987). The water composition for 16 different aquifers is reported in Water Atlas of Saudi Arabia (MAW 1985). The details of different aquifers are given in Table 9.2. The groundwater from the aquifers were analyzed and sodium adsorption ratio (SAR), adjusted sodium adsorption ratio (adj RNa), adjusted sodium adsorption ratio (adj SAR) exchangeable sodium percentage (ESP), calcium/magnesium ratio (Ca^{++}/ Mg^{++}), and chloride/sulfate ratio (Cl^{-}/SO_4^{--}) were calculated from analytical data.

The chemical composition of the water samples of Riyadh region is presented by plotting on a Piper trilinear diagram (Piper 1944) (Fig. 9.6). The Piper diagram provides a convenient method to classify groundwater types, based on the ionic composition of different water samples. This diagram shows the main minerals present in water, calculated on the basis of the major concentrations of ions in Riyadh region which is rich in calcium - magnesium sulphate - chloride water type.

Table 9.2 Most important features of different aquifers

Aquifers	Water Depth (m)	Discharge (Ls^{-1})	Location (Region of Saudi Arabia)
Al-Saq	150–1500	100	Central – North
Wajid	150–900	40–80	Southern
Tabuk	60–2500	15–20	Central – North
Minjur	1200–2000	60–120	Central
Dhruma	100	60–120	Central
Biyadh	30–200	25–50	Northern
Wasia	100–800	85–110	Central – East
Umm-ER-Radhuma	160–200	50–100	Eastern
Dammam	50–100	7–22	Eastern
Neogene	NA	50–100	Eastern
Jilh	NA	10–18	West – Riyadh
Khuff	NA	7–23	East – El-Qawayh
Aruma	NA	30–32	Central – East
Jubalia	NA	NA	Central
Basalt	NA	NA	Western
Aluuvial	NA	50	Western - Costal

NA data not available
Source: MAW (1985)

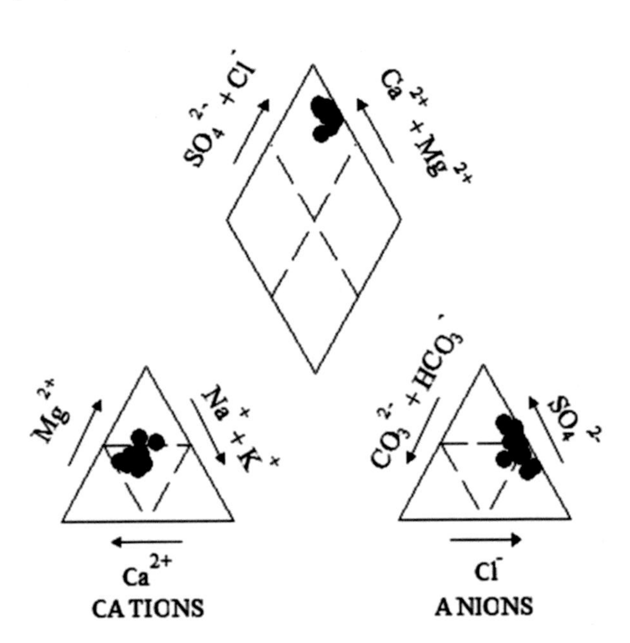

Fig. 9.6 Piper – tri-linear diagram showing the major ionic composition of Riyadh region groundwater. (Source: Al-Omran et al. 2005)

Table 9.3 Yield potential of date palm with varying soil salinity (EC_e) and irrigation water salinity (EC_w)

	Yield Potential				
	100%	90%	75%	50%	maximum EC_e
EC_e (dSm^{-1}	4	6.8	10.9	17.9	32
EC_w (dSm^{-1})	2.7	4.5	7.3	12.0	–

Source: Ayers and Westcot (1985)

9.2.6 *Salinity and Date Palm Production*

Mass and Hoffman (1977) provide an extensive list of salinity coefficients for a number of field, vegetable, forage and fruit crops. These coefficients consist of a threshold and the rate yield declines with increasing salinity (slope). The salinity threshold (a) is the maximum average soil salinity (ECe) the crop can tolerate in root zone without the decline in yield. Using these coefficients, the yield potential (% yield) can be estimated from the following equation:

$$Y = 100 - b\left(EC_e - a\right) \quad (9.1)$$

Where

Y = relative yield
b = the rate yield declines with increasing salinity.
a = threshold salinity value for date palm (4 dsm^{-1})
EC_e = electrical conductivity of root zone

Table 9.3 shows the yield reduction of date palm at different values of soil salinity (ECe) and irrigation water salinity (ECw) as reported by Mass and Hoffman (1977). Although, the date palm is a fruit plant, it is one of the most resistant (tolerant) plants to salinity where ECe can be 4 dsm^{-1} without losing any yield. The table provides irrigation water salinity (ECw) that, if used continuously to achieve LF of 15–20% (Ayers and Westcot 1985), would result in yield potential of 100, 90, 75, and 50%. The ECw values at 100% yield potential for date palm are 2.7 dsm^{-1} which represent the poorest water quality that, if used continuously, will produce and ECe level of 4.0 dsm^{-1} which is equal to salinity threshold value. If the average values of ECe at the root zone throughout the season was 10.9 dsm^{-1}, or ECw with leaching fraction of 15–20% is 7.3 dsm^{-1}, then the yield potential of date palm is 75%.

9.2.7 *Estimation Evapotranspiration*

9.2.7.1 Estimating Crop ET

Since long, different methods were employed to estimate the water requirements of different crops. As a result, numerous methods have been developed and adopted for different crops. Some of these methods are more accurate than others and some more convenient to use than others, because of the availability of information of metrological data for the date palm trees. Since the direct measurement of crop evapotranspiration (ET_c) is expensive, time consuming and laborious, it is usually estimated from more easily available climatic data. This approach involves the estimation of a meteorological related reference evapotranspiration (ET_r) and a set of ET crop coefficients (K_c) to determine crop water requirement.

Various procedures have been used to obtain the necessary reference ET_r data and several types of crop coefficients curves have been published (Wright and Jensen 1972). However, all the existing methods of estimating crop ET from climatic data involve some empirical relationships and assumptions, hence local or regional verification or calibration is necessary to gain higher reliability in obtaining practical utility of ET equation. The recent development of computerized weather stations provides hourly-integrated measurement of meteorological variables and a means to estimate ET on short-term basis and with more accuracy.

9.2.7.2 Estimating Reference Evapotranspiration (ET_r)

The equations for estimating ET_r may be broadly classified as those based on combination theory, humidity data, radiation data and miscellaneous methods that involve multiple correlation for ET and various climatic data. The Technical Committee on Irrigation Water Requirements presents comprehensive details of these methods, American Society of Civil Engineers (Jensen et al., 1990). The most commonly used procedures to estimate….are the Penman, Jensen- Haise, Blaney-Criddle, Thornthwaite and pan evaporation methods. These methods were calibrated under local conditions by measuring the evapotranspiration for two years from alfalfa grown in lysimeters and obtaining the climatic data from weather station in the area (Al-Omran et al. 2004).

9.2.7.3 Crop Coefficients

Many evapotranspiration estimating methods result in an ET estimate for a reference surface of water or reference crop of grass or alfalfa. Extensive research has been conducted on reference ET methods and crop coefficients because of their use in irrigation scheduling and water resources allocation, management, and planning. The available methods for estimating reference ET when properly used with reliable crop coefficients permit furnish crop ET within the accuracy of most field-irrigation systems to deliver water (Jensen and Wright 1978). Various procedures have been used during the past three decades to obtain the experimental crop and reference data needed to develop ET crop coefficient. Several sets of curves derived from these data have been published (Wright and Jensen 1978). Although crop coefficients have been suggested for the arid and semiarid climates yet haven't been tested under severe hot and arid climate of Saudi Arabia.

9.3 A Case Study of Date Palm Water Requirement

The aim of present study was to determine the date palm water requirements of eight regions of Saudi Arabia taking into consideration the shaded area of the tree and irrigation water quality and to compare it with the actual water added by farmers in each region of Saudi Arabia.

The agricultural sector consumed more than 85% of water consumption which reached to more than 23 billion m^3 in 2012 (Ministry of Electricity and Water 2014). As the demand of water increases, an effective and accurate evaluation of crop water requirement (CWR) is essential for planning, designing, operating, managing farm irrigation systems. The efficient use of water resources for irrigation can be done through estimation of CWR. Evapotranspiration (ET) plays a major role in irrigation water management. Allen et al. (1998) reported that many factors may play a role in limiting crop development, which are: water availability, soil salinity, poor land fertility, poor soil and water management, plant density and soil water contents. In Saudi Arabia, the limiting factor in agricultural development is water availability to irrigate the increasing number of date palm trees. Based on the recent statistical reports (General Authority for Statistics 2015), the total number of date palm trees are 28.5 million on 54000 ha drip irrigation and 53200 ha surface irrigation fields.

In Saudi Arabia, estimation of water requirement of date palm has been reported by many researchers. These estimates differ between 6200–55000 m^3/ha. Alazba (2001) estimates water requirement to be between 15000–55000 m^3/ha, depending upon irrigation system or leaching requirement. Al-Ghobari (2000) has estimated the total annual amount of water required by one date palm tree as 136 m^3 in Najran of south western region. Kassem (2007) monitored water requirements in Qassem region. Using soil water balance method, he determined the annual water use with drip irrigation as 16400 m^3/ha, with a density of 100 tree/ha. Al-Amoud et al. (2012) estimated the actual water use in the range between 21360–28290 m^3/ha, for density of 100 tree/ha. A study conducted by Ismail et al. (2014) in the western part of Saudi Arabia calculated water requirement based on Penman-Montieth equation for ET_o, Kc ranged from 0.8–1.0, and the evapotranspiration area (23 m^2/tree), to be 7300 m^3/ha, for density of 100 tree/ha. Recently, Dewidar et al. (2015) estimated water requirement of date palm using non-weighing lysimeter. They reported that volumetric palm water requirement per day fell between 87–297 L/day, with daily average of 182 L and crop coefficient ranged from 0.74 to 0.91. In Kuwait, the date palm water requirement using drainage type lysimeters through water balance budget was ranged between 23392–27251 m^3/ha (Bhat et al. 2012). In Algeria, Mihoub et al. (2015) reported that the annual water requirement is 17411 m^3/ha, for a density of 120 tree/ha by drip irrigation compared to 26117 m^3/ha by surface irrigation. In Jordan, Jordan valley, Mazahrih et al. (2012) reported that the amount of applied irrigation water per date palm tree were 27, 40, 53 and 67 m^3 for the irrigation treatments 50, 75, 100 and 125% ET_c respectively.

9.3.1 Experimental Sites

This study was conducted on eight different regions of Saudi Arabia to estimate monthly and annual irrigation water requirements of date palm (*Phoenix dactylifera L.*) of Klayas variety. Field measurements and determination of ET_c were taken during one year starting Oct. 2013–Sept. 2014 on complete grown tree (more than 10 years old). Fields that have been selected are located in regions of the Medina (Al Ula), Tabuk (Teimaa), Makkah (Al Jumum), Al Jouf (Sakakah), Riyadh (Sodos), Qassim (Riyad Al Khabra, Hail (AL Kaedh), East Region (Al Ahsa) (Fig. 9.1). The characterization of the soil and irrigation water are shown in Tables 9.4 and 9.5 (Al-Shemeri 2016).

9.3.2 Meteorological Data

Small weather stations were installed at each site of the study to monitor the changes in meteorological parameters during the study period. The meteorological data recorded were: net radiation (MJ/ m^2day), wind speed (m/hr), air temperature (°C), relative humidity (%) and rainfall (mm). The air water vapour pressure deficit (kPa) was calculated using daily and hourly average temperatures and relative humidity. Finally, the reference evapotranspiration (ET_r, mm/day) was calculated according to the Penman-Monteith (PM) equation as specified by the FAO protocol (Allen et al. 1998).

Table 9.4 The physical and mechanical analyses of the soil

	Mechanical analysis				Soil constants			Ca		Hydraulic
Sites	Sand %	Silt %	Clay %	Soil Texture	W.P %	F.C %	SP %	CO_3 %	O.M %	conductivity cm/h
Medina	88.9	5	6.1	Sandy	5.08	10.15	20.3	3.82	0.1	6.5
Tabuk	81.4	7.5	11.1	Sandy Loam	7.5	15	30	6.34	0.72	5
Makkah	71.4	17.5	11.1	Loam Sandy	8	16	32	3.56	1.78	3.1
Al Jouf	90.7	5	4.3	Sandy	5.8	11.6	23.2	2.3	0.11	6.25
Riyadh	65.7	17.5	16.8	Loam Sandy	6.95	13.9	27.8	36.9	0.58	2.9
Qassim	47.1	30	22.9	Loam	8.9	17.8	35.6	7.93	2.88	2.2
Hail	83.9	7.5	8.6	Sandy Loam	7.7	15.4	30.8	2.7	0.15	3.55
East Region	67.5	17.5	15	Sandy Loam	8.3	16.6	33.2	14.1	1.88	4.2
Max	**90.7**	**30**	**22.9**	–	**8.9**	**17.8**	**35.6**	**36.9**	**2.88**	**6.25**
Min	**47.1**	**5**	**4.3**	–	**5.08**	**10.15**	**20.3**	**2.3**	**0.1**	**2.2**

Table 9.5 The analyses of the irrigation water

				Cation (meq/L)				Anion, meq/l		
Sites	TDS (mg/L)	SAR	pH	EC (dS/m)	Mg^{++}	Na^{+}	K^{+}	$CO_3^{=}$	HCO_3^{-}	Cl^{-}	$SO_4^{=}$
Medina	544	3.35	7.63	0.85	1.1	4.8	0.15	0	1.6	5.0	2.5
Tabuk	390	2.23	7.6	0.61	1.9	3.11	0.12	0	1.5	3.0	2.5
Makkah	1004	3.31	7.5	1.57	2.9	6.62	0.21	0	2.0	7.5	6.0
Al Jouf	960	4.15	7.4	1.5	2.6	7.81	0.15	0	2.0	8.1	5.2
Riyadh	646	3.42	7.61	1.01	1.98	5.4	0.18	0	2.1	5.1	4.8
Qassim	1536	8.44	7.8	2.4	3.0	16.99	0.30	0	2.76	16.88	5.12
Hail	601	3.45	7.6	0.94	1.0	5.0	0.20	0	2.1	5.1	2.6
East Region	998	3.66	7.5	1.56	3.0	7.5	0.25	0	2.6	7.0	6.5

9.3.3 *Estimation Method of ET*

9.3.3.1 Penman Montieth Method

Using the Penman Monteith equations (9.2, 9.3, 9.4, 9.5 and 9.6) based on climate data on the farm as part of the national project of the rationalization of the irrigating water in agriculture (RIWA), Ministry of Environment, Water, Agriculture to estimate the water needs. Then, calculate the total irrigation water requirements based on the quality of irrigation water and soil salinity, taking into account the values of crop coefficient Kc for each month, irrigation efficiency and shaded area of date palm. The combined FAO Penman-Monteith method was used to calculate ET_o through the following equation:

$$ET_O \frac{0.408\Delta(Rn-G)+\gamma\left(\frac{900}{T+273}\right)U_2(e_s-e_a)}{\Delta+\gamma(1+0.34U_2)} \tag{9.2}$$

Where:

ET_o = Reference evapotranspiration (mm/day)
Rn = Net radiation at the crop surface (MJ/m^2 per day)
G = Soil heat flux density (MJ/m^2 per day)
T = Mean daily air temperature at 2 m height (°C)
U_2 = Wind speed at 2 m height (m/sec)
e_s = Saturation vapour pressure (kPa)
e_a = Actual vapour pressure (kPa)
$e_s - e_a$ = Saturation vapour pressure deficit (kPa)
Δ = Slope of saturation vapour pressure curve at temperature T (kPa/°C)
γ = Psychrometric constant (kPa/°C)

As crop evapotranspiration ETc can be calculated as:

$$ET_c = K_c \times ET_r \tag{9.3}$$

Where

K_c = crop coefficient ranged from 0.8 – 1.0 depend on the month of year as noted in (Allen et al. 1998)
$ET_r = ET_o$ = Reference crop evapotranspiration (mm/day)
ET_c = Crop evapotranspiration (mm/day)

The percentage of evapotranspiration area (S_e) was calculated from actual shaded are at noon in June to the actual area to each tree from the following equation as described by Hellman (2010) for grape:

$$S_e = \frac{Shaded\,area\,per\,tree}{Actual\,area} x100 = \frac{\pi R^2}{10m\,x\,10m} \tag{9.4}$$

Where

S_e = The percentage of evapotranspiration area.
R = radius of tree (m).
Shaded area = Area of the shade of one tree measured at noon.

Leaching requirements were calculated using the following equation (Doorenbos and Pruitt 1977).

$$LR = \frac{EC_{iw}}{2MaxEC_e} \times \frac{1}{Eff} \tag{9.5}$$

Where

LR = The fraction of the water to be applied that passes through the entire root zone depth and percolates below
EC_{iw} = Electrical conductivity of irrigation water (dsm^{-1})
EC_e = Electrical conductivity of the soil saturation extract for a given crop appropriate to the tolerable degree of yield reduction (dsm^{-1}).
Max EC_e = Maximum tolerable electrical conductivity of the soil saturation extract for a given crop (dsm^{-1}).
Eff = Leaching efficiency (90% for sandy and loamy sands).

Calculating the Grass Water Requirements (*GWR*)

$$GWR = \frac{ET_c \times S_e}{(1 - LR) \times Effir} \tag{9.6}$$

Where

GWR= Gross Water Requirement (m^3/ha).
ET_c = Crop Evapotranspiration (m^3/ha).
Effir = Efficiency (%), 90%.
LR = Leaching Requirements.
S_e = The percentage of evapotranspiration area.

9.3.3.2 Water Balance Method

Water balance method by difference of soil moisture contents between two irrigations period by measuring changes in moisture contents after and before irrigation at the root zone using a device to measure moisture (Terra Sen Dacom Sensor) at depths of 10 – 120 cm all year. After verifying the accuracy of moisture sensitive, calibrated sensors with direct method (gravimetric laboratory method) with data from the sensors for a period of two months for three sites. The total amount of irrigation for one year calculated by the following equation:

$$\mathrm{ET} = \mathrm{P} + \mathrm{I} - \mathrm{Dr} \pm \Delta S \quad (9.7)$$

Where

ET = Consumptive use (in mm)
P = Precipitation (in mm)
I = Irrigation added (in mm)
Dr = Drinage (in mm) and
ΔS = change in soil water content (in mm)

9.3.3.3 The Amount of Applied Irrigation Water

a) The study site: The amount of applied irrigation water throughout the year by readings of flow meter (actually added) in the field experiment using soil moisture and data of meteorological stations.
b) Farmers fields: The amount of applied irrigation water throughout the year by flow meter added by farmers (actually added to the fields by farmers adjacent to the field of study).

9.3.4 Results and Discussions

9.3.4.1 Climatic Conditions at the Experimental Sites

The observed average values of the climatic variables at the eight sites are presented in Table 9.6. The data revealed that the highest maximum temperature during the year in the Makkah and East Region were 49.9 and 47.5 °C, while the lowest

Table 9.6 The observed average values of the climatic variables eight sites

Sites	Stat	T-Mean °C	T-max °C	T-min °	Rainfall mm	Radiation MJ/m^2	RH-min %	W.speed m/s	ET_0 mm	Kc	ETc mm
Medina	Min.	9.30	13.50	2.80	0.00	2.05	6.00	0.40	2.23	0.80	1.78
	Max.	34.40	40.10	30.30	0.60	11.22	44.00	5.10	12.21	1.00	12.21
	Ave.	24.34	30.51	17.81	0.00	8.19	16.51	2.53	7.11	0.91	6.63
Tabuk	Min.	5.60	9.50	-3.30	0.00	0.79	6.00	0.20	1.18	0.80	0.94
	Max.	33.20	41.80	26.30	8.00	9.76	75.00	4.20	10.57	1.00	10.57
	Ave.	21.67	28.72	14.35	0.11	7.23	18.47	1.95	5.72	0.91	5.32
Makkah	Min.	20.20	23.90	13.10	0.00	1.81	6.00	0.50	2.43	0.80	1.94
	Max.	39.50	49.90	32.00	12.80	9.15	58.00	2.80	8.40	1.00	8.40
	Ave.	30.36	37.71	23.77	0.04	6.67	23.27	1.36	5.46	0.91	5.03
Al Jouf	Min.	4.10	7.00	-1.60	0.00	0.46	4.00	0.80	0.75	0.80	0.60
	Max.	37.50	43.60	32.30	42.00	10.89	92.00	6.50	15.65	1.00	15.65
	Ave.	22.47	28.62	15.93	0.38	7.45	22.05	2.72	6.59	0.91	6.19
Riyadh	Min.	7.40	10.70	-0.70	0.00	1.17	5.00	0.50	1.22	0.80	0.98
	Max.	37.10	44.10	30.70	16.60	10.00	87.00	4.30	12.72	1.00	12.72
	Ave.	24.90	31.81	16.95	0.26	7.52	17.73	1.91	6.29	0.91	5.86
Qassim	Min.	6.00	9.90	-1.90	0.00	1.41	5.00	0.70	1.51	0.80	1.21
	Max.	38.00	45.70	30.50	13.40	9.08	76.00	5.00	13.34	1.00	12.74
	Ave.	25.12	32.80	16.82	0.08	6.90	17.58	2.23	6.50	0.91	6.05
Hail	Min.	4.80	7.30	-2.40	0.00	0.76	7.00	0.40	0.76	0.80	0.61
	Max.	37.10	43.80	31.20	30.40	10.81	95.00	4.80	10.43	1.00	10.43
	Ave.	22.31	29.50	15.08	0.54	7.51	22.32	2.14	5.90	0.91	5.50
East Region	Min.	9.10	14.40	0.50	0.00	0.88	6.00	0.60	1.16	0.80	0.90
	Max.	39.20	47.50	32.70	26.80	10.28	89.00	4.80	12.69	1.00	12.70
	Ave.	25.60	34.28	18.81	0.40	6.98	20.96	2.08	6.30	0.91	5.88

minimum temperature during the year in the Tabouk and Hail were −3.3 and −2.4 °C. The highest maximum net radiation during the year in the Madinah and Al-Jouf were 11.22 and 10.89 MJ/m^2 while the lowest minimum net radiation during the year in the Al-Jouf and Hail were 0.46 and 0.76 MJ/m^2. The highest maximum relative humidity during the year in the Hail and Al-Jouf were 95 and 92 % while the lowest minimum relative humidity during the year in the Al-Jouf wase 4 %. The highest maximum wind speed during the year in the Al-Jouf and Madinah were 6.5 and 5.1 m/s while the lowest minimum wind speed during the year in the Tabouk and Madinah were 0.2 and 0.4 m/s. The results of the study showed that the crop evapotranspiration, ETc (mm/year) of the sites in, Medina, Tabuk, Makkah, Al Jouf, Riyadh, Qassim, Hail, East Region were 2418.75, 1940.51, 1837.76, 2259.03, 2139.23, 2207.41, 2032.09, 2144.87 mm/year, respectively. These results indicate that the estimation of ETc at the different sites of Saudi Arabia is affected by weather conditions. The highest value of ETc was in Medina field site, which could be attributed to the highest net radiation and temperatures.

9.3.4.2 Date Palm Water Requirement in the Experimental Sites

a) *Using the Penman Monteith equation (56) based on climate data*

The results of the study in Table 9.8 showed that the irrigation water requirements (m^3/ha) after taking into account the proportion of cultivated area for each tree of the sites in Medina, Tabuk, Makkah, Al Jouf, Riyadh, Qassim, Hail, East Region were 9495.24, 7340.18, 7298.93, 8913.59, 8614.96, 8568.68, 7996.99, 8510.72 m^3/ha, respectively, with the 100 Palm trees/ha. The total annual irrigation water requirements (m^3/tree) in these sites were: 95, 73.4, 73, 89, 86, 85.7, 80, 85 m^3, respectively, as the radius of shaded area per tree is 3.5 m with effective diameter of 90%, and the rate of leaching were: 12, 8, 13, 12, 14, 11, 13, 13%, respectively. With an irrigation efficiency of 90%, it was found that the on an average, overall irrigation water requirements at all the sites was 8342.41 m^3/ha/year with 100 (trees/ha). These values of ETc and CRW are attributed to the metrological conditions of each site. However, the reduction in the estimated CWR to an average of 8342 m^3/ha compared to overall average of 20000 m^3/ha as reported by many researchers (Al-Amoud et al. 2012; Ismail et al. 2014; Mihoub et al. 2015; Dewidar et al. 2015) is mainly attributed to the percentage of vegetative cover or shaded area (Se) of the tree. As we calculated the Se values as (0.33) of the actual area of the tree. Therefore, the practice distance of 10 m x 10 m between trees on the farms of Saudi Arabia does not seem appropriate at all the sites. This area of 100m^2 for each tree is an overestimate of the crop water requirements and therefore, it must be change to 7 m x 7 m in order to have a higher vegetative cover on the date palm farms.

b) *Water balance method*

The results of water balance method indicate the linear relationship with $r^2 = 0.90 - 0.93$ between the data of Terra Sen Dacom sensors and direct method (Gravimetric Method) as depicted in Fig. 9.7) for a period of two months on the three sites. The

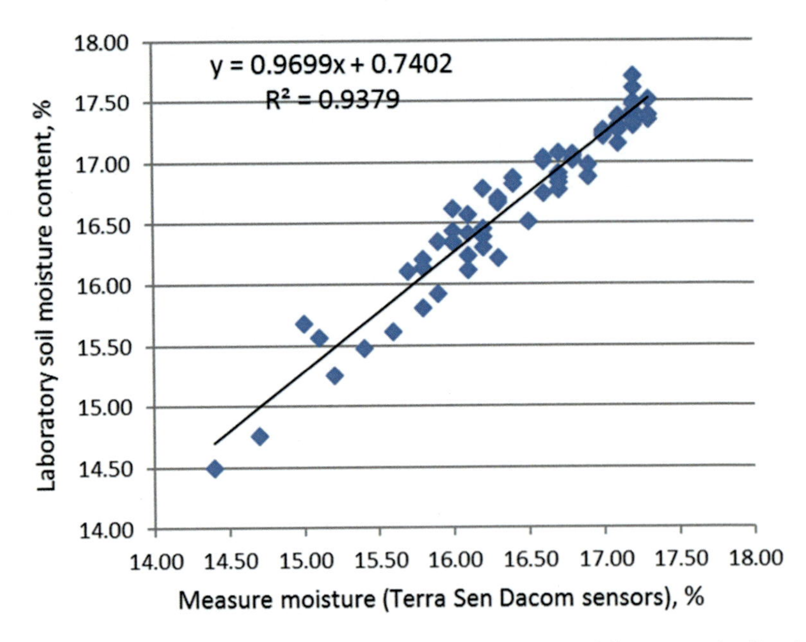

Fig. 9.7 Relationship between gravimetric soil moisture contents and the measured soil moisture contents (Terra Sen Dacom sensors)

results presented in Table 9.4 show that the volume of water consumed was 3604.31, 3515.25 m^3/ha/year for Qassim and Al Jouf, respectively. The amount of rainfall for Qassim and Al Jouf during the season were 92.85, 434.99 m^3/ha/year, respectively. The water balance methods showed that the water consumption for the two sites were very low as compared to ETc estimation by P-M or water added to field. This reduction in total amount of water consumption is mainly due to short depth of the sensor installed in the site (120 cm). It seems that about 50% of water added to date palm trees goes to waste and is being lost through percolation.

c) *The amount of applied irrigation water in study sites*

Table 9.7 shows that the amount of irrigation water actually added by a flow meter of all study sites, of the Medina, Tabuk, Makkah, Al Jouf, Riyadh, Qassim, Hail, East Region were 11305.0, 9463.9, 9692.0, 11252.75, 1007.4, 10035.0, 10272.5, 10082.8 m^3/ha/year, respectively. While these volumes added by the farmers in adjacent, farms were: 13717, 12277, 12220, 13340, 12050, 12880, 12620, 12610 m^3/ha/year, respectively. The increases of the amount of irrigation in adjacent farms by the farmers are mainly due to poor knowledge on irrigation requirements. Before installing the irrigation water monitoring system on the study sites, the farmers used to add three times higher than this volume and that could reach to 35000 m^3/ha.

Table 9.7 Compared the amount water applied in the different methods sites and increase water ratio (%) compared to Penman-Monteith Method

Sites	Water Requirements of Different Methods (m^3/ha/year)				The Increase Water Ratio, (%) Compared to Penman-Monteith Method	
	Penman-Monteith method	Water balance method	Applied Irrigation Water			
			Field Study	Farmer Adjacent	Field Study	Farmer Adjacent
Medina	9495.24	–	11305.0	13717.00	16.0	30.8
Tabuk	7340.18	–	9463.9	12277.00	22.4	40.2
Makkah	7298.93	–	9692.0	12220.00	24.7	40.3
Al Jouf	8913.59	3515.25	11252.8	13340.00	20.8	33.2
Riyadh	8614.96	–	10007.4	12050.00	13.9	28.5
Qassim	8568.68	3604.31	10035.0	12880.00	14.6	33.5
Hail	7996.99	–	10272.5	12620.00	21.2	36.6
East Region	8510.72	–	10082.8	12610.00	15.6	32.5

Table 9.8 Water use efficiency Kg/m^3, Yield Kg/ha and water saving, % in the field study as compared to the adjacent farmer's field

Sites	Field Study			Farmer Adjacent			Water Saving (%)	**EC_e**	Yield %
	Water applied, m^3/ha/year	Yield, Kg/ha	Water use, Kg/m^3	Water applied, m^3/ha/year	Yield, Kg/ha	Water use, Kg/m^3			
Medina	11305.00	7482	0.66	13717.00	7374	0.54	17.58	1.000	100.00
Tabuk	9463.90	6240	0.66	12277.00	6170	0.50	22.91	0.935	100.00
Makkah	9692.00	5406	0.56	12220.00	5324	0.44	20.69	4.600	97.84
Al Jouf	11252.75	6215	0.55	13340.00	6150	0.46	15.65	4.840	96.98
Riyadh	10007.40	7620	0.76	12050.00	7520	0.62	16.95	2.050	100.00
Qassim	10035.00	6742	0.67	12880.00	6531	0.51	22.09	10.950	74.98
Hail	10272.50	6908	0.67	12620.00	6708	0.53	18.60	2.600	100.00
East Region	10082.80	8400	0.83	12610.00	8520	0.68	20.04	6.030	92.69

9.3.5 *Water Use Efficiency Kg/m^3, Yield Kg/ha and Water Saving*

Table 9.8 shows that the productivity per hectare ranged between 5406 kg.ha^{-1} in Makkah and 8400 kg.ha^{-1} Al Ahasa. Water use efficiency (WUE) of palm in Medina, Tabuk, Makkah, Al Jouf, Riyadh, Qassim, Hail and East region in study sites were: 0.66, 0.66, 0.56, 0.55, 0.76, 0.67, 0.67, 0.83 kg.m^{-3}, respectively, while in the neighboring fields, these values were: 0.54, 0.50, 0.44, 0.46, 0.62, 0.51, 0.53, 0.68 kg/m^3, respectively. The water savings were: 17.58, 22.91, 20.69, 15.65, 16.95, 22.09, 18.60, 20.04 %, respectively.

Based on the equation given by Mass and Hoffman (1977) (Yield % = 100 – b (EC_e - a), on the reduction of yield using saline water on all sites of the study. For the date palm trees, the threshold salinity values (a) are 4.0 dsm^{-1} and (b) as 3.6%. As revealed by Table 9.8 the date palm production was affected by salinity in Al-Qassim site with a reduction 25% followed by East Region farm at 7.31%. The rest areas were not affected by salinity.

9.4 Summary and Conclusions

For the improvement of date palm cultivation in the Kingdom of Saudi Arabia, new innovative solutions are needed that in addition to the irrigation water issues could also address constraints like: pest and diseases control, draught, water shortage, soil and water salinization, water quality production, and socio economic and institutional constraints. Integrated efforts by irrigation researches, technicians, managers, and the farmers have to be put together in order to conserve water used in date palm cultivation. No single technology can solve all the water quantity and quality problems confronting irrigation of date palm and other crops. Many technologies, such as irrigation scheduling, advanced irrigation systems, limited irrigation methods, soil moisture management, wastewater irrigation, can be used to use less water for date palm irrigation. An Improved water management is required at all levels of irrigation including planning and design, project implementation, and operation and maintenance. These management improvements require comprehensive changes in institutions and organizations, water policy and law, rehabilitation or introduction of new irrigation systems, education and training of the farmers and extension workers, and researchers and development priorities.

The determination of annual evapotranspiration and water requirements of eight different regions of Saudi Arabia is an offer to improve irrigation water management of date palm. Regions that have been selected for the study include: Medina (Al Ula), Tabuk (Teimaa), Makkah (Al Jumum), Al Jouf (Sakakah), Riyadh (Sodos), Qassim (Riyad Al Khabra, Hail (Al Kaedh), East Region (Al Ahsa). The results of the study showed that the crop evapotranspiration, ETc (mm/year) without taking shaded area per tree, in the regions of Medina, Tabuk, Al Jouf, Riyadh, Qassim, Hail, Al Ahsa were 2418.75, 1940.51, 1837.76, 2259.03, 2139.23, 2207.41, 2032.09, 2144.87 mm/year, respectively. Irrigation water requirements (m^3/ha) after taking into account the proportion of cultivated area for each year are 9495.24, 7340.18, 7298.93, 8913.59, 8614.96, 8568.68, 7996.99, 8510.72 m^3/ha, respectively, for the 100 Palm/ha. The annual total irrigation water requirements in these regions were found to be 95, 73.4, 73, 89, 86, 85.7, 80, 85 $m^{3/}$tree, respectively as the radius of shaded area per tree is 3.5 m. The decrease of the CRW in all sites of study to around 8000 m^3/ha is mainly attributed to percentage of shaded area of date palm tree. The study suggests maintaining the tree-to-tree distance of 7 m x 7 m, instead of practicing 10 m x 10 m in order to reduce the estimation of CRW of date palm trees. The water balance methods showed that water consumption for the two sites were very low compare to ETc estimation by P-M or water added to field. This reduction in total amount of water consumption is mainly due to short depth of the sensor installed in the site (120 cm). It seems that 50% of water added to date palm tree goes to waste as leaching water.

References

Al-Amoud, A. I., Mohammed, F. S., Saad, A. A., & Alabdulkader, A. M. (2012). Reference evapotranspiration and date palm water use in the Kingdom of Saudi Arabia. *International Research Journal of Agricultural Science and Soil Science, 2*(4), 155–169.

Alazba, A. A., (2001). *Theretical estimate of palm water requirements using Penman-Monteith model*. A paper number 12100, 2001 ASAE annual meeting.

Al-Barrak, S. A. (1990). Characteristics of some soils under date palm in Al-Hassa Eastern oasis, Saudi Arabia. *Journal of King Saud University of Agricultural Science, 2*(1), 115–130.

Al-Barrak, S. A. (1997). Characteristics and classification of some coastal soils of Al-Hassa, Saudi Arabia. *Journal of King Saud University of Agricultural Science, 9*(2), 319–333.

Al-Ghobari, H. M. (2000). Estimation of reference evapotranspiration for southern region of Saudi Arabia. *Irrigation Science, 19*, 81–86.

Al-Jaloud, A. A., & Hussain, G. (1992). Water quality of different aguifers in Saudi Arabia and its predictive effects on soil properties. *Arid Soil Research and Rehabilitation, 7*, 85–101.

Allen, R.G., Pereira L.S., Raes D., & Smith M. (1998) Crop Evapotranspiration. FAO irrigation and drainage paper No.56, FAO, Rome., Italy, 300 pp.

Al-Omran, A. M., Choudhary, M. I., Shalaby, A. A., & Mursi, M. M. (2002). Impact of natural clay deposits on water movement in calcareous sandy soil. *Journal of Arid Land Research and Management, 16*, 185–193.

Al-Omran, A. M., Mohammad, F. S., Alghobari, H. M., & Alazba, A. A. (2004). Determination of evapotranspiration of tomato and squash using lysimeters in central Saudi Arabia. *International Agricultural Engineering Journal, 13*(1&2), 27–36.

Al-Omran, A. M., Falatah, A. M., & Al-Matrood, S. S. (2005). Evaluation of irrigation well water quality in Riyadh region, Saudi Arabia. *Journal of King Abdulaziz University, 16*(2), 23–40.

Al-Shemeri, F. 2016. *Estimation of Date Palm Water Requirement in Saudi Arabia*. M.Sc. Thesis. King Saud University, Riyadh. Saudi Arabia. In Arabic.

Ayers, A.S. & D. W. Westcot. (1985). *Water quality for agriculture*. FAO #29, Rome, 1985.

Bashour, I. I., Al-Mashhady, A. S., Prasad, J. D., Miller, T., & Mazroa, M. (1983). Morphology and composition of some soils under cultivation in Saudi Arabia. *Geoderma, 29*, 326–340.

Bhat, N. R., Lekha, V. S., Suleiman, M. K., Thomas, B., Ali, S. I., George, P., & Al-Mulla, L. (2012). Estimation of Water Requirements for Young Date Palms Under Arid Climatic Conditions of Kuwait. *World Journal of Agricultural Sciences, 8*(5), 448–452.

Dewidar, A. Z., Ben, A. A., Al-Fuhaid, Y., & Essafi, B. (2015). Lysimeter based water requirements and crop coefficient of surface drip-irrigated date palm in Saudi Arabia. *International Research Journal of Agricultural Science and Soil Science, 5*(7), 173–183.

Doorenbos, J., & Pruitt, W. O. (1977). Guidelines for predicting crop-water requirements. In *Irrigation and drainage paper No. 24* (2nd ed., pp. 1–107). Rome: FAO, United Nations.

General Authority for Statistics. (2015). *Detailed results of the Agriculture Census*. Kingdom of Saudia Arabia, www.stats.gov.sa

Hamza, A. G., Abu-Mustaf, A. M., & Hassan, M. M. (1975). The investigation of diuretic water of Abu-Hamata well. *Bulletin, Faculty of Science, Rriyad Universtiy., 7*, 269–271.

Heakal, M. S., & AlAwajy, M. H. (1989). Long-Term effects of irrigation and date palm production on Torripsamments, Saudi Arabia. *Geoderma, 44*, 261–273.

Hellman, E. 2010. *Irrigation scheduling of grapevines with evapotranspiration data*. Texas A&M University, Texas AgriLife Extension Service: College Station, Texas. http://winegrapes.tamu.edu/grow/irrigationscheduling.pdf
Ismail, S. M., Al-Qurashi, A. D., & Awad, A. A. (2014). Optimization of irrigation water use, yield, and quality of Nabbut-Saif Date Palm under dry land conditions. *Irrigation and Drainage, 63*, 29–37.
Jahangir, M., Al-Salim, S. A., Al-Mishal, M. I., Faruq, I. M., Al-Zahrani, Y., & Al-Sharif, A. S. (1987). Chemical profiling of ground water of Al-Kharj, Saudi Arabia. *Pakistan Journal of Scientific and Industrial Research, 30*, 9–13.
Jensen, M. E., & Wright, J. L. (1978). The role of evapotranspiration models in irrigation scheduling. *Transactions of the American Society of Mechanical Engineers, 21*(1), 82–87.
Jensen, M. E., R. D. Burman, & R. G. Allen, eds. (1990) "Evapotranspiration and Irrigation Water Requirements", A.S.C.E. Manual, New York, p. 332.
Kassem, M. A. (2007). Water requirement and crop coefficient of date palm trees «Sukariah CV». *Misr Journal of Agricultural Engineering, 24*, 339–359.
Maas, E. V., & Hoffman, G. J. (1977). Crop salt tolerance\-current assessment. *Journal of the Irrigation and Drainage Division, 103*(2), 115–134.
MAW. (1985). *Ministry of Agriculture and Water annual statistical reports*. Riyadh, Saudi Arabia.
Mazahrih, N. T., Al-Zu'bi, Y., Ghnaim, H., Lababdeh, L., Ghananeem, M., & Abu-Ahmadeh, H. (2012). Determination actual evapotranspiration and crop coefficients of date palm trees (Phoenix dactylifera) in the Jordan Valley. *American-Eurasian Journal of Agriculture and Environmental Science, 12*(4), 434–443.
Mee, J. M. (1983). Saudi groundwater chemistry and significance. *The Arab Gulf Journal of Scientific Research, 1*(1), 113–120.
Mihoub, A., Samia, H., Sakher, M., El-Hafed, K., Naoma, K., Kawther, L., Tidjani, B., Abdesselam, B., Mohamed, L., Yamina, K., & Amor, H. (2015). Date Palm (Phoenix dactyllifera L.) irrigation water requirements as affected by sanlity in Oued Righ conditions, North eastern Sahara, Algeria. *Asian Journal of Crop Science, 7*(3), 174–185.
Ministry of Agriculture. (2000). *Ministry of Agriculture annual statistical reports*. Riyadh, Saudi Arabia
Ministry of Electricity and Water. (2014). *Annual reports*. Riyadh. Saudi Arabia.
Piper, A. M. (1944). A graphic procedure in the geochemical interpretation of water-analyses. *Eos, Transactions American Geophysical Union, 25*(6), 914–928.
Richards, L. A. (1954). *Diagnosis and improvement of saline and alkali soils* (Vol. 78, No. 2, p. 154). Washington, DC.
Wright, J. L., & Jensen, M. E. (1972). Peak water requirement of rops in southern Idaho. *Proceedings of the American Society of Civil Engineers, Journal of the Irrigation and Drainage Division, 98*(IR2), 192–201.
Wright, J. L., & Jensen, M. E. (1978). Development and evaluation of evapotranspiration models for irrigation scheduling. *Transactions of ASAE, 21*(1), 88–91.
Zaid, A. (2002). *Date palm 2002*. Date Palm Cultivation. FAO.156, Rome.

Dr. Abdulrasoul Al-Omran is a Professor of Soil Sciences and Water Management at the King Saud University (K. S. U). He grew up in Al-Hassa, Eastern Region, Saudi Arabia, where he completed his early schooling as well as four years of College of Agriculture, KSU. 1975.

In 1979, he entered M.Sc. at University of California, Davis, USA in Water Science (Irrigation), and PhD from Oregon State University, Corvallis, USA, 1984, in Soil Science. He is Editor-in-chief of Journal of Saudi society for Agricultural Sciences (JSSAS) since 2003, and associate editor for other journals such as Arid Land Research and Management, USA (2003-present). He published over 100 articles in national and international journals in the field of crop water requirements, water quality and conservations. He wrote and translated more than five books from English to Arabic.

Mr. Fahad Alshammari is a Postgraduate Student. He entered M.Sc. at King Saud University in Date Palm Water Requirements. He works at Ministry of Environment, Water and Agriculture, Riyadh, Saudi Arabia.

Dr. Samir Eid is a Researcher Assistant at the Ministry of Environment, Water and Agriculture, Riyadh, Saudi Arabia and a researcher at agricultural Engineering research institute, Aricultural Research Center, Egypt.

Mr. Mahmoud Nadeem is a Scientific Researcher at Soil Science Department, Desert Research Center (DRC), Matarya, Cairo, Egypt (1980–1987). Now at Soil Science Department, King Saud University, Riyadh, Saudi Arabia (1987–2017). Where he completed his early schooling as well as four years of College of Agriculture, Ain Shams University, Cairo Egypt, in Soil Sciences. He entered M.Sc. at Ain Shams University, Cairo Egypt, 1980, in Soil Sciences. He published over 30 articles in national and international journals in the field of Soil Chemistry and Crop Water Requirements.

Chapter 10
The Contribution of CSR to Water Protection in the Maghreb Region: Engineering a New Approach to Assure Water Security

Abdelhafid Aimar

10.1 Introduction

The Maghreb is a region which is currently facing many challenges in the environmental sphere and in the exploitation of natural resources. Today, the threats of water scarcity and insecurity are imminent in the region. Population growth, urbanization, climate change, poor water management, and soil contamination are severely affecting water supplies and endangering community livelihoods. Water is an essential natural resource, it is vital for economic and social development, as well as for sustained environmental management. Growing demand for water and the increasing scarcity and deterioration of water quality mean that its use and management have become of central interests to governments and businesses in the region and elsewhere.

Environmental degradation in the Maghreb region is essentially of natural origins, but has been accelerated by human activities during the last decades. While it is caused by soil degradation, desertification, droughts, floods, and water scarcity (Tabet-Aoul 2011), it has been aggravated by agricultural runoffs and domestic and industrial discharges. Indeed, pollution is presently deteriorating the quality of water and reducing freshwater availability in the region. The engineering of a new water approach appears to be crucial to reduce social vulnerabilities, secure safe water supplies, and ensure water security in the region. But this does not seem to be solely the job of governments. In an already ecologically fragile Maghreb region, a collaborative action is urgently required to address water scarcity and contamination risks. Any action to protect water resources in the region seems to necessitate the involvement of both private and public businesses as they are using about 10% of water resources and are partly responsible for water contamination.

A. Aimar (✉)
Faculty of Economics, Commerce and Management, University of Jijel, Jijel, Algeria

M. Behnassi et al. (eds.), *Climate Change, Food Security and Natural Resource Management*, https://doi.org/10.1007/978-3-319-97091-2_10

Although the industrial business sector use less water compared to the agricultural sector, the impact of their externalities on the environment is far-reaching. It contaminates surface and groundwater, causes serious damages to soil and air, and affects both biodiversity and human health. Industrial businesses in the region need to redirect their action to strike a balance between business development on the one hand, and water protection and risk reduction on the other. Indeed, the risks emanating from industrial activities can be devastating. Corporate social responsibility (CSR) is about risk management. It deals with various industrial risks with the purpose of minimizing or eliminating the negative impact of industrial operations on both community and environment. This paper illustrates the water risks with which the Maghreb region is concerned and draws attention to the contribution of CSR to water conservation and pollution risk reduction. It examines whether the involvement of business sector can abate the threats of water deterioration in the region. It further seeks to know how CSR could allocate resources to positively and effectively impact the use of valuable water resources, control water contamination risks and sustain water resource management. The aim of this paper is to formulate a new approach capable of combating water scarcity and degradation, containing future water risks, and ensuring water security for the region's populations.

10.2 The State of Water in the Region

The Maghreb is one of the driest and ecologically vulnerable regions in the world. More than 70% of the land is dry, desertification is increasing, rainfall is dropping and water resources are scarce and unevenly distributed. Reports – such as the 2009 United Nations Development Program (UNDP) and 2010 Arab Forum for Environment and Development (AFED) reports – are warning that countries in the region are under severe water scarcity (in some areas less than 500 cubic meters per capita per year), which is far below the global average in terms of water availability, estimated at 6000 m^3 (El-Ashry et al. 2010; UNDP 2011). This dangerous situation is forecast to deteriorate further in the foreseeable future because of many factors such as the fast population growth, the increasing economic growth, climate change, pollution, and low water use efficiency. Moreover, the region suffers from a weak water governance structure. Water institutions are less efficient and lacking adequate technical capabilities and coordination (Chatila 2010). Moreover, large public sectors subsidies and deficient water policies have limited participation and created irresponsible practices (El-Ashry et al. 2010). This has led to unequal distribution of water resources, water wastage, increasing pollution, lack of transparency, and inefficient water services throughout the Maghreb region. This water scarcity constitutes a serious challenge to the region's human security (mainly food and human health security and development).

In addition to the hard ecological conditions, agricultural production in the region is deteriorating the imbalance between water demand and supply as it consumes more than 80% of total water withdrawals compared to an average of 14% and 6% for domestic and industry use respectively. But the share of agriculture in

Table 10.1 Water availability and use in the Maghreb region

Country	Natural renewable resource (bn cubic meters)	Per capita renewable availability (cubic meters)			Annual water usage		% use by sector		
		2006	2015[a]	2025[a]	Bn cubic meters	As a % of total water resources	Agric.	Domestic	Indust.
Algeria	11.50	350	620	558	4.59	40	60	25	15
Morocco	20.00	940	297	261	16.84	84	95	–	05
Tunisia	03.35	450	405	373	2.53	72	84	12	04

Source: Barghouti (2010)
[a]Projections

Morocco is much higher, about 95 as indicated in Table 10.1. Besides, irrigated agriculture is expanding in the region to boost food production and reduce imports, especially after the last global food crisis. For example, the irrigated area in Algeria was reported to have increased to 1.6 million hectares in 2014 against 1.1 million in 2012 (Ruitenburg 2012). This is likely to raise the agriculture's share of water use in the long run. This imbalance in water use between different sectors is exacerbating the problem of water shortages. For example, in Tunisia it is estimated that the quantity of water consumed daily by an irrigated area of 1000 ha is equal to the amount of water used by a city with one million people (Molle 2011). Thus, a reallocation of water resources between different users is vital. Water transfers should be done from agriculture to urban and industrial areas.

Furthermore, the Maghreb region suffers from water overuse, wastage and inefficiency. This is particularly driven by low prices and the low levels of people's consciousness. According to some statistics, the price of water could be as much as 35% of the cost of production, and this percentage is only 10% for desalinated water (El-Ashry et al. 2010). In this respect, it is important to note that more than 50% of the quantities of water used in agriculture are wasted because of old irrigation systems (El-Ashry et al. 2010). Undoubtedly, agricultural production could be raised significantly with less water and millions of cubic meters could be saved to be reallocated to other socio-economic activities.

10.3 Water Deterioration in the Maghreb Region

Many global reports on water resources (such as WPP 2011) have made it blatantly clear that countries worldwide will not be able to meet their most pressing goals in the areas of human development (sanitation, access to water supply), food security (agricultural water management), energy security (hydropower, cooling water), and urban development (protection from droughts and floods) without a major shift in the way they manage their water resources. In the Maghreb region, water resources are scarce and are being undermined by changes in rainfall and temperature, as well as by anthropogenic activities. The region is projected to witness more impacts in the foreseeable future (Ben Abdelfadel and Driouech 2008).

Today, water deterioration has become a serious global threat as it is a principal cause and a basic component of environmental degradation. Water deterioration is the depletion of freshwater resources through overexploitation and misuse (Wisler 2014). Overexploitation and pollution worldwide have been affecting the environment negatively over the last few centuries (Patterson n.d.). But today, the overuse and pollution of water resources have been exacerbated by population growth, increased urbanization, higher standards of living, and climate change (Climate Institute 2007–2010). In the Maghreb region, water deterioration has been reducing freshwater availability and impacting soils, ecosystems and human health. If this trend continues, economic and social development in the region will likely be obstructed and human health will be in serious danger in the decades to come.

10.3.1 Water Pollution

Pollution is one of the biggest problems in the world today. It is not just the result of natural phenomena, such as floods and droughts; rather it is also caused by human activities. It occurs at different levels and it affects all the elements of an ecosystem, including air, water, and soil. It impacts immensely all species, including mankind, and destroys wildlife and natural habitats. Pollution equally reduces freshwater availability and causes illnesses and deaths. While efforts are being made to prevent a global environmental disaster, pollution remains a serious threat to life on the planet. According to some estimates, over 80% of used water worldwide is not collected or treated (Corcoran et al. 2010). In developing countries, up to 90% of wastewater flow untreated into rivers, lakes and highly productive coastal zones, threatening health, food security, and access to safe drinking water (UNwater 2013).

There are several sources of water pollution ranging from sewage and fertilizers to soil erosion. It is primarily due to:

- *Agricultural practices related to the use of high levels of agro-chemicals.* Chemical fertilizers, insecticides and herbicides used to increase production actually pollute the air, soil and water (Patterson n.d.). Fertilizers enter both human and livestock waste streams that eventually enter groundwater, while nitrogen, phosphorus, and other chemicals from fertilizer can acidify both soils and water (Tilman et al. 2001). Likewise, irrigation increases salt and nutrient content in soils and damages streams and rivers from damming and removal of water (Tilman et al. 2001).
- *Increasing inflows of domestic and industrial chemicals and toxic waste into water bodies.* Different kinds of industries discharge chemicals daily which pollute streams and rivers and damage fragile ecosystems. All pollutants have a negative, often devastating, impact on vegetation and aquatic ecosystems (Painter n.d.).
- *Acid rain can also have negative impacts on water, soil, and the whole environment.* It occurs when carbon monoxide and sulfur dioxide from industrial plant emissions combines with moisture present in the air. A chemical reaction creates this acid precipitation. Acid rain can acidify and pollute lakes and streams. It can acidify the water or soil to a point where no life can be sustained (Skye n.d.-a).

- *The lack of clean sanitation.* It is estimated that about 1 billion people worldwide still lack access to sanitation services and some 2.4 billion people remain without access to improved sanitation facilities. In rural areas, only 51% are using improved sanitation compared to 82% in urban areas (UNICEF 2015).
- *Desalination of seawater.* Brine effluents are discharged back into the sea making seawater increasingly saltier and marine life difficult.

In the Maghreb region, urbanization and industrial activities, as well as irrigation, are expanding fast, generating increasing volumes of harmful waste. Domestic sewage, industrial chemicals, fertilizers, pesticides and other various contaminants are being discharged daily into waterways. It is estimated that the MENA countries generate a total of about 10 km^3 of wastewater per year, of which 5.7 km^3 is treated (El-Ashry et al. 2010). In Tunisia alone, the World Bank recorded more than 750 sources of water pollution and 155 million cubic meters of waste every year (Croitoru and Sarraf 2010). These increasing wastes deteriorate the quality of water and pose serious threats to public health in the whole region. As the figures indicate in Table 10.2, the discharge of water pollutants in the region is increasing immensely, and Algeria is by large the most generator of organic water pollutants because of accelerated industrialization and urbanization. Pollution reduces the availability and safety of freshwater resources in the region which are already scarce.

More importantly, contaminated water has drastically negative impacts on vegetation, aquatic ecosystems and humans. Pollutants raise the temperature of the water and increase the rate of algal blooms, forcing fish to leave to much cooler places or causing massive fish die-offs (Skye n.d.-b). Furthermore, polluted water is a contributor to a wide range of health problems and illnesses. It causes typhoid, diarrhea, cholera and other potential toxins which are great in numbers. A case in point of water pollution in the region caused by industrial chemicals is the privately-owned leather treatment plant located at El-Milia region in eastern Algeria. This plant pumps the water it uses into its leather treatment processes from Oued El Kebir valley and releases huge quantities of toxic chemicals, such as chromium, into the water stream without proper treatment. Undoubtedly, the leather industrial effluents are extremely harmful to surface water bodies and pose serious threats to human health and the environment (Chowdhury et al. 2013). Moreover, such irresponsible practices reduce freshwater availability in the region and impact drastically the whole environment.

Table 10.2 Water pollution levels from organic pollutants in the Maghreb region (1990–2003)

Country	Emission of organic water pollutants (metric tons daily in 1990)	Emission of organic water pollutants (metric tons daily in 2003)	Emission of organic water pollutants (kg per worker daily 1990)	Emission of organic water pollutants (kg per worker daily 2003)
Algeria	107.0	–	0.25	–
Morocco	41.7	72.1	0.14	0.16
Tunisia	44.6	55.8	0.18	0.14

Source: Constructed from World Bank (2007)

Moreover, over-extraction of groundwater in the Maghreb region is exposing groundwater aquifers to saline water, particularly in coastal areas. According to some estimates, Algeria, Morocco and Tunisia are extracting renewable groundwater resources at a rate ranging from 20 to 60%. In some regions, like the Souss-Massa plain in Morocco, aquifers are being overexploited at notorious rates (Molle 2011). Likewise, the central areas in Tunisia suffer a serious drop in the level of groundwater and an increase in water salinity (Elloumi 2011). Thus, over-extraction depletes aquifer reserves in the region, affects the quality of water, raises salinity in the soil and causes water supply irregularities, particularly in dry seasons. In addition, over-extraction affects forests, pastoral resources and ecosystems which are already exposed to sand movement from the south and rising temperature.

10.3.2 Fast Population Growth and Urban Development

The population in the Maghreb region is growing fast. Table 10.3 clearly indicates that the total population in Algeria, Morocco, and Tunisia rose to nearly 78 millions people in 2010 compared to about 35 millions in 1980. The population in the region has grown to 85 millions in 2015 and is projected to further grow to about 100 millions by 2050 (World Bank 2016). This growth will lead to further increases in withdrawals from the water supply for agricultural, domestic, and industrial uses. This is expected to further pressurize the region's scarce water resources, something which will likely exacerbate environmental degradation and water deterioration.

Moreover, the urban population in the Maghreb region is forecast to increase to more than 60% of the total population in the years to come (UNDESA 2014). This will concentrate the demand for water in urban areas, and puts stress on the fresh water supply from industrial and human contaminants (Climate Institute 2007–2010). Urbanization causes overcrowding and increases unsanitary living conditions, which will in turn expose a growing number of people to illnesses (Abdul Rahman and Suppian 2010). Besides, a large percentage of the population in the region still lacks access to sanitary water and sewer systems. This will continue causing diseases and deaths from contaminated water in the future.

Indeed, total water withdrawals in the Maghreb region are increasing fast because of population growth and urbanization. Table 10.4 clearly shows that water withdrawals in the region have increased by nearly 26% in 15 years. But in Algeria, this figure

Table 10.3 Estimated and projected population growth in the Maghreb region 1950–2050 (mn)

Country	1950	1980	2010	2020	2025	2050
Algeria	8.753	18.811	35.468	40.180	42.043	46.522
Morocco	8.953	19.567	31.951	35.078	36.406	39.200
Tunisia	3.530	6.457	10.481	11.518	11.921	12.649
Total	**21.236**	**34.835**	**77.900**	**86.776**	**90.370**	**98.371**

Source: Zeitoon (2012)

Table 10.4 Increases in the Maghreb region's water demand

Country	Population 1985 (mn)	Population 2000 (mn)	Total water withdrawals 1985 (10 m³)	Total water withdrawals 2000 (10 m³)
Algeria	21.86	30.46	3.50	6.07
Morocco	22.10	27.84	11.00	12.61
Tunisia	7.10	9.56	2.48	2.70
Total	51.06	67.86	16.98	21.38
	32.90% increase in population		**25.91% increase in water use**	

Source: Barghouti (2010)

has nearly doubled over the same period. Thus, population growth and associated demand for water have accelerated the use of water resources; a situation which exacerbates the groundwater depletion and, therefore, further environmental degradation.

Thus, fast population growth in the Maghreb region is projected to make excessive demands on water, agriculture and livestock and other natural resources. This will likely result in drastic impacts such as, inter alia (Patterson n.d.):

- The use of chemical fertilizers, insecticides and herbicides to increase agricultural production pollutes the air, soil and water with toxic chemicals. Fertilizer run-offs cause toxic algal blooms that kill aquatic animals.
- Removing trees and other plants to increase areas of cultivation causes habitat loss and threatens the survival of numerous species of animals and plants.
- Monoculture keeps cost of production low, but it reduces biodiversity and negatively impacts the soil.
- Large-scale farming of animals increases their susceptibility to diseases. Waste generated in the farms and meat-processing plants can affect the water quality in the area.
- The greater distance food items have to travel to reach the consumer, the greater the transportation's impact is on the environment (carbon emissions, atmospheric pollution, food waste, etc.).

10.3.3 Climate Change

Increased climate variability is posing new challenges to water resource management at an international scale because the climate and hydrological cycle are strongly linked (World Bank 2009). Climate change is particularly affecting precipitation patterns, river flows, groundwater, land temperature, and soil humidity. Extreme climate variations are causing floods, droughts, desertification, major storms, and wild fires. Droughts and floods are expected to become frequent in different areas at various times. Rises in temperature will change rainfall patterns and increase the rate of evaporation. These intensified climate events are degrading the environment worldwide and deeply affecting economies, food crops, human health, livelihoods, and livestock. Poor and arid countries are the most vulnerable.

Although Maghreb countries are not responsible for global warming (e.g., Morocco and Algeria's share in global emissions of greenhouse gases is only 0,8% and 0.36% respectively) (Abdelbari 2015), climate change variations are forecast to exacerbate the water scarcity in the region. According to leading Global Climate Models (GCMs), temperature is expected to rise by an average of 2 °C, evaporation rates by 25%, and rainfall is likely to drop by 25% by the turn of this century (Assaf 2010). These projected climatic changes will affect the amount of water available to replenish groundwater and surface water, including dam levels. Moreover, the increase in temperature will cause a sea-level rise, which may affect the fresh water supply of coastal areas as well. The intrusion of saltwater underground results in an increase of salinity in reservoirs and aquifers. An increase in water temperature can also affect ecosystems greatly in the region because of species' sensitivity to temperature.

Thus, the aridity of the Maghreb region is expected to intensify and water availability will likely decrease drastically in the decades to come. The region will be exposed to recurrent and extended droughts, crop failures, and livestock losses. Table 10.5 clearly shows that millions of people in the region will be suffering from water shortages in the coming decades. Rain-fed agriculture is also projected to drop by as much as 40% in Algeria and Morocco by the end of this century (Barghouti 2010).

From this perspective, the cost of adaptation to the effects of climate change could be astronomical in the region. According to the World Bank report (2010), the adapting global cost to a 2 °C global increase in temperature could reach as far as US$ 100 billions per annum between 2020 and 2050. Thus, the Maghreb region's governments are urgently required to raise preparedness to face the serious threats and uncertainty of climate change. This can particularly be done through investing largely in water infrastructures and their staff, as well as raising water use efficiency. If achieved properly, such efforts could mitigate the impact of climate change in the region.

Table 10.5 Future scenarios of climate change impacts on water in the Maghreb region and elsewhere

Type of change	Effects on human security	Affected area
1 °C rise in earth temperature	Reduced water runoff in Ouergha watershed by 10%	Morocco
2 °C rise in earth temperature	1–1.6 billion people will be affected by water shortages	Africa, MENA countries, Southern Europe, parts of South and Central America
3 °C rise in earth temperature	Increased water stress for additional 155–600 millions people	North Africa

Source: Constructed from UNDP (2009)

10.3.4 Water Wastage

Deficient water management causes wastage in water use and contributes to water shortages. This happens because of misallocation of water supplies, insufficient law enforcement and/or leaky water pipes. Agriculture can be a major source of water wastage when old irrigation systems are used. In the Maghreb region, agricultural production is consuming more than 80% of total water withdrawals. Undoubtedly, this percentage can be reduced considerably to save millions of cubic meters yearly if modern irrigation systems, such as sprinkler or dripping systems, are widely adopted by farmers in the region.

Likewise, weak law enforcement in the region often leads to illegal drilling into groundwater aquifers. This contributes to overexploitation of water resources in the region. For instance, farmers in Algeria pump large amounts of water from water wells or streams in the summer season to flood melon and water melon fields without much control, monitoring or care. Such practices contribute to the depletion of the region's water resources and cause further soil and water deterioration.

Moreover, water leaks are common in the Maghreb region, similarly to other Arab countries. These are due to aging pipe infrastructure and inefficient distribution networks. The region loses millions of cubic meters due to this problem every year. According to some estimates, water leaks in the Arab region are as much as 30% of water consumption as compared to 5% in Europe (Bishara 2014). Water leaks are expected to increase in the future because water pipelines will continue aging, and hence existing water supply infrastructure will be unable to support the growing burden of population growth and industry in the region (Water Academy 2004). So, unless tackled seriously, water leakages will likely exacerbate the problem of water scarcity in the Maghreb region.

10.4 Business Organization, the Environment and Society

Business organizations are part of the community in which they operate. They use the community's human and material resources (inputs) to generate products (outputs) which the community needs. While doing so, they search to maximize the economic earnings in order to enhance their competitiveness and ensure their permanent existence. However, while carrying out their operations, business organizations may negatively affect the environment and workers' health and safety. More significantly, they generate numerous negative impacts on both environment and society (air and water pollution, soil degradation, violations of human rights at work, and generating harms to surrounding communities...).

Therefore, it is vital for business organizations to cooperate with public authorities, scientific community, and people to mitigate the negative effects of their activities. Their management strategies need to embrace the environmental and social dimension and share both the social return and cost with the community. Thus, environmental and social responsibility of business organizations is a matter of self-preservation, and an integral part of sound resource management practices (Skye n.d.-a). In this sense, the contribution to the protection of the environment is beneficial to businesses as it enables them to face the criticism and pressures emanating from the community (Chauveau and Rosé 2003) and respond to stakeholders' interests (individuals or groups whose interests could be affected by the decisions and activities of the organization).

10.4.1 Corporate Social Responsibility (CSR): Meaning and Aspects

There is no clear-cut definition of CSR. It still derives its strength, acceptance and deployment of the voluntary nature. CSR depends very much on market, the scope and forms of the business activity, and the human and financial capabilities of the corporation. The World Bank views CSR as "the commitment of business to contribute to sustainable economic development, working with employees, their families, the local community and society at large to improve quality of life, in ways that are both good for business and good for development" (Petkoski and Twose 2003). The World Business Council dor Sustainable Development (WBCSD) also defines CSR as "the continuing commitment by business to behave ethically and contribute to economic development while improving the quality of life of the workforce and their families, as well as of the local community and society at large" (WBCSD 1998). The European Commission further views CSR as "a concept whereby companies integrate social and environmental concerns in their business operations and in their interaction with their stakeholders on a voluntary basis" (European Commission 2002).

Although they differ, CSR definitions have some common elements:

- Commitment to contribute to sustainable development.
- Voluntary character of social responsibility, social responsibility becomes part of the organizational culture and it is not subject to legal or contractual obligations.
- Consensus on the important role that could be played by the business organization in the community, being one of its components.
- Commitment by the business organization to make social responsibility one of the pillars of its strategy.
- Transparency in information gathering and diffusion inside and outside the business organization, allowing authentication and demarcation of the good practices of the business organization and keep track of the progress taking place.
- Ability to involve other parties in the operations of the business organization and to have relations with many members of the community in which it operates (such as civil society organizations, medias, labor unions, local authorities, etc.).

In fact, CSR is more than philanthropy giving and volunteering. Since the early 1950s, calls have constantly emerged regarding the necessity of business organizations' commitment to the community in which it operates instead of focusing only on making the maximum possible profits. Today, businesses play a basic developmental role, and philanthropy giving for development has become an integral part of their activities. The business management became responsible not only for the economic efficiency of corporate's activities expressed by profitability index, but also for what it should play towards environmental and social concerns arising from the performance of those activities. In other words, companies have become obliged to assume the environmental and social responsibility, together with the economic responsibility. The aim is to help achieve sustainable development. Table 10.6 clearly illustrates the economic responsibilities model on the one hand, and the environmental and social responsibilities model on the other hand.

Protecting the environment from the negative impact of operations is a core responsibility. Besides their legal obligations, which differ according to region and country, companies are seen to have a broad responsibility to protect the physical environment throughout their supply chains. They should commit to continuous improvements in eco-efficiency (doing more with less), and managing the full life-cycle of their products. In this respect, the WBCSD (1998) noted in its dialogue session that companies should be: proactive on the environment and seek solutions that can lead to competitive advantage; and responsible and leading companies pave the way for others to follow suit.

10.4.2 CSR-Associated Concepts

The concept of CSR is connected to a set of concepts and issues. They all related to environmental issues, such as resource use and preservation. This section focuses on five basic terms: sustainability, triple bottom line, ISO 26000, eco-efficiency, and ISO 14046.

Table 10.6 Economic and social models of enterprises

Economic responsibility model	Social responsibility model
Production	Quality of life
Exploitation of natural resources	Conservation of natural resources
Increase of production	Increase of productivity and efficiency
Internal decisions based on market conditions	Decisions based on market conditions under the community's control
Economic return (profit)	Balance between economic returns and social returns
Interest of the organization or the manager or owners	Interest of the enterprise and the community
A limited role for the government	An active role for the government

10.4.2.1 Sustainability

CSR is tightly linked to sustainability which essentially relates to human use and protection of natural resources. Sustainability has been given a wide range of definitions but most of them focus on two basic aspects: the aspect of time and the aspect of human activity (Zwahlen 1995). Such definitions have been put forward by a number of scholars in the 1980s and 1990s like Allen (1980), Pearce (1988), Mâler (1991), Pearce and Atkinson (1993), and Titilola (1994). In their views, sustainability is the use of resources either for a very long time or forever. Viederman (1993) discussed 'sustainable society' rather than 'sustainable use', without excluding the aspects of time and human activities in his definition of sustainability. Jowsey and kellett (1995) further raised the 'ratio of demand to resource base' and 'environmental impact'. From his side, Zwahlen (1995) spoke strictly of the 'sustainability of the use of resources' rather than the 'sustainability of a resource'. Thus, the aspect of 'resource use' is a basic element in the definitions which have been given to sustainability.

Sustainability has also been advanced as improving the quality of life without increasing the use of natural resources in order to sustain Earth's capacities for regeneration and waste absorption. In this context, Elkington (1997) noted that a sustainable society needs to meet three conditions: its rates of use of renewable resources should not exceed their rates of regeneration; its rates of use of non-renewable resources should not exceed the rate at which sustainable renewable substitutes are developed; and its rates of pollution of emission should not exceed the assimilative capacity of the environment. Hart and Milstein (2003) further defined sustainability as the expectations of improving the social and environmental performance of the present generation without comprising the ability of future generations to meet their social and environmental needs. Thus, the notion of sustainability raises the problem of human over-exploitation of resources, and stresses the need for the protection of natural resources and ecosystems.

Later on, the concept of 'sustainability' became attached to further environmental aspects in the wider range of human activities, such as energy production, flood control, irrigation, land reclamation, urbanization, and industrialization. The concepts 'reduce, reuse and recycle' become the basis of sustainability. 'Reduce' means reducing the use of resources by providing energy efficient systems and water conservation measures. 'Reuse' signifies reusing resources in order to restrict demand on Earth's resources. 'Recycle' is the use of recycled materials with the purpose of decreasing new resource demand and protecting the environment (UNCRD 2011).

10.4.2.2 Triple Bottom Line

The triple bottom line (TBL) is a concept composed of three Ps (profit, people and planet). *People* relates to fair labor practices, to the community and region where the business operates. *Planet* refers to sustainable environmental practices. *Profit* is the economic value generated by the business organization while carrying out its operations (Kanj and Chopra 2010). This concept was first put forward by John Elkington in 1994 (Elkington 1994, 2004). It essentially aims to measure the financial, social, and environmental performance of the business organization over a period of time (Alhaddi 2015). In fact, TBL seeks to measure the overall cost of the performance of the business organization and its success through the use of the economic, social, and environmental aspects. Thus, TBL seeks to evaluate CSR. The more a business organization measures social and environmental impact the more it becomes responsible towards the community and the environment in which it operates.

However, it should be noted that the three 'Ps' do not have the same measures. While it is easy to measure the bottom line and the flows of money (profit), it is difficult to measure the 'planet' and 'people' in terms of cash. So, environmental variables (planet) can be better measured by natural resources and their sustainability (e.g., energy consumption, water consumption, air quality and land use). As to the social dimensions of TBL (people), a number of measurements can be used, such as education, equity, access to social resources, health and well-being, quality of life, and social capital (Slaper and Hall 2011). Today, TBL is common practice in non-profit and for-profit organizations, as well as in governmental institutions. The aim is to evaluate their performance and impact on both the community and the environment. Nestlé is a case in point. It is not only interested in the creation of value for its shareholder, but also in the creation of value for society. It believes that healthy and sustainable economic activity creates jobs. Nestlé equally focuses on water as it is particularly important in its production and marketing activities (supplies of agricultural raw materials and manufacturing processes, as well as for consumers to prepare its products) (Bulcke 2012).

10.4.2.3 ISO 26000:2010

ISO 26000 has been introduced in 2010 and today it is viewed as the international standard of CSR. It explains the meaning and application of CSR at the level of business organization and introduces its seven principles. These are (ISO 2010):

- **Accountability**: an organization should be accountable for its impacts on society, the economy and the environment.
- **Transparency**: an organization should be transparent in its decisions and activities that impact on society and the environment.
- **Ethical behaviour**: an organization should behave ethically. In other words, behaviour should be based on the values of honesty, equity, and integrity.
- **Respect for stakeholders' interests**: an organization should respect, consider and respond to the interests of its stakeholders.
- **Respect for the rule of law**: an organization should accept that respect for the rule of law is mandatory.
- **Respect for international norms of behaviour**: an organization should respect international norms of behaviour, while adhering to the principle of respect for the rule of law.

- **Respect for human rights**: an organization should respect human rights and recognize both their importance and their universality.
- While addressing the seven principles of CSR, ISO 26000 further puts forward four principal clauses (ISO 2014):
- **The environment**: prevention of pollution, sustainable resource use, climate change mitigation and adaptation, protection of the environment, biodiversity, and restoration of natural habitats.
- **Fair operating practices**: promoting social responsibility in the value chain.
- **Consumer issues**: protecting consumers' health and safety, and sustainable consumption.
- **Community involvement and development.**

Therefore, the aim of ISO 26000 is to assist organizations to carry out operations in a socially responsible manner. In so doing, business organizations contribute to the spreading of the benefits of CSR and the promotion of sustainable development.

10.4.2.4 Eco-efficiency

The concept of eco-efficiency was first put forward in 1990 by Schaltegger and Sturm as a "business link to sustainable development" (ESCAP 2009). It was seen as a practical tool to change the way humans produce and consume resources. As far as business organizations are concerned, eco-efficiency was intended to be a basic approach to promote economic and environmental benefits and to contribute to sustainable development. This can be achieved through more efficient uses of resources and lower pollution (ESCAP 2009). In this context, the WBCSD recommends for business organization to adopt this approach to be able to reduce consumption and impacts on the natural environment, as well as to increase the value of products. Thus, the aim of eco-efficiency is to reduce the use of resources to achieve both economic and environmental benefits. However, this aim can be reached only if this approach is adopted at both the micro and macro levels.

With the purpose of promoting the use of this approach, the United Nations Economic and Social Commission for Asia and the Pacific (ESCAP) developed a set of eco-efficiency indicators to measure the progress of eco-efficiency in a country's economy. These eco-efficiency indicators are intended to respond to the different challenges of sustainability. These challenges include (ESCAP 2009):

- Impacts of economic activity on the environment (e.g. resource consumption, pollution emissions, waste).
- Effects of resource productivity on the economy (e.g. economic efficiency).
- Impacts of environmental degradation on economic productivity (e.g. reduction in absorptive capacity, loss of forest cover).
- Effects of environmental improvement on society (e.g. congestion costs, improvement in wellbeing, social costs).

Thus, the concept of eco-efficiency correlates economic benefits and environmental performance. It particularly seeks to reduce both the use of resources and the release of toxic materials. It aims to enhance the quality and value of products and to reduce their environmental impact.

10.4.2.5 ISO 14046:2014

Established in 2014 by the International Organization for Standardization, ISO 14046 is the new corporate water footprint. It can be defined as "the total volume of freshwater that is used directly and indirectly to run the business" (Hoekstra et al. 2011). It specifies the principles, requirements and guidelines of assessing and reporting water footprints (Humbert 2009). It further seeks to provide consistent assessment techniques for calculating and reporting a water footprint as a stand-alone assessment (only water related impacts). It is expected to benefit organizations, governments and interested parties worldwide by providing transparency, consistency and credibility to the task of assessing water footprint and reporting the water footprint results of products, processes or organizations (ISO 2013). Water footprint assessment identifies potential environmental impacts related to water. It also identifies the quantity of water used and the changes in water quality.

Therefore, ISO 14046 presents a number of benefits (ISO 2013):

- Assess and prepare for the future risks of water use;
- Identify ways to reduce the environmental impacts of water use;
- Improve efficiency at product, process and organizational levels;
- Share knowledge and best practice with industry and government; and
- Meet customer expectations of increased environmental responsibility.

10.5 How Can CSR Contribute to Water Protection in the Maghreb Region?

The business organization's contribution to the protection of water resources involves a set of processes. In most cases, environmental contributions are often compulsory because the related legislations and directives define the pollution norms and objectives, taking into account the absorptive capacity of the environment. Failure to comply with existing laws and directives, the business organization may cause serious damages to the environmental resources and ecosystems' balance and, therefore, to the community. When the business organization abides by the defined pollution norms and objectives, this is called compulsory contribution or 'legal social responsibility'. When the business organization achieves better pollution norms and objectives, then the additional costs borne by the organization are called voluntary environmental and social contribution. In both cases, the environmental contributions made by the business organization are considered positive. But when the business organization fails to respect the pollution norms, then its contribution is negative. This can be the result of:

- A lack of pollution's monitoring by the business organization;
- A partial control of pollution;
- The location of business, for instance in an industrial area, where pollutants accumulate in quantities higher than the defined standards;
- Non-respect or partial respect of the standards and specifications for the quality or safety of the products.

The mainstreaming of CSR approach in business operations is still very weak in the Maghreb region despite the governments and civil society's efforts to promote the protection of water resources and the environment. Indeed, business organizations in the region are extracting and using the volumes of water and other natural resources which their operations require without much control, monitoring or care. They only pay for the cost of the water they use, and some of them use the water they need for free, and they do not usually care much for it. Operating responsibly is by no means a shared practice at the organizational level. Therefore, the Maghreb region is lagging behind in terms of CSR, and is required to find means and ways to encourage business organizations to mainstream CSR principles and guidelines in their management strategies in order to reduce the water they use in their production processes and to release less waste. This is vital to conserve the ecosystems and achieve environmental and social sustainability.

10.5.1 Measures to Enhance Water Protection and Efficiency

Concerned public institutions in the Maghreb region, particularly water authorities, are required to enhance water protection and restore water quality in order to enhance water security. This operation can be achieved through:

- Adopting strict and effective regulations to curb the levels of water pollution and to reduce effluents at the source (Chatila 2010). Businesses should be monitored to know what they do and how they do it.
- Imposing pollution charges to drive businesses to reduce industrial waste.
- Obliging businesses to disclose information on the discharge of industrial pollution to the public. This enables the community to compare businesses in terms of social and environmental responsibility before making any choices.
- Elaborating and enforcing discharge wastewater quality guidelines, standards and norms to minimize industrial pollution and its impact on the environment, public health, and water availability.
- Ensuring compliance with water use regulations in the business sector. Concerned institutions should insist on the separation of toxic pollutants.

- Introducing efficient water pricing systems to rationalize the use of water by business organizations. This will help reduce water wastage.
- Controlling mining runoff that is harmful to water resources with the purpose of reducing it.
- Setting product quality standards to reduce industrial waste and debris.
- Upgrading environmental, health, safety, and labor standards.
- Forcing suppliers of manufacturing requirements to the business sector to abide by the required environmental specifications and standards included in the business contracts. Businesses should also cooperate with their suppliers to preserve and reduce the water used in producing the inputs needed for their manufacturing processes.
- Empowering workers in the business sector to become tools for water protection inside and outside their workplace. They should be encouraged and sponsored to carry out activities geared to environmental protection.
- Taking bold steps to reduce the environmental impact of the products and services. Business organizations should analyze the complete lifecycle of their products to reduce pollution. Colossal quantities of water in the region could be saved annually if such practices are adopted.[1] Moreover, businesses should be committed with external stakeholders to reduce impacts of materials and consumer use of products, including the amounts of water and energy they use throughout the product cycle and, therefore, the impact on climate change.
- Conducting research to gather more information about the environmental impact of their products. Business organizations should aim in their manufacturing processes at reducing greenhouse gases through maximizing energy efficiency and shifting to renewable energy. Governments in the Maghreb region should elaborate climate policies and laws to help businesses develop energy transition and foster energy efficiency.
- Identifying practical, low-cost and cost-saving opportunities. The aim is to raise operational efficiencies, while using less materials, water and energy, as well as releasing less waste and emissions.

10.5.2 Incentives to Promote CSR

To promote CSR strategies at the organizational level, governments in the region are required to provide a set of incentives to the business sector. In this regard, the following measures can be effective:

- Economic incentives (such as tax reduction) to businesses mainstreaming social and environmental sustainability, and saving energy and water in their management strategies and processes.

[1] It was reported that, through innovating its apparel making process, Levi Strauss & Co in the USA managed to reduce the amount of water it uses in some of its processes by 96%, thus saving more than 1 billion liters of water in manufacturing since 2011(Levistrauss 2015).

- Substantial financial support to less pollutant manufacturers to help reduce their social and environmental costs.
- Community participation to enhance water protection. Consumers should be committed to assist socially responsible corporations by purchasing and consuming their products.
- Governmental financial support to business organizations to reduce waste loads to cut down waste treatment costs. This could encourage businesses to produce and supply quality products and services in the most sustainable way. They should shift as much as possible to the production of recyclable products.
- Encouraging business organizations provide grants to environmental associations where workers can volunteer in their spare time.
- Encouraging private businesses to engage in recycling and reuse of unwanted items, such as bottles, cans, paper and textiles in their production processes. Organizations should encourage consumers to collect and not to throw away used material. The aim is to put to a new use what has been used and to reduce industrial and consumer discharge.
- Stimulating businesses to invest in water recycling to be reused in their processes or allocated to other uses. Using recycled water instead of freshwater in industrial processes should be promoted in the region.[2]
- Granting SCR awards for best quality and eco-friendly products and services, as well as for businesses using less water, energy and chemicals, and producing less waste. They should be encouraged to achieve zero discharge of toxic chemicals. The aim is to reduce their overall potential impact on workers, consumers and the environment.
- Providing assistance in the development, implementation and disclosure of water sustainability policies and practices. Businesses must be assisted to adopt integrated water management systems.
- Promoting eco-friendly knowledge and ethical business practices in the manufacturing sector. Collaboration between business organizations operating in the same sector of activity with the purpose of achieving zero discharge in manufacturing processes.

Collaboration between business organizations and research bodies should be consolidated to induce technological innovation and promote the use of clean production processes. This can help the Maghreb region to reduce the industrial waste and emissions, cut the energy and water costs, protect the scarce water resources, and ultimately sustain the business organizations in the region.

[2] In this respect, the project Tenorio in San Luis Potosi (Mexico), which uses water it buys from a nearby wastewater treatment plant for cooling towers, could be a good example to follow. It should be noted that this project enabled the area of San Luis Potosi to reduce groundwater extraction of at least 48 million cubic meters and increased aquifer sustainability. Moreover, by using recycled water instead of freshwater the plant managed to cut water costs by 33% (Rex and Foster 2014).

10.6 Conclusion

The Maghreb region is projected to become drier and face serious water problems in the decades to come. Fast population growth, urbanization, pollution, climate change, and climate change are undermining the scarce water resources and posing serious threats to the region's economic, social and human development. More attention to the water sector is desperately needed. Non-conventional water resources and sustainable practices appear to be vital to help the region deal with future water challenges.

The Maghreb region should consider the use of water and other resources in the future with much care. The Maghreb region needs to promote sustainable practices in all activities, including the private ones. If the region continues to use the dwindling scarce water resources inefficiently and fails to protect them, it will experience severe water shortages in the foreseeable future. This will likely make governments and business organizations in the region unable to sustain industrial production and to foster development.

Although the business organizations in the Maghreb region are using about 10% of total water withdrawals, they are discharging tonnes of toxic chemicals daily into water bodies, often without proper treatment. It is estimated that the Maghreb region with the other MENA countries generate a total of about 10 km^3 of wastewater per year, of which 5.7 km^3 is treated. This is increasing water pollution, reducing freshwater availability and threatening the safety of the region's populations. Private corporations should be aware that wasting and polluting the scarce water resources directly impact the environment, and consequently affect the community in which they operate. On the contrary, when they cooperate to protect the environment and conserve the scare water resources for the populations of the region, they, in fact, conserve this precious resource to sustain their production and secure their existence. Through the adoption of CSR practices and operate responsibly, the business sector effectively contribute to water, health, and environmental security in the region.

References

Abdelbari, T. (2015). Changement climatiques: l'Algérie réclame une aide financière et technologique [Climate change: Algeria calls for financial and technological assistance], tsa-algérie. 17 Sept 2015. Available from: http://www.tsa-algerie.com/20150917/changements-climatiques-lalgerie-reclame-une-aide-financiere-et-technologique/

Abdul Rahman, H., & Suppian, R. (2010). Environmental degradation and human disease. *Health and the Environment Journal, 1*(2). Available from: http://www.hej.kk.usm.my/pdf/HEJVol.1No.2/Article12.pdf

Alhaddi H. (2015). Triple bottom line and sustainability: A literature review. *Business and Management Studies, 1*(2), 6–10. September, Redfame Publishing.

Allen, R. (1980). *How to save the world*. London: Kogan Page.

Assaf, H. (2010). Water resources and climate change. In M. El-Ashry et al. (Eds.), *Arab environment, water, sustainable management of a scarce resource*. Beirut: AFED Report.

Barghouti, S. (2010). Water sector overview. In M. El-Ashry et al. (Eds.), *Arab environment, water, sustainable management of a scarce resource*. Beirut: AFED Report.
Ben Abdelfadel, A., & Driouech, F. (2008). Climate change and its impacts on water resources in the Maghreb Region. Arab Water Council. Available from: http://www.arabwatercouncil.org/administrator/Modules/Events/IWRA%20Morocco%20Paper.pdf
Bishara, A. (2014). The dangers of death…from thirst (Arabic). El-Watan Newspaper. 18/05/201. Available from: http://alwatan.kuwait.tt/articledetails.aspx?Id=359039
Bulcke, P. (2012). Nestlé CEO on the need for international standards. 25 Jul 2012. Available from: http://www.iso.org/iso/home/news_index/news_archive/news.htm?refid=Ref1628
Chatila, J. G. (2010). Municipal and industrial water management. In M. El-Ashry et al. (Eds.), *Arab environment, water, sustainable management of a scarce resource*. Beirut: AFED Report.
Cheauvau, A., & Rosé, J. J. (2003). *L'entreprise responsable [The responsible enterprise]*. Paris: Editions d'Organisation.
Chowdhury, M. et al. (2013). Treatment of leather industrial effluents by filtration and coagulation processes. *Water Resources and Industry, 3*. Sept. Available from: http://www.sciencedirect.com/science/article/pii/S2212371713000085
Climate Institute. (2007–2010). Water and climate change. Available from: http://www.climate.org/topics/water.html
Corcoran, E., Nellemann, C., Baker, E., Bos, R., Osborn, D., & Savelli, H. (2010). *Sick water? The central role of wastewater management in sustainable development. A rapid response assessment*. Arendal: United Nations Environment Programme, UN-HABITAT, GRID.
Croitoru, L., & Sarraf, M. (2010). The cost of environmental degradation. Case studies from the Middle East and North Africa. The World bank.
El-Ashry, M., Saab, N., & Zeitoon, B. (2010). *Arab environment, water, sustainable management of a scarce resource*. Beirut: AFED Report.
Elkington J. (1994). Towards the sustainable corporation: Win-win-win business strategies for sustainable development. *California Management Review, 36*(2), 90–100. Available from: https://doi.org/10.2307/41165746.
Elkington, J. (1997). *Cannibals with forks – triple bottom line of 21st century business*. Stony Creek: New Society Publishers.
Elkington J. (2004). Enter the triple bottom line, A. Henriques & J. Richardson (Eds.), *The triple bottom line: Does it all add up?*, London: Earthscan. The chapter available from: http://www.johnelkington.com/archive/TBL-elkington-chapter.pdf
Elloumi, M. (2011). Pour une gestion durable des ressources naturelles, les limites du cadre institutionnel tunisien. In T. Dahou et al. (Eds), *Pouvoirs, sociétés et nature au sud de la Méditerranée*. Éditions Karthala.
ESCAP. (2009). *Eco-efficiency indicators: Measuring resource-use efficiency and the impact of economic activities on the environment, the United Nations economic and social commission for Asia and the Pacific*. Bangkok: United Nations publications.
European Commission. (2002). Green paper – promoting a European framework for corporate social responsibility. Available from: http://eurlex.europa.eu/smartapi/cgi/sga_doc?smartapi!celexplus!prod!DocNumber&lg=en&typedoc=COMfinal&an_doc=2001&nu_doc=366
Hart, S. L., & Milstein, M. B. (2003). Creating sustainable value. *Academy of Management Executive, 17*(2), 56–67.
Hoekstra, A. Y., Chapagain, A. K., Aldaya, M. M., & Mekonnen, M. M. (2011). *The water footprint assessment manual: Setting the global standard*. London: Earthscan.
Humbert, S. (2009). *ISO standard on water footprint: Principles, requirements, and guidance*. Paris: UNEP/CEO Water Mandate.
International Organization for Standardization (ISO). (2010). *Guidance on social responsibility* (1st ed.). Geneva: ISO 26000.
International Organization for Standardization (ISO). (2013). No more waste – tracking water footprints. Geneva. Available from: http://www.iso.org/iso/home/news_index/news_archive/news.htm?refid=Ref1760

International Organization for Standardization (ISO). (2014). ISO guidance on social responsibility. In *Discovering ISO 26000*. Geneva.

Jowsey, E., & Kellett, J. (1995). The comparative sustainability of resources. *International Journal of Sustainable Development and World Ecology, 2*(2), 77–85.

Kanj, G. K., & Chopra, P. K. (2010). Corporate social responsibility in a global economy. *Total Quality Management & Business Excellence, 21*(2), 119–143.

Levistrauss (2015). Sustainability, products: Waterless. Available from: http://levistrauss.com/sustainability/products/waterless/

Mäler, K.-G. (1991). National accounts and environmental resources. *Environmental and Resource Economics, 1*(1), 1–15.

Molle, F. (2011). Politiques agraires et surexploitation de l'eau au Maghreb et au Machrek. In T. Dahou et al. (Eds), *Pouvoirs, sociétés et nature au sud de la Méditerranée*. Editions Karthala.

Painter S. (n.d.). Environmental issues of river and lake pollution. Available from: http://greenliving.lovetoknow.com/environmental-issues/environmental-issues-river-lake-pollution

Patterson, S. (n.d.). How do humans affect the environment. Available from: http://greenliving.lovetoknow.com/How_Do_Humans_Affect_the_Environment

Pearce, D. (1988). The sustainable use of natural resources in developing countries. In R. K. Turner (Ed.), *Sustainable environmental management principles and practice*. Boulder: Westview Press.

Pearce, D., & Atkinson, G. (1993). Capital theory and the measurement of sustainable development: An indicator of weak sustainability. *Ecological Economics, 8*(2), 103–108.

Petkoski, D., & Twose, N. (2003). *Public policy for corporate social responsibility, July 7–25*. Washington, DC: The World Bank. Available from: http://info.worldbank.org/etools/docs/library/57434/publicpolicy_econference.pdf

Rex, W., & Foster, V. (2014). Ahead of world water day: Let's talk about…energy? Available from: http://blogs.worldbank.org/water/ahead-world-water-day-let-s-talk-about-energy

Ruitenberg R. (2012). Algeria to boost irrigated area to 1.million Hectares in 2014. Available from: http://www.bloomberg.com/news/articles/2012-07-24/algeria-to-boost-irrigated-area-to-1-6-million-hectares-in-2014

Skye J. (n.d.-a). Causes of environmental degradation. Available from: http://greenliving.lovetoknow.com/Causes_of_Environmental_Degradation

Skye J. (n.d.-b). Effects of water pollution. Available from: http://greenliving.lovetoknow.com/Effects_of_Water_Pollution

Slaper, T.f., & Hall, T.J. (2011). The triple bottom line what is it and how does it work. Indiana Business Research Centre, Indiana University, Kelley School of Business. Available from: http://www.ibrc.indiana.edu/ibr/2011/spring/article2.html

Tabet-Aoul, M. (2011). Innovating ways to face the effects of environmental degradation. Hydrométéorological Institute. Feb 10. Available from: http://www.mei.edu/content/innovating-ways-face-effects-environmental-degradation

Tilman D., et al. (2001). Forecasting agriculturally driven global environmental change, *Science, 292*, 281–284. Available from: http://cedarcreek.umn.edu/biblio/fulltext/t1791.pdf

Titilola, T. (1994). Indigenous knowledge systems and sustainable agricultural development in Africa: Essential linkages. *Indigenous Knowledge and Development Monitor, 2*(2).

UNICEF. (2015). Progress on sanitation and drinking water. Available from: http://www.unicef.org/publications/files/Progress_on_Sanitation_and_Drinking_Water_2015_Update_.pdf

United Nations Development Program (UNDP). (2009). *Arab human development report 2009: Challenges to human security in the Arab countries*. New York: Regional Bureau for Arab States.

United Nations Development Programme (UNDP). (2011). *Arab development challenges report 2011*. Cairo: Regional Centre for Arab States.

United Nations for Regional Development (UNCRD). (2011). reduce, reuse and recycle (the 3Rs) and resource efficiency as the basis for sustainable waste management. 9 May. New York.

United Nations-Department of Economic and Social Affairs (UN-DESA). (2014). World urbanization prospects: The 2014 revision. Available from: https://esa.un.org/unpd/wup/publications/files/wup2014-highlights.Pdf

UNwater. (2013). Facts and figures. Available from: http://www.unwater.org/water-cooperation-2013

Viederman, S. (1993). *The economics and economy of sustainability: Five capitals and three pillars. Talk delivered to delaware estuary program*. New York: Jessie Smith Noyes Foundation.

Water Academy. (2004). The cost of meeting the Johannesburg targets for drinking water. Available from: Water-academy.org. 22/06/2004.

Wisler, A. (2014). Impacts of environmental degradation. Available from: http://greenliving.lovetoknow.com/Impacts_of_Environmental_Degradation

World Bank. (2007). *Making the most of scarcity: Accountability for better water management results in the middle east and north Africa*. Washington, DC: the World Bank.

World Bank. (2009). *Water and climate change: understanding the risks and making climate-smart investment decisions*. Washington, DC: The World Bank.

World Bank. (2010). *The cost to developing countries of adapting to climate change, new methods and estimates, the global report of the economics of adaptation to climate change study, consultation draft*. Washington, DC: The World Bank.

World Bank. (2016). *Population, total*, Washington, DC: The World Bank. Available from: http://data.worldbank.org/indicator/SP.POP.TOTL?locations=1A

World Business Council dor Sustainable Development (WBCSD). (1998). Corporate social responsibility, meeting changing expectations. Available from: http://www.wbcsd.org/Pages/Adm/Download.aspx?ID=108&ObjectTypeId=7

WPP. (2011). *Water partnership programme, driving change in water*. 2010 annual report. Washington, DC.

Zeitoon, B. M. (2012). Population, consumption, and sustainability options. In N. Saab (Ed.), *Arab environment, survival options*. Beirut: AFED Report.

Zwahlen, R. (1995). The sustainability of resources versus the sustainability of use: A comment. *International Journal of Sustainable Development, 2*(4). World Ecology 2, Electrowatt Engineering Services, Ecology Department, Zurich.

Abdelhafid Aimar is a full professor at the Faculty of Economics, Commerce and Management, University of Jijel, Algeria. He has been head of Economics and Management Departments (1996–2005), vice-dean of postgraduate studies (2008–2012) at the Faculty, and now his is head of the teaching staff of the Faculty and president of the scientic committee of the Commerce Department. He has been conducting an important number of research projects in the fields of global economy, finance, develompent, and climate change. At present, he is head of the Faculty's research team working on sustainable management of water resources in the Maghreb region.

Chapter 11
The Impact of Soil Degradation on Agricultural Production in Africa

Olaf Pollmann and Szilárd Podruzsik

11.1 Introduction

The soil degradation has serious impacts on many related fields, especially in Africa. Soil is not only producing food, soil is also filtering rainwater, regulating the climate, sinking and storing carbon. In a small part of earth, an entire population of millions of organisms are living and using the soil to survive, to establish peoples and to keep the soil as fertile as it is. Without fertile soil, loss of biodiversity cannot be stopped, global warming cannot be kept at 2 °C nor will access to adequate food be assured for everybody on earth. Globally, the building of cities and roads damage the soil surface irreparably. In order to further the goal of soil protection, the United Nations considered 2015 as the international year of soils. Some facts illustrate the growing acuteness and importance of soil (from Soil Atlas 2015):

- Land and soil have a multitude of social, ecological, cultural, spiritual, and economic functions worldwide.
- Fertile soil is vital and it forms just a thin layer on the earth surface. It takes 2000 years to create 10 cm of topsoil.
- Millions of hectares of land are lost every year through inappropriate farming techniques, for the construction of cities and roads, and through deforestation. Cities eat into fields, and fields expand at the expense of forest and pastureland.

O. Pollmann (✉)
SCENSO – Scientific Environmental Solutions, Sankt Augustin, Germany
e-mail: o.pollmann@scenso.de

S. Podruzsik
Department of Agricultural Economics and Rural Development, Corvinus University of Budapest, Budapest, Hungary
e-mail: szilard.podruzsik@uni-corvinus.hu

M. Behnassi et al. (eds.), *Climate Change, Food Security and Natural Resource Management*, https://doi.org/10.1007/978-3-319-97091-2_11

- Without protecting the soil, it will be impossible to feed a growing world population, keep global warming below 2 °C, or halt the loss of biodiversity.
- Land ownership is distributed inequitably – even more so than income. Access to land is fundamental in the fight against hunger and poverty. In many countries, women are disadvantaged compared to men.
- Land prices are rising almost everywhere. If individual or communal rights are not assured, local people are forced off the land.
- Competition for land is growing. The causes include the spread of fodder crops, and the growing use of crops to produce 'green' biofuels.
- Global trade has turned arable land into a mobile resource. Developed and emerging economies are exporting their hunger for land to the developing world. They import land in the form of products grown abroad.
- Despite the fact that chemical fertilizers are being used, yields are not increasing as rapid as expected. Organic farming stimulates soil organisms and improves soil fertility in the long term – something that mineral fertilizers fail to do.
- Modern city planning must include soil conservation. Infrastructure and housing must use less fertile land, especially in countries with declining populations.
- An international regulatory framework based on human rights must ensure that the distribution of land is equitable and that fertile soils are not monopolized by the rich.[1]
- Protecting the soil is a global task. But individuals can make a significant contribution by purchasing local products and eating less meat.

With this background, the impact of soil degradation was investigated on examples of agricultural production in Africa. The importance to understand the market relevance and key markets in the agricultural sector is increasingly essential for the success of agricultural production in Africa.

11.2 Indicators of Soil Quality

Soil quality and efficiency are dependent on the use of the soil. Most of the African soil is used for agricultural purposes, e.g. to produce food for human consumption. Therefore, the quality of soil also defines the kind of crops to be planted.

Half of the soil usually contains mineral particles such as sand and clay. About 20% is water, 20% is air, and the rest are plant roots and soil organic matter like

[1] Since 2008, the term 'land grabbing' gained notoriety around the globe. It refers to large-scale land acquisitions mainly by private investors but also by public investors and agribusiness that buy farmland or lease it on a long-term basis to produce agricultural commodities. These international investors, as well as the public, semi-public or private sellers, often operate in legal grey areas and in a no man's land between traditional land rights and modern forms of property. In many cases of land grabbing, one could speak of a land reform from above, or of the establishment of new colonial relationships imposed by the private sector (http://www.globalagriculture.org/report-topics/land-grabbing.html). For more information, see for instance Behnassi and Yaya (2011).

organisms and humus. Humus stores nutrients and water and gives the soil a stable structure with plenty of pores. Humus also contains the carbon plants absorbed from the air as the greenhouse gas carbon dioxide. Beside the topsoil, it is also important to have deep subsoil that allows roots to penetrate and extract water even when the top soil has run dry. These layers regulate the soil fertility.

Soils that are especially fertile are good for growing crops, while less-fertile soils are more suitable for meadows, pastures, and forests. Therefore, even less fertile soil can also be valuable for ecological reasons. If a soil is used too intensively or in an appropriate way, its function declines and it starts to degrade (Fig. 11.1). An estimated 20–25% of soils worldwide are already affected by serious degradation and more than 8 million hectares degrade each year (Soil Atlas 2015).

Because of extensively farming, soils are losing significant amounts of organic matter including humus and soil organisms. Due to these factors, the natural fertility of soils is declining. Techniques such as high-yielding seeds, fertilizers, pesticides, monoculture and irrigation have led to sharp rises in yields. The farm production almost tripled in the last 50 years worldwide while the area of agriculture land expanded by only 12%. This results in changing type of farming (Table 11.1) or even the entire use of agricultural fields.

Agronomic measures involve changing how the crops are grown. Ploughing and planting along contour instead of up and down the slope for example can reduce erosion. Intercropping or rotating cereals with legumes restores soil fertility and reduces the need for nitrogen fertilizer.

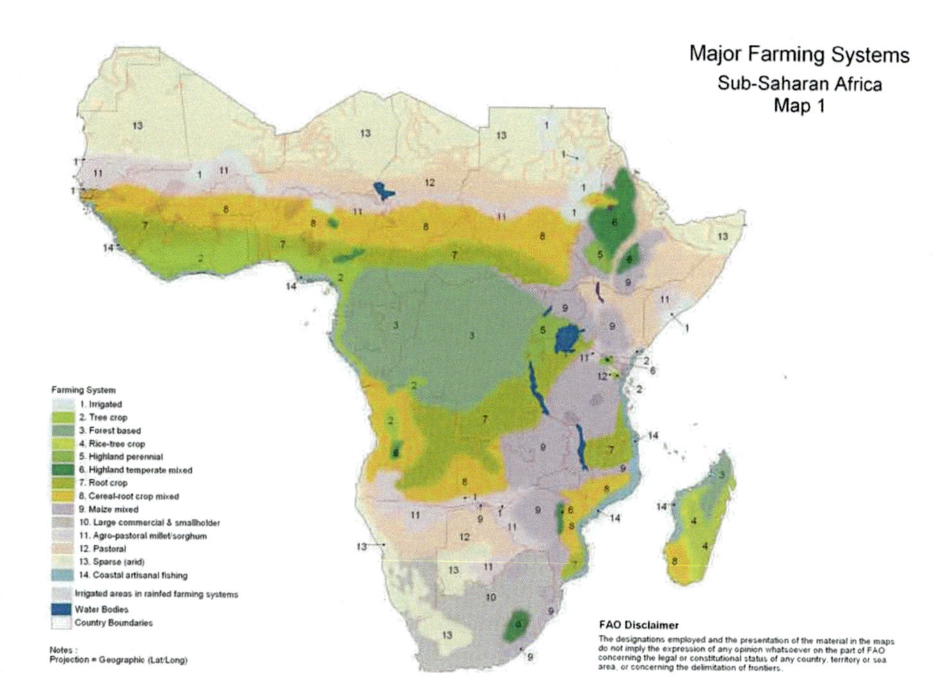

Fig. 11.1 Sub-Sahara Africa farming systems. (Source: Dixon et al. 2001)

Table 11.1 Major farming systems of Sub-Saharan Africa

Farming systems	Land area (% of region)	Agric. Popn. (% of region)	Principal livelihoods	Prevalence of poverty
Irrigated	1	2	Rice, cotton, vegetables, rainfed crops, cattle, poultry	Limited
Tree crop	3	6	Cocoa, coffee, oil palm, rubber, yams, maize, off-farm work	Limited-moderate
Forest based	11	7	Cassava, maize, beans, cocoyams	Extensive
Rice-tree crop	1	2	Rice, banana, coffee, maize, cassava, legumes, livestock, off-farm work	Moderate
Highland perennial	1	8	Banana, plantain, enset, coffee, cassava, sweet potato, beans, cereals, livestock, poultry, off-farm work	Extensive
Highland temperate mixed	2	7	Wheat barley, tef, peas, lentils, broadbeans, rape, potatoes, sheep, goats, livestock, poultry, off-farm work	Moderate-extensive
Root crop	11	11	Yams, cassava, legumes, off-farm work	Limited-moderate
Cereal-root crop mixed	13	15	Maize, sorghum, millet, cassava, yams, legumes, cattle	Limited
Maize mixed	10	15	Maize, tobacco, cotton, cattle, goats, poultry, off-farm work	Moderate
Large commercial and smallholder	5	4	Maize, pulses, sunflower, cattle, sheep, goats, remittances	Moderate
Agro-pastoral millet/sorghum	8	8	Sorghum, pearl millet, pulses, sesame, cattle, sheep, goats, poultry, off-farm work	Extensive
Pastoral	14	7	Cattle, camels, sheep, goats, remittances	Extensive
Sparse (arid)	17	1	Irrigated maize, vegetables, date palms, cattle, off-farm work	Extensive
Coastal artisanal Fishing	2	3	Marine fish, coconuts, cashew, banana, yams, fruit, goats, poultry, off-farm work	Moderate
Urban based	Little	3	Fruit, vegetables, dairy, cattle, goats, poultry, off-farm work	Moderate

Source: Dixon et al. (2001)

In Africa, smallholder farmers sow and weed by hand or using special animal-drawn implements that disturb the soil as little as possible. Changing the techniques, the farmers have to learn new skills, change probably the crops they use, invest in new equipment, and put more effort into controlling the weeds.

A 2007 study (Badgley et al. 2007) has shown that in developing countries the yields from organic farming methods were an average of 92% of those using con-

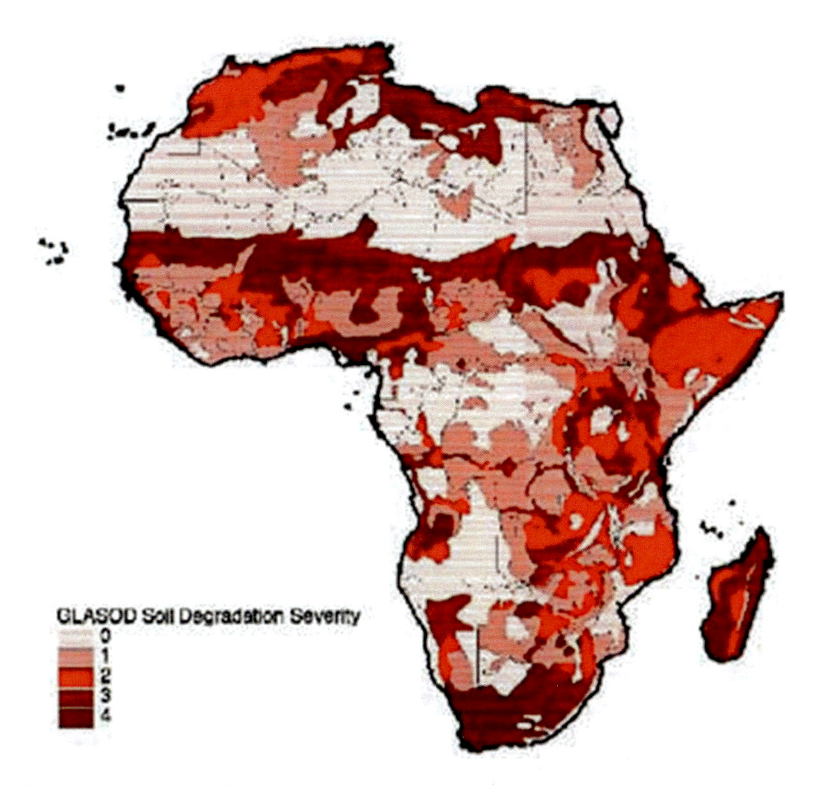

Fig. 11.2 Soil degradation in Africa. (Source: FAO 2001)

ventional methods, and organic systems produce 80% more than conventional farms in developing countries, because the materials needed for organic farming are more accessible than synthetic farming materials to farmers in some poor countries. In the tropics, an analysis showed that organic farming boosted yields by up to 74% without depleting the long-term soil fertility (Soil Atlas 2015).

With these mentioned indicators and additional soil indicators like soil wetness, pH, water-holding capacity, etc. the productivity of different soils can be evaluated (Nkonya et al. 2013). The soil quality has dropped over the last centuries (Fig. 11.2). With this kind of evaluation it will be possible to estimate the economic efficiency and therefore concrete market potentials of the soils.

11.3 Importance of Soil Quality and Related Economy for Africa

The soil quality in agricultural driven countries is directly related and essential for the economic growth. On the first sight, it could be the most effective way to increase the yields. But the increase of yields in the short run has a negative impact on soil quality in the future and therefore a drop in soil fertility, soil quality and also resultant in limited economic benefits from soil degradation.

Some African countries have understood the relations between quick agricultural yields and sustainable economic growth (Fig. 11.3). With the 'green economy', poverty, creating employment, and improving the overall well-being of the population will be addressed. The green economy must be built from the cultural traditions and roots in Africa that are deeply tied to environmental stewardship; it must be inclusive and engage women, youth, and the spirit of future generations to be truly transformative.

Beside, mismanagement and negative circumstances in agriculture malnourishment in rural areas are an additional serious problem in Africa and therefore also for the African economy. Malnourishment affects roughly 223 million people in sub-Saharan Africa – about one quarter of the region's total population – according to estimates by the Food and Agriculture Organization (FAO).

Fig. 11.3 Countries with specific green economy strategies. (Source: UNEP 2015)

One important fact to increase economic aspects combined with increased yield and nourishment is to use organic fertilizers such as manure, compost, and plant residues to increase total productivity. African farmers need a local adapted solution to increase the needed economy. All outside solutions have high potential to fail.

11.4 Conclusion

The study describes the potential of soil quality and soil efficiency to create the linkage between agricultural soil treatment and local economic success on the African market. Small-scale farming has shown a very positive success in constant yields, and therefore constant soil fertility even by using organic farming techniques.

To raise the yields, chemical fertilizers are added to the still fertile soil only to raise yields in the short run. With this change of the entire soil structure and chemical composition, the economic advantage cannot be kept up.

For antagonizing this process, it is important for each African country to establish land use concepts to decouple organic farming of small farmers from bigger commercial communities. This decoupling process could keep the soil fertility for small farming communities while other profit-yielding agriculture will support the regional economy.

Farming should always be a field for securing food and nutrients to the local people in the first sight before it gets to a profit-yielding agriculture of competing counties. Many samples in Africa has shown that this approach can be achieved but still new techniques of organic farming have to be established and accepted on the basic agricultural level of application.

References

Badgley, C., et al. (2007). Organic agriculture and the global food supply. *Renewable Agriculture and Food Systems, 22*(2), 86. https://doi.org/10.1017/S1742170507001640 Lay summary – New Scientist (July 12, 2007).

Behnassi, M., & Yaya, S. (2011). Land resource governance from a sustainability and rural development perspective. In M. Behnassi, S. A. Shabbir, & J. D'Silva (Eds.), *Sustainable agricultural development: Recent approaches in resources management and environmentally-balanced production enhancement* (pp. 3–23). Netherlands: Springer.

Dixon, J., Gulliver, A., & Gibbon, D. (2001). *Farming systems and poverty – improving farmers' livelihoods in a changing world*. FAO and World Bank. http://www.fao.org/docrep/003/Y1860E/y1860e04.htm.

FAO. (2001). *Two essays on socio-economic aspects of soil degradation, economic and social development* (Paper 149). ISBN: 9251046298.

Nkonya, E., et al. (2013). *Economics of land degradation initiative: Methods and approach for global and national assessments* (ZEF-discussion papers on development policy, No. 183).

Soil Atlas. (2015). The soil atlas 2015, jointly published by the Heinrich Böll Foundation, Berlin, Germany, and the Institute for Advanced Sustainability Studies, Potsdam, Germany. Published: January 2015 (1st edn), p. 68. Licence: CC BY-SA 3.0.
UNEP. (2015). *Building inclusive green economies in Africa experience and lessons learned 2010–2015 United Nations Environment Programme*.

Olaf Pollmann is a civil engineering and natural scientist. He is holding a Doctorate (Dr.-Ing./PhD) in the field of environmental-informatics from the Technical University of Darmstadt, Germany and a second Doctorate (Dr. rer. nat./PhD) in the field of sustainable resource management from the North-West University, South Africa. Olaf Pollmann is a visiting scientist and extraordinary senior lecturer at the North-West University and CEO of the company SCENSO – Scientific Environmental Solutions in Germany. He is also deputy head of the section "African Service Centers" in West (WASCAL) and Southern Africa (SASSCAL) on behalf of the Federal Ministry of Education and Research (BMBF).

Dr. Szilárd Podruzsik holds a PhD in economics from the University of Economic Sciences and Public Administration in Budapest. He works for the Corvinus University of Budapest as a senior lecturer. His research fields cover the areas of agriculture and food industry. Currently, his research focuses on the food consumer welfare, food logistics and its process optimisation. In his research, he applies different models to help stimulate, estimate and evaluate the relevant sectors.

Chapter 12
Effect of Treated Wastewater on Plant, Soil and Leachate for Golf Grass Irrigation

Hind Mouhanni and Abdelaziz Bendou

12.1 Introduction

Treated wastewater reuse has been a common alternative for irrigation in many countries which are characterized by arid climate sand water shortage. Application of reclaimed wastewater reuse for irrigation has been expanded due to many reasons such as: its potential in terms nutrients inputs; socio-economic implications; reduction of environmental pollution; and enhancement of the quality of water resources.

In Mediterranean countries, treated wastewater is exponentially used for irrigating ornamental plants in areas suffering from water scarcity. This is perceived as an economic way to decrease pollution of surface waters and provide groundwater recharge for other agricultural uses. A great number of publications have recognized the benefits of this practice. In this perspective, this work was conducted with the aim to investigate the effects of irrigating golf grass with municipal reclaimed water, which contains higher concentration of soluble salts compared to ground water, on leachate and soil (Mouhanni et al. 2008, 2011, 2012; Mouhanni and Bendou 2011).

Similar results were obtained in United States by Gregory et al. (2010) who concluded that the turf grass irrigated by reclaimed wastewater are moderately or highly salt tolerant when fully established. However, continuous irrigation with reclaimed water poses a potential soil Na accumulation problem. Turf grass assimilates a large amount of N and P with minimal potential losses to ground water. In Spain, Salgot

H. Mouhanni (✉)
Département de Chimie, Centre des Sciences et Techniques, Campus Universitaire Ait Melloul, Ait Melloul, Morocco

Équipe des Matériaux, mécanique et Génie civil – ENSA- UIZ, Agadir, Morocco

Équipe d' Électrochimie, Catalyse et Environnement- FS-UIZ, Agadir, Morocco

A. Bendou
Ibn Zohr University, Agadir, Morocco

M. Behnassi et al. (eds.), *Climate Change, Food Security and Natural Resource Management*, https://doi.org/10.1007/978-3-319-97091-2_12

et al. (2006) studied wastewater reuse and concluded that reclaimed wastewater can be reused for different applications depending on specific water quality categories.

Biological and chemical parameters have to indicate all potential pathogens and chemical intoxications in relation to the origin of sewage. Therefore, it is necessary to find adequate indicators which can be performed by chemical as well as biological quantitative risk assessment. In USA, Zalesny et al. (2008) studied the sodium and chloride accumulation in leaf, woody and root tissue of Populus after irrigation with Landfill leachate and fertilized well water (control). The monitoring started from 2005 to 2006 in Rinelander, Wisconsin. The results showed that the leachate irrigated soils at harvest had the greatest Na^+ and Cl^- levels. The soil Na^+ concentration was nearly 24 times than the control. The leachate soil Cl^- concentration was three times greater than the control. Across all genotypes, Na^+ levels were greatest in the leaves. Similarly, woody sequestered high amounts of Na^+ and Cl^-. As a conclusion, human activities have increased the salts in areas dedicated to plant growth.

In India, Jalali et al. (2008) studied the effects of irrigation with wastewater on soil and groundwater quality in two soil column. The addition of the wastewater resulted in increased exchangeable Na^+ on the exchange complex at the expense of exchangeable Ca^{2+}, Mg^{2+}, and K^+. Hence, the ESP of both soils increased. No adverse effects on soil structure were observed, provided that wastewater is continuously used for irrigation. A change from wastewater to better water quality may well invoke soil structural damage. The use of wastewater for irrigation also increased Mg^{2+} and K^+ losses from the soils.

If wastewater is applied to soil for a long period, leaching concentrations will reach groundwater at high levels (as part of irrigation water evapotranspiration). A rough assessment may thus help to confirm whether or not wastewater use is acceptable.

In our study, we dealt with the case of Agadir (south of Morocco: altitudes between 30 and 31 °N) (Fig. 12.1). The Agadir region is an agricultural area that is characterized by an arid climate, very limited water resources, and poor nutrient soils. The agricultural

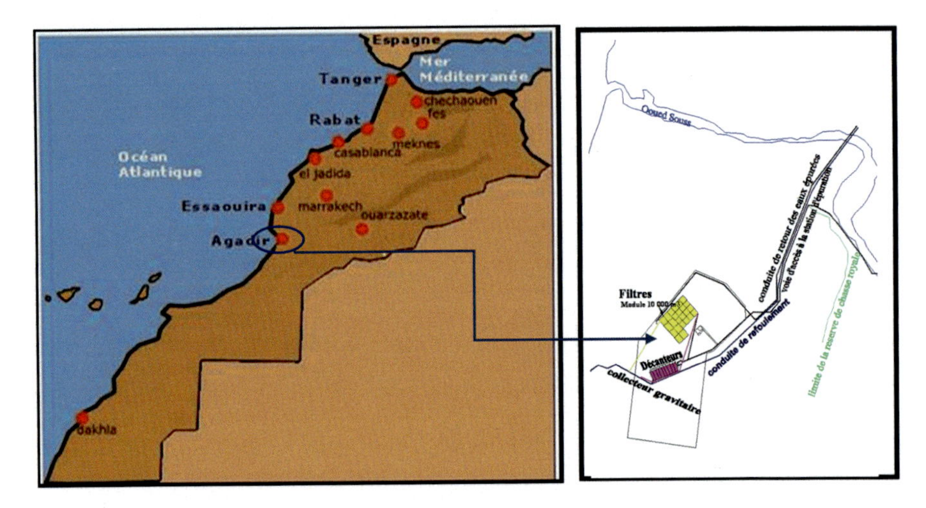

Fig. 12.1 Localization of the M'zar plant of Agadir

sector is the largest water consumer. Thus, the use of treated wastewater in agriculture is a good alternative that will help preserve water resources in the region. Moreover, given the nutritional wealth of the treated wastewater, this solution will permit a recycling of these items and reduce the abusive misuse of fertilizers (Chenini et al. 2002).

The current potential of wastewater treated by the Agadir M'zar plant (Fig. 12.1), which might be used for unrestricted irrigation (category A WHO standards), is 10,000 m^3/day and will reach 50,000 m^3/day in the medium term. A feasibility study on the reuse of the Agadir M'zar plant wastewater was launched by the Water Supply Service of Agadir (RAMSA[1]). In this context, the total surface of green spaces of Agadir city is estimated to be 878×10^4 m^2 with a need of water for irrigation reaching 8106 m^3/year. With a daily flow of 50,000 m^3/day, the treated wastewater of the M'zar plant will completely fill this need. The golf grass alone occupy 30.5% (268×10^4 m^2) of the total area of green space in Agadir (878×10^4 m^2), with a water consumption estimated to be 3216,103 m^3/year (Gregory et al. 2010; Mouhanni et al. 2012).

This study focuses on the reuse of treated wastewater for golf grass irrigation. It provides the planning, protocol and results of the tests that were carried out to evaluate the effects of the reuse of treated wastewater for golf grass irrigation. Particular attention was given to the monitoring of parameters of germination and growth of grass plants irrigated with treated wastewater compared to those irrigated by groundwater. In addition, the analysis of soils in three different depths (20 cm, 40 cm and 60 cm) has been done to demonstrate the interaction between the soil and the two types of water for irrigation.

12.2 Material and Methods

12.2.1 Experimental Site

The in situ tests had been performed on the site of the wastewater treatment plant of M'zar Ait Melloul (Fig. 12.1) where two zones of land have been managed: one for irrigation tests using treated waste water released by the plant; and the other for irrigation tests using the groundwater drawn from a well located in the wastewater treatment plant area.

12.2.2 Tests Scheduling

In order to study the feasibility and evaluate the impact of treated waste water use for irrigation of golf grass, three varieties of golf grass (V1, V2, and V3) had been used on three parcels of land (P1, P2, and P3). For comparison purpose, the same tests are reproduced in the same conditions while using groundwater (drawn from the well). Each parcel had a dimension of 25 m^2 and was subdivided into two

[1] Régie Autonome Multiservices d'Agadir.

parcels of 12,5 m² in order to perform a repetition of the conditions of each test. The subdivision of every parcel was assured by the pose of a plastic insulator with a depth of 0,5 m in order to prevent infiltration between the subdivisions. Every parcel contains a layer of 20 cm of soil composed of 75% of the plant earth and 25% of sand.

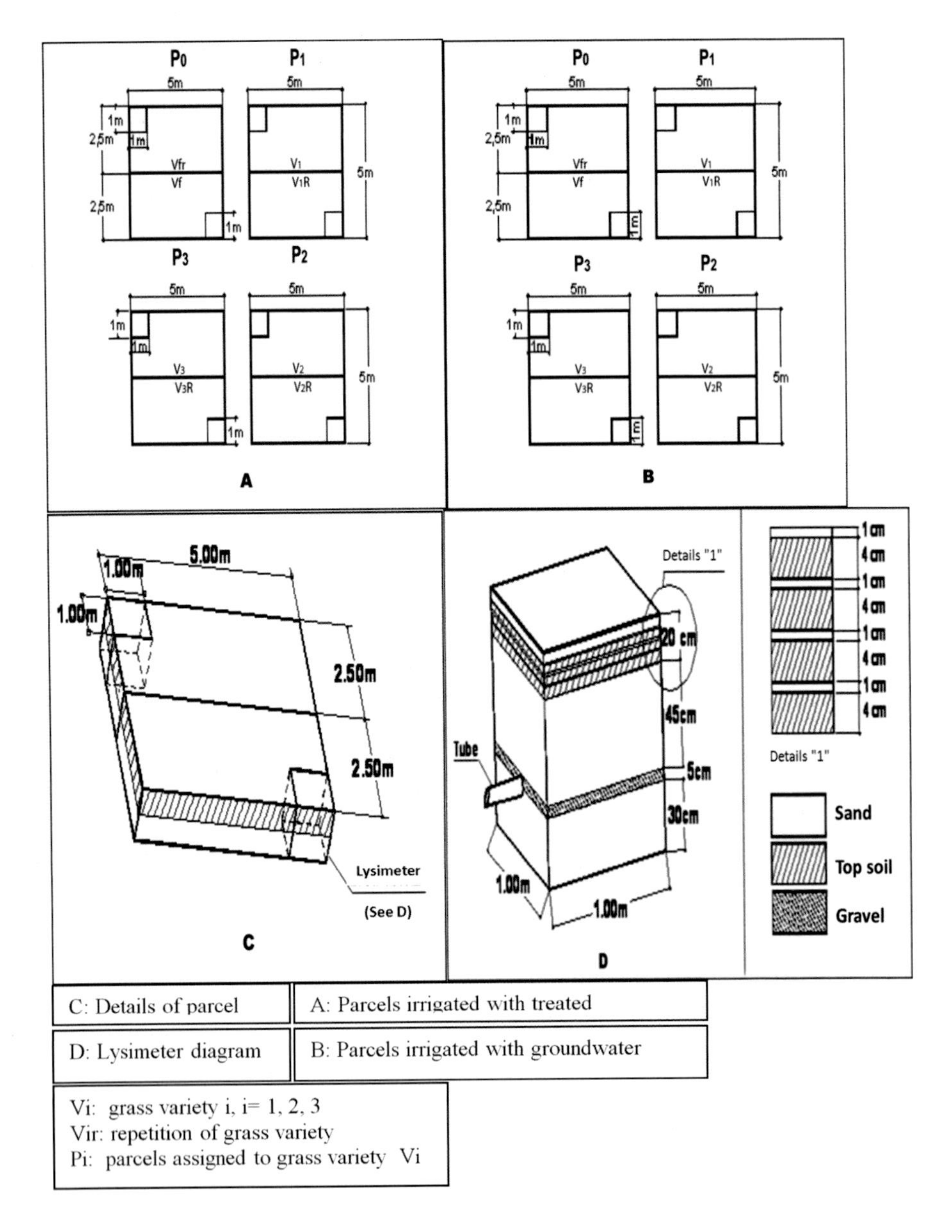

Fig. 12.2 Description of the experimental design and explanatory Diagram giving the disposition of the parcels and lysimeters with the assigned varieties of grass to each parcel

Table 12.1 Physico-chemical parameters of soil and sand

Parameters	Texture	pH	Organic matter (%)	Total nitrogen (%)	Total limestone (%)	EC 1/5 (dS/m)	Soluble salts (g/kg)	P_2O_5 assimilable (ppm)	K_2O exchangeable (ppm)
Topsoil	LSC [a]	8.70	1.85	0.15	5.20	0.12	0.21	14.56	128.70
Sand	S	9.6	0.05	0.01	36.7	0.045	0.16	1	24.9

[a]*LSC* loamy Sandy Clay

A lysimeter was managed in each parcel opposite corners (Fig. 12.2). The lysimeters had a volume of 1 m^3 and reproduced the conditions of soil and the variety of grass sowed in the concerned parcel; they were conceived with a good water tightness permitting the recuperation of leaking water from soil after irrigation. The Fig. 12.2 presents a diagram that shows the disposition of the parcels and lysimeters with the assigned varieties of grass to every parcel, and the repetitions and dimensions of the different characteristics. On this diagram, a fourth parcel noted Po was planned for tests using the composted sludge as organic amendment.

12.2.3 *Characteristics of Soils and Irrigation Waters*

The soils of the parcels are composed of 75% of plant top soil and 25% of sand. The pedological analyses of the soil constituents are presented in the Table 12.1.

According to the results summarized in Table 12.1, the soil presents an alluvial texture with few of clay and little sand, what proves that its amendment to the sand is going to improve its water and nutriments retention capacity. It is slightly provided in major nutriments and organic matter: the total nitrogen, the assimilated phosphor as well as in exchangeable potash. Besides, the sands are very poor in any fertilizing element. The saltiness of the two types of soil is very low; otherwise the alkalinity of the sands is higher than that of the plant earth. Therefore, soil cannot provoke any prejudicial risk to the cultures as it doesn't provide them with any nutriment which imposes to the grass to take advantage solely from fertilizing elements contained in waters or brought by possible amendments.

Waters used for the irrigation of the experimental parcels are of two types: the groundwater of Souss plain, drawn from the well located in the wastewater plant zone; and the treated wastewater of the plant M'zar Ait Melloul, which uses the infiltration percolation process on bed sands. The main features of waters used for irrigation are indicated below in the Table 12.2.

The assessment of the analyses of the treated wastewater and ground water permits their classification as water for irrigation according to the United States Department of Agriculture (USDA). The USDA classification is based on the values of the electrical conductivity (EC) and the one of the Sodium Adsorption Ratio (SAR) factor (Ayers and Westcot (1988):

- The ground waters are classified as C2S1, they can be used on any type of soil with a minimal risk of sodium accumulation. The saltiness of waters

Table 12.2 Ionic balance of treated wastewater and groundwater used for irrigation

	Groundwater		Treated wastewater	
Measured parameters	mg/L	($\times 10^{-3}$) mole/L	mg/L	($\times 10^{-3}$) mole/L
pH	7.4		7.1	
EC at 25 °C dS/m	0.58		3.15	
SAR	0.84		6.66	
Cl^-	128	3.61	720	20.31
K^+	2.89	0.07	43	1.1
Na^+	30.52	1.32	487	21.21
Ca^{2+}	47.8	1.19	294.8	7.35
Mg^{2+}	32.1	1.32	65.9	2.71
Total Nitrogen	1.5	0.1	44.6	3.18

Table 12.3 The varieties of grass sowed in the different parcels

Parcels	P1	P2	P3
Species	Agrostides	Ray grass	Mixture
Sowed variety (Vi)	Penccross (**V1**)	Ray grass Anglais (**V2**)	Ray grass Anglais 60%
			Red fescue 40% (**V3**)

(CF = 0,520 dS/m) is close to the limit affecting the growth of the grass without applying special treatments for saltiness reduction (limit situated at EC = 0,750 dS/m).

- The treated wastewaters are classified as C4S1, they are very saline (EC at 25 °C is of 3,15 dS/m). However, they can be applied on any type of soil with a SAR value less than 10.

One notes that the treated wastewaters have an important ionic load owing to the contents of the chlorides, sodium and nitrates, able to affect the absorption of other cations like magnesium (Mg^{2+}).

12.2.4 Protocol of Irrigation, Seed And Follow-Up

The parcels irrigated by the treated wastewaters were distant from those irrigated by the groundwater (drawn from the well) to avoid any contamination. Their irrigation was done by the same system which consisted s on an aspersion by a gun having a constant flow rate of 828 l/h, thus ensuring the constancy of the same volume of irrigation water. Every parcel of 25 m^2 is irrigated three times a day and received s a total of 90 l of water per day (Table 12.3).

One of the objectives of this study aims to compare the parameters of grass growth (tillering and evolution of leave length) of the different species of golf grass: the Ray-Grass, the agrostide and the fescue, irrigated with the treated wastewaters

Table 12.4 Means with standard deviation of pH, EC and ionic composition in mmol/l of leachate

	Treated wastewater			Groundwater		
Parameters	V1	V2	V 3	V1	V2	V 3
pH	7,88 ± 0,40	7,99 ± 0,44	7,92 ± 0,38	8,00 ± 0,30	7,92 ± 0,13	7,86 ± 0,25
EC	11,26 ± 4,74	10,01 ± 4,22	9,61 ± 3,17	0,66 ± 0,21	0,82 ± 0,30	0,92 ± 0,23
Na+	99,82 ± 27,43	102,98 ± 28,73	66,72 ± 16,44	3,06 ± 0,59	3,06 ± 0,63	3,16 ± 0,50
Cl-	62,51 ± 31,63	58,46	51,71	3,89	4,39	5,80
		29,67	19,34	1,37	1,01	1,86
Ca++	2,54 ± 0,92	2,41	2,30	0,46	0,48	0,51
		0,86	0,68	0,17	0,14	0,18
Mg++	4,30 ± 1,65	3,83	3,43	0,71	0,85	0,79
		1,36	0,90	0,16	0,27	0,14
K+	0,80 ± 0,31	0,68	0,82	0,12	0,13	0,08
		0,23	0,35	0,04	0,02	0,06
SO4--	1,70 ± 0,79	1,52	1,21	0	0	0
		0,54	0,56	0	0	0
HCO3-	4,21 ± 2,10	4,04	4,47	2,73	3,72	3,66
		1,97	1,53	0,78	1,29	1,95
SAR	17,19 ± 7,82	16,38	16,58	0,99	1,11	1,37
		9,36	5,76	0,45	0,56	0,45

and with the ground water. The Table 12.4 presents the different seeds (V1, V2, and V3) composed of pure or mixed species as well as the parcels assigned to each seed.

The first grass plants germinated after 3 days from the sow. The parameters of grass growth (number of talles (shoots) and length of the leaves) have been monitored during the first 41 days from the apparition of the first plantations on the different parcels (P1, P2, and P3) sowed by the different varieties of grass (V1, V2, and V3) and irrigated by the two qualities of water (groundwater and treated wastewater).

The samplings of soil and leachate are done every 6 days from the first day of germination. This follow-up lasted 60 days resulting in 10 sampling operations. For every sampling operation, 16 samples of leachate and composite samples of soil are collected by auger on the diagonal of each plot and it conserved for analysis.

12.3 Results and Discussion

12.3.1 Influence of the Quality of Irrigation Waters on the Germination of the Seeds

To study the influence of the quality of groundwater and treated wastewater on the germination of the grass seeds, tests of germination concerning the three varieties of grass (V1, V2, and V3) had been performed on the site of the treatment plant. The soil used in the trays of alveolar had the same composition as the one in the parcels.

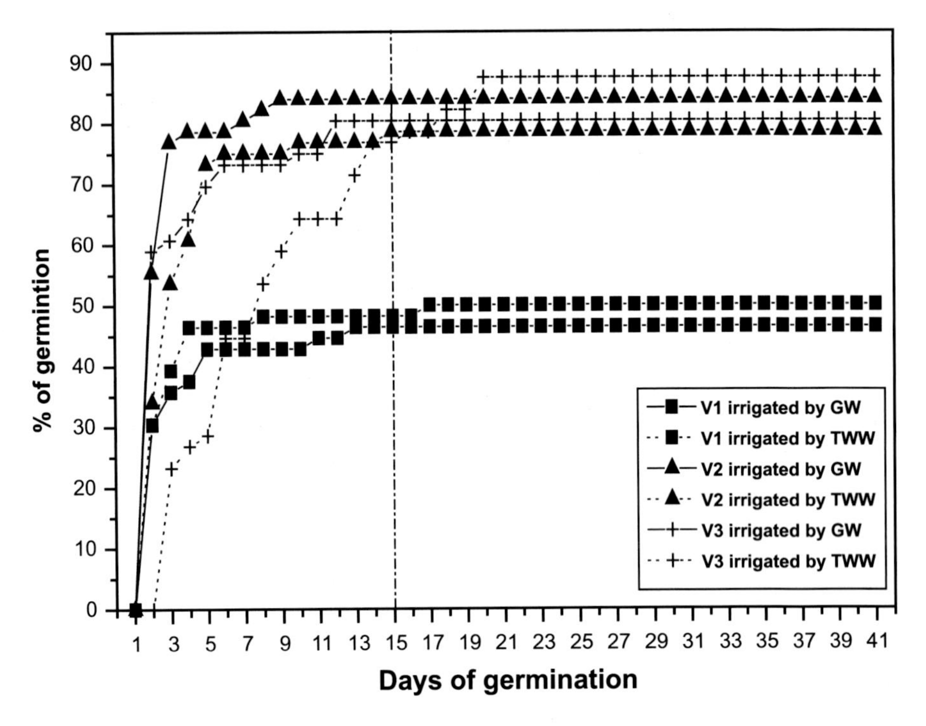

Fig. 12.3 Evolution of the percentage of germination of the different varieties irrigated by the two water qualities: *GW* ground water and *TWW* treated wastewater

For every variety of grass, two trays with 56 alveolars were sowed (a seed by alveolar): one was irrigated by the ground water, and the other by the treated wastewater. The germination of the seeds in the six trays was monitored during 41 days.

The results of these tests are represented in Fig. 12.3 that shows the evolution of the percentage of germination with the time for every variety of grass and for the two types of irrigation water: groundwater and treated wastewater.

The evolution of the percentage of germination in Fig. 12.3 shows that the germination of the varieties V2 and V3 was favored by the irrigation with the groundwater during the first 15 days. The germination of these two varieties accused a time of adaptation to the saltiness of the treated wastewater during this phase. The % of germination of the variety V1 is slightly favored by the treated wastewater. The germination of this variety shows a better tolerance to the saltiness since the first days.

After the 15th day, the percentage of germination remained steady and reached a maximum of 84% for the varieties V2 and V3, whereas the % maximal doesn't exceed 48% for V1. The influence of the water quality on the maximal % of germination reached remained weak for all varieties and fluctuates around ±5%.

12.3.2 *Influence of the Quality of the Water Irrigation on the Parameters of Grass Growth*

The evolution of the number of talles, counted every 2 days, in the different parcels is represented in Fig. 12.4.

The talles are the supplementary shoots produced from a seed mother by stolon or rhizome. The proliferation of talles (tillering) is an important phenomenon that conditions the density of the grass carpet. The increase of the number of talles before the maximal germination rate was reached due to the contribution of germination and to the phenomenon of tillering. According to the evolution of the % of germination discussed above, the maximal germination was reached after 15 days of seedling that corresponds to the 13th day of growth. The increase of the number of talles after the 13th day of growth was exclusively due the phenomenon of tillering.

The irrigation by the groundwater favored the tillering in a first time, but the gap recorded in relation with the irrigation by treated wastewater reduced quickly and annulled itself on the 21st day of growth. According to this tendency, one can expect that the tillering will be favored by the irrigation with treated wastewater after the 21st day.

The length of the leaves provided information on the speed of growth of the aerial part of the plant (grass). A sample of leaf by parcel was collected every 2 days. The length of the leaves of every sample was measured on graph paper. The evolution of this length, for the different varieties irrigated by the two types of water, is represented in Fig. 12.5 for the 21 days of follow-up.

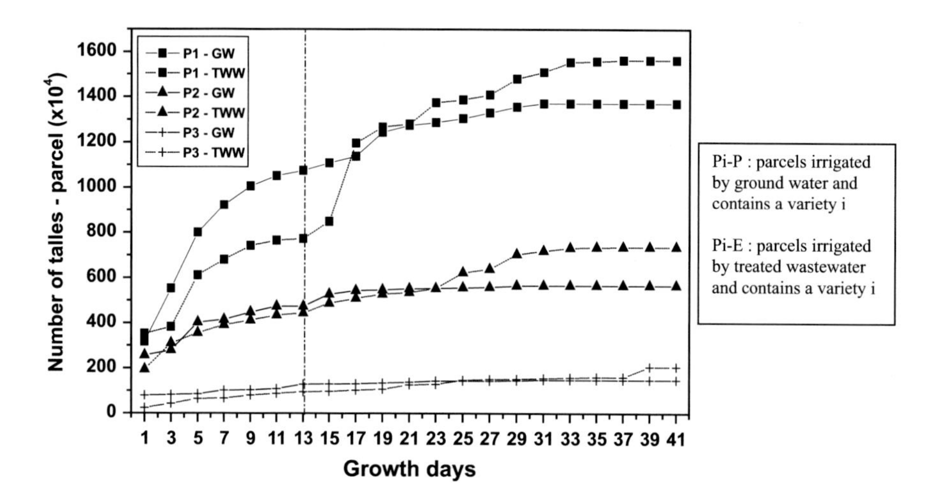

Fig. 12.4 Evolution of the number of talles in the parcels irrigated by the two qualities of water: *GW* groundwater and *TWW* treated wastewater

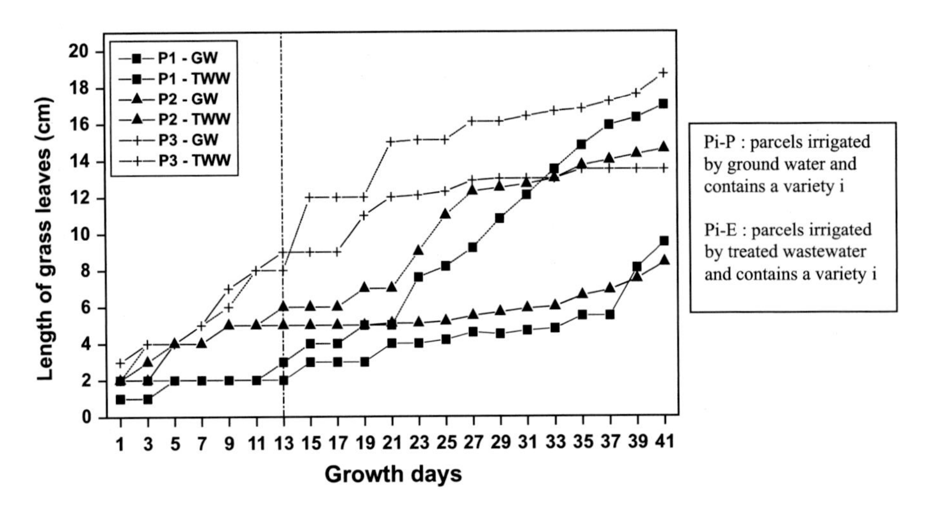

Fig. 12.5 Evolution of leaves length in the parcels irrigated by two water qualities: *GW* groundwater and *TWW* treated wastewater

Before the maximal germination rate is reached, we could not decide about the effect of the irrigation by treated wastewater on the evolution of the leaves length before the 13th day of growth indicated on the Fig. 12.5. However, after the 13th day of growth, it was obvious that the irrigation by the treated wastewaters favored the increase of the leaves length compared to the groundwater irrigation, and it was observed for all varieties.

12.3.3 Leachate Characteristics

As shown in Table 12.4, the ionic content in leachate, in case of irrigation with treated wastewater, is very important. Leachate presents an increase in electric conductivity, Na+, Cl-, and SAR. This increase can be explained by leaching of soluble salts from the soil to the groundwater through irrigation water. The concentration of soluble salt in soil, sand, and the mixture (75% soil and 25% sand) are respectively 0.41, 0.16 and 0.34 g/kg which confirm their leaching (Al-Hamaiedeh and Bino 2010; Francesco and Alarcon 2009).

The high electrical conductivity and SAR in leachate is strongly correlated to the concentration of sodium chloride and reflects a salinization of soil and groundwater in the case of irrigation with treated wastewater. Furthermore, the ionic content in leachate in the case of irrigation with groundwater is very low and it didn't have any negative effect on soil and ground water.

Table 12.5 Ionic Analysis of soils irrigated with reclaimed wastewater

Variety	Soil deep cm	pH	EC dS/m	SS g/l	Na+	Ca++	Mg++	Cl-	K+	SO4–	HCO3-
					mmole/l						
V1	0–20	8.5	1.44	1.02	24.97	8.73	9.95	50.07	0.42	2.32	20.11
	20–40	8.18	1.42	1.00	15.61	5.37	7.82	32.19	0.24	1.32	17.23
	40–60	8	1.05	0.74	14.27	4.28	5.36	30.81	0.22	0.85	15.47
V2	0–20	8.21	1.4	0.99	22.17	5.7	7.11	50.07	0.48	5.74	28.72
	20–40	8.11	1.17	0.83	8.09	4.99	5.69	32.19	0.24	0.72	14.36
	40–60	7.68	0.68	0.43	7.74	2.85	5.01	28.61	0.20	0.42	11.36
V3	0–20	8.37	1.69	1.2	13.03	6.91	7.66	34.66	0.46	0.84	20.11
	20–40	8.17	1.17	0.83	11.56	5.82	6.64	29.21	0.28	0.63	15.47
	40–60	7.88	1.06	0.75	10,21	3,74	4.98	25,04	0.26	0.44	13.08

12.3.4 Ionic Analysis of Soil

The results in Table 12.5 showed a gradual decrease of parameters of salinity (soluble salts, sodium, chloride, potassium, calcium, and magnesium) as a function of depth. This decrease might be explained due to the following reasons: the plants consumed some of these constituent elements (Ca^{2+}, K^{+}, Mg^{++}), while the sodium and chloride provide the phenomenon of ionic exchange with soil aggregates.

The sulfates and bicarbonates prove the formation of precipitates of calcium carbonate and magnesium by increasing concentrations of these elements resulting the leaching in irrigation water (Mouhanni et al. 2008, 2012; Tomas 2008; Vuokko et al. 2010). Therefore, many parameters of salinity are concentrated in the top 20 cm of soil and decrease in the 40 and the 60 cm of soil respectively.

According to the study of Al-Hamaiedeh and Bino (2010), which concerned the effect of treated GW reuse on the properties of soil and irrigated plants at Al-Amer villages, Jordan, the results showed that salinity, sodium adsorption ratio (SAR), and organic content of soil increased as a function of time; therefore leaching of soil with fresh water was highly recommended. The same results had been shown by Lucila et al. (2007) in their study which assessed the soil and groundwater impacts by treated urban wastewater reuse in the irrigation of golf course. They observed that NaO_2 increased of more than 1000 mg kg^{-1} in the top of soil, while Cl^{-} concentration in the aquifer reached up to 1200 mgl^{-1} 10 month after stating irrigation.

12.4 Conclusion

The monitoring of the leachate quality and soil irrigated turf plots by treated wastewater have allowed us to conclude that this kind of irrigation water presents a risk of salinization for the groundwater, and especially in the case of heavy textured soils. Therefore, caution in the management of irrigation intakes to prevent the accumulation of salts in the rhizosphere and increased concentrations of sodium chloride in the water must be taken into consideration. Indeed, the purified water can be valued on coarse textured soils – like sandy or well-drained soils – to ensure leaching and reduce the phenomenon of ion exchange with the soil aggregates. The sulfate ions and bicarbonate of treated wastewater do not bring about concentrations that can present a risk to the water. However, precipitates carbonates of calcium or magnesium may be provided. This will cause the problem of salinization which is a major handicap in the reuse of the treated wastewater for irrigation.

Acknowledgements Our thanks go to RAMSA, Regional Office of Agricultural Development of the Souss Massa Daraa, National Institute of Agronomic Research of Agadir, and the Water and Chemistry Laboratory in Franche Comté University in Besançon town in France for their assistance during the conduct of this study.

References

Al-Hamaiedeh, H., & Bino, M. (2010). Effect of treated grey water reuse in irrigation on soil and plants. *Desalination, 256*, 115–119.

Ayers, R.S. and Westcot, D. (1988). The quality of water in agriculture. FAO Irrigation and Drainage. pages? Edt?

Chenini, F., Trad M., Rejeb, S., Chaabouni, Z., Xanthoulis, D., (2002).Optimization and durability of the treatment and reuse of wastewater in agriculture. Report Ministry of Agriculture, Environment and Water Resources, National Institute for Research in Agricultural Engineering, Forestry, Tunisia.

Gregory, E., Ervin, R., & Zhang, X. (2010). Reclaimed water for turfgrass irrigation. *Water, 2*, 685–701.

Jalali, M., Merikhpour, H., Kaledhoukar, M. J., & der Zee Seatm, V. (2008). Effects of wastewater irrigation on soil sodicity and nutrient leaching in soils calcareus. *Agricultural Water Management, 95*, 143–153.

Lucila, C., Fabregat, S., Josa, A., Suriol, J., Vigues, N., & Mas, J. (2007). Assessment of soil and groundwater impacts by treated urban wastewater reuse. A case study: Application in a golf course (Girona, Spain). *Science of the Total Environment, 374*, 26–35.

Mouhanni, H. et Bendou, A. (2011). Impact de la réutilisation des eaux usées épurées en irrigation sur la croissance de la plante gazon. *Revue Internationale d'héliotechnique-énergie-environnement* N° 43 14–21.

Mouhanni, H., Hamdy, H., Bendou, À., & Benzine, L. (2008). Réutilisation des eaux usées épurées pour l'irrigation du gazon des golfs: Impact sur la germination et la croissance du gazon. *Revue Internationale d'hélio technique-énergie-environnement, 38*, 27–33.

Mouhanni, H.; Hamdy, H.; Bendou, A., Cavalli, E.; Benzine, L. (2011). Impact de la réutilisation des eaux usées épurées en irrigation: Analyse ionique des sols. *Revue Déchets Sciences et Techniques* N° 59 mars.

Mouhanni, H., Hamdy, H., Bendou, A., Cavalli, E., & Benzine, L. (2012). Impact de la réutilisation des eaux usées épurées en irrigation: Analyse ionique des lyxiviats. *Revue des Sciences de l'Eau, 25*, 69–73.

Pedrero, F., & Alarcon, J. J. (2009). Effect of treated wastewater irrigation on lemon trees. *Desalination, 246*, 631–639.

Salgot, M., Huertas, E., Weber, S., Dott, W., & Hollender, J. (2006). Wastewater reuse and risk: Definition of key objectives. *Desalination, 187*, 29–40.

Tomas, C. (2008). Water reuse of South Barcelona's wastewater reclamation plant. *Desalination, 218*, 43–51.

Vuokko, O. K., Tapio, S., & Pekka, E. P. (2010). Bilateral collaboration in municipal water and wastewater services in Finland. *Water, 2*, 815–825.
Zalesny, J. A., Zalesny, R. S., Jr., Wiese, A. H., Sexton, B., & Hall, R. B. (2008). Sodium and chloride accumulation in leaf, woody, and root tissue of populus after irrigation with landfill leachate. *Environmental Pollution, 155*, 72–80.

Hind Mouhanni Specialist in Food technology and attached to the faculty of applied sciences, Chemistry department in Ait Melloul Campus, I'm a ex-head of professional licensing "Processing and valorization fishery products" in the Higher Institute of Marine Fisheries. I have generally 11 years of professional experience of working on various R & D organizations and area of expertise: microbiology, chemistry, treatment of water, processing food products. I had several formations under my specialty as well as publications of my research work in different measures. My work has been presented in several forums, conferences, workshops and international scientific.

Dr. Abdelaziz Bendou is a Researcher Professor and Director of Higher School of Business and Management of Agadir and Vice-President for Cooperation.at Ibn Zohr University of Agadir, Morocco. Bendou has a large experience in training, research, and higher education institutions such as the Higher Shool of Technology, the National Shool of Ingineering Sciences and the Higher Shool of Business and Management (Ibn Zohr University). His core scientific interests cover the areas of themodynamic, solar drying, energie, food process, management, and environment.

Chapter 13
Ecoservice Role of Earthworm (*Lumbricidae*) Casts in Grow of Soil Buffering Capacity of Remediated Lands Within Steppe Zone, Ukraine

Sergiy Nazimov, Iryna Loza, Yurii Kul'bachko, Oleg Didur, Oleksandr Pakhomov, Angelina Kryuchkova, Maria Shulman, and Tatiana Zamesova

13.1 Introduction

Protecting environment, managing natural resource, and ensuring environmental safety of human life are essential conditions for the sustainable economic and social development of European countries (Sklenicka et al. 2004; Pecharová et al. 2011; Behnassi et al. 2014). In this regard, solving the environmental problems of anthropogenic disturbed areas within Eastern Europe is on the front burner (Thassitou and Arvanitoyannis 2001; Pecharová and Hrabankova 2006; Böhm et al. 2009). Coal industry activity is considered to be one of the most powerful factors leading to deterioration of natural landscapes variety (Strzyszcz 1996).

Soil contamination with heavy metals affects primarily the soil biological and ecological conditions; it can change the conservative soil signs such as humus status, structure, acidity, and other characteristics. Soil contamination leads to partial and, in some cases, complete loss of soil fertility that reduces soil economic cost (Orlov 1994; Orlov et al. 2005; Truskavetskiy 2003). While investigating the man-made contamination, such contamination brings up the challenge of soil protective capacity under the elevation of heavy metals concentrations. The better soil protective properties, the more heavy metals can be insolubilized and no plant-available. As a result, over amounts of chemicals in food chains are restricted, and their migration to surrounded environment ecosystems is limited (Ilyin 1995; Cooke and Johnson 2002).

As a result of coal mining, the lands intended for economic purpose are withdrawn from agricultural use. Thus, such lands are replaced by man-made landscapes, i.e. dumps and open-cuts, which are characterized by subsidence, highly

S. Nazimov · I. Loza (✉) · Y. Kul'bachko · O. Didur · O. Pakhomov · A. Kryuchkova
M. Shulman · T. Zamesova
Laboratory of Biological Monitoring, Biology Research Institute, Dnipropetrovsk National University, Dnipropetrovsk, Ukraine

M. Behnassi et al. (eds.), *Climate Change, Food Security and Natural Resource Management*, https://doi.org/10.1007/978-3-319-97091-2_13

mineralized groundwater table rise, acid mine drainage, toxic contamination (Jachimko 2012; Pecharová et al. 2001). These factors affect species wealth and biological diversity within territories disturbed by production activities (Ripl et al. 1994). Disrupted lands developed during coal mining can be partially restored by remediation (Wang et al. 2001; Pecharová et al. 2001; Cooke et al. 2002).

Main agro-chemical characteristics determining remediated lands productivity and suitability for biota living are actual acidity (pH) and salinity (Arranz-González 2011). Scientists notice that mine spoil often has high density, low coefficient of structure, as well as high salinity of water extracts. Such characteristics determine extremely low suitability of such substrates for biota existence (Skousen et al. 1998; Cooke and Johnson 2002). The next remediation stage is covering with topsoil, such as compost mass or humus layer. Humus topsoil of ordinary chernozem or humus-free subsoil is usually used as remediated layers under conditions of Ukrainian Steppe, particularly within Western Donbass. Final stage of damaged lands remediation is a biological stage. The most using type is phyto-remediation by herbal, arboreal and shrubbery plantations (Böhm et al. 2011; Singh and Singh 2006; Pakhomov et al. 2009).

Among soil invertebrates, earthworms have a leading role in formation of stability mechanisms in arboreal (forest) plantations. As a result of their life activity, earthworms make a significant ecological contribution to transformation of soil characteristics and properties (Lavelle et al. 2006; Didur et al. 2011; Aira et al. 2003; Choosai et al. 2010; Bottinelli et al. 2010; Kul'bachko et al. 2011). Earthworms so called 'ecosystem engineers' are the animals effecting on environment and communities of soil invertebrates, and resulting in ecosystem successions (Jones et al. 1994; Aira et al. 2010; Eisenhauer 2010). Such earthworm environmental-forming function does not duplicate by any groups of living organisms; it plays a crucial role in protecting and improving the fertility in remediated soils (Didur et al. 2013; Kul'bachko et al. 2014).

Tropho-metabolic activity of saprophages is considered to be an important element in the formation of numerous environmental properties (Brygadyrenko 2016); that causes maintaining of buffer properties in artificial soil against copper contamination within remediated areas. Copper among such metals as zinc, molybdenum, cadmium, and lead – is one of the main man-made environmental pollutants (Pokarzhevsky 1985; Safonov 2005).

Mainly, buffering capacity is a soil capability to resist changes under impact of various factors. It may be acid-alkaline buffering capacity (pH-buffering capacity) (Truskavetskiy 2003), or buffering capacity to contamination by heavy metals (Ilyin 1995; Pampura et al. 1993). Some studies examining the buffering capacity of Ukrainian soils are agriculturally orientated (Truskavetskiy 2003); they are devoted to study of buffering capacity of different soil genetic types, and not associated with a soil-zoological component (Gamkalo 2005, 2001). There is no information about the participation of soil saprophages in buffering properties formation and maintenance of remediated lands under arboreal plantations.

The study goal is determining the tropho-metabolic effect of earthworm (*Lumbricidae*, as zoogenic component of ecosystem services) to maintenance stability of remediated soil against some negative effects such as soil solutions with acid pH values and copper contamination.

13.2 Materials and Methods

The present research was carried out within Steppe zone, in Western Donbass as a part of Donetsk Coal Basin (Fig. 13.1). The western part of this basin located in Dnipropetrovsk region is called Western Donbass. In the basin area, it was found about 40 layers with a working capacity 0.6–1.6 m, which underlies 400–1800 m depth. Coal production is performed by open-cut mining. Flat and conical piles at eleven mines of Western Donbass contain more than 70 million tons of phytotoxic sulfur-containing carbon-bearing and argillaceous shale, argillite with a high content of pyrite (FeS_2), troilite (FeS), chalcopyrite ($FeCuS$), etc. When deeply buried deposits of Cretaceous period are moved onto surface, it is initiated the processes of physical weathering, oxidation, dissolution, hydrolysis, and burning. A number of other negative factors are also determined, such as high concentration of soluble toxic salts, alkalinity level rise, low absorbency and permeability, high spoil density, low carbon and plant-available nitrogen (Novitskiy 2011; Mtui et al. 2006).

Soil samples are selected within remediated areas in Norway maple plantations, on two different remediated sites that differ in stratigraphic structure (location of layers from top to bottom): Variant 1: loess loam 0–50 cm, tertiary sand: 50–100 cm;

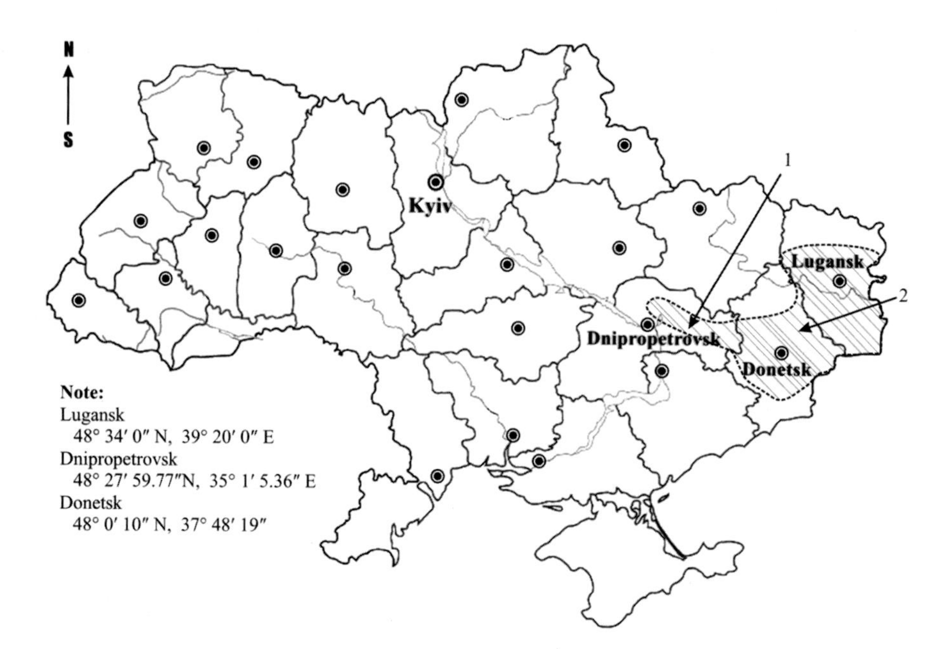

Fig. 13.1 Location of Donbass Coal Basin in Ukraine: (1) Western Donbass; (2) Central Donbass

mine spoil: 100–700 cm; Variant 2: humic layer of ordinary black soil: 0–50 cm, loess loam: 50–100 cm, tertiary sand: 100–150 cm, mine spoil: 150–700 cm.

Samples of fresh cast of earthworm *Aporrectodea caliginosa* (Savigny 1826) are selected from abovementioned remediated lands. It is the only species of worm found on the investigated remediated site. *Aporrectodea caliginosa* (Savigny 1826) is saprophage, secondary destructor, humificator (Milcu et al. 2006; Striganova 1980). According Bouché classification it belongs to endogeic morpho-ecological group of earthworms (Butt and Lowe 2011).

Sampling plots in the study site were selected on the basis of occurrence of fresh worm casts. The fresh earthworm casts were represented by small soil conglomerates weighting 0.8–13.4 g. Collected fresh worm casts were combined into a common sample for each study variant weighing approximately 1 kg. The samples were taken to the laboratory and weighed, air-dried, powdered with mortar and pestle, and passed through a fine sieve (1 mm-mesh). For each test, the sample of 10 g was taken from the combined sample in triplicate. For each experiment was calculated mean and standard error.

Differences in the physico-chemical properties of worm casts and adjacent soil were tested by Student's t-test. All statistics were calculated using Statistic software Statistica version 6.0.

13.2.1 Evaluation Method of Earthworm Casts and Soil Buffering Capacity Against Different pH Values

To assess the buffering capacity of earthworm casts, soil and subsoil, we used an approach based on determination of changes in their pH level after adding of acids or alkalis solutions. The experimental design for acid and alkaline exposure ranges is shown in Table 13.1.

A series of weights of soil samples, earthworm casts and heat-treated sand (reference) were prepared for analysis. Fixed amounts of acid and alkali solutions were added to all samples, while the total amount of solution (water-acid or water-alkali) was constant (25 mL). Ratio soil (or earthworm cast): solution was amounted to 1:2.5. Results of actual acidity measuring were plotted. X-axis indicates the number

Table 13.1 Design of experimental study in determination of pH-buffering capacity in soil samples and earthworm casts

Reagent	Volume of reagent added, mL					
	Acid exposure range					
0.1 M HC1	–	1.5	3	4.5	6	7.5	9
H_2O	25	23.5	22	20.5	19	17.5	16
	Alkaline exposure range						
0.1 M NaOH	–	1.5	3	4.5	6	7.5	9
H_2O	25	23.5	22	20.5	19	17.5	16

of milliliters of acid (or alkali) added, and Y-axis indicates the corresponding pH values. Obtained plotted curves allow buffering capacity of samples by the buffering area calculation to be estimated in acid and alkaline ranges. Buffering area is defined as area between titration curves of sample and reference (quartz sand), and expressed in conditional units.

To calculate buffering capacity, method of numerical integration was applied. The task was solved by means of Simpson formula (Atkinson 1989; Chapra 2012):

$$\int_a^b f(x)dx = \frac{b-a}{6n}\begin{bmatrix}(y_0 + y_{2n}) + 2(y_2 + y_4 + \dots y_{2n-2}) + \\ +4(y_1 + y_3 + \dots + y_{2n-1})\end{bmatrix},$$

where

a and *b* are the lower and upper integration limits, respectively;

n is a number of pairs of parabolic trapezia, corresponding to a number of ordinates (points) excluding one, taken by half;

y are integration variables (values of a function at the corresponding points).

In square brackets, the extreme ordinates (y_0 and y_{2n}) are taken with a coefficient of "1"; other ordinates with even indices are taken with a coefficient of "2", and odd indices are taken with a coefficient of "4". To calculate area by Simpson formula an odd number of ordinates is required. If there are five ordinates, then $n = 2$ because they form only two pairs of curved (parabolic) trapezia (the first pair: $[y_0,y_1]$, $[y_1,y_2]$, the second pair: $[y_2,y_3]$, $[y_3,y_4]$); if there are seven ordinates, then $n = 3$, etc. Measurement of actual acidity (pH) in samples was performed in triplicate. Mean, its standard error and significant difference of the mean were calculated (Van Emden 2008; Zar 2010; Weiss 2012).

13.2.2 Evaluation Method of Earthworm Casts and Soil Buffering Capacity Against Different Levels of Copper Contamination

To determine zoogenic environmental-forming function in soil resistance formation against copper contamination, we studied immobilization (immobility) – mobilization (mobility) of copper in earthworm casts and bulk soils, and participation of earthworm casts in formation of resistance against contamination with copper. Previously prepared air-dry soil and casts samples were weight, transferred in 150 mL flasks and added copper sulphate pentahydrate solution in 1:10 ratio, which contains copper at concentrations 5–40 mg/L with 5 mg/L grade. Flasks were shaking at 2 h and leaved during a day. Then a suspension was stirred and mixed again, only the funnels with filter content were used in further assay. Matter on the filter was transferred in glass weighing bottle and air-dried. Sample weights were analyzed on moving forms of copper compounds by atomic absorption method.

To assess the impact of earthworms' tropho-metabolic activity for maintaining resistance of their habitats to copper pollution we used effect and toxicant immobilization efficiency. Effect (E. inf.) is a difference of areas between reference and sample buffering capacity in nominal units. Effect value indicates the soil material stability to contaminant. Than it is higher, soil is the more stable to such effect. Effectiveness of toxicant immobilization (E. imm.) (binding) is calculated as effect to reference ratio. Reference is presented as a curve made by points relevant to logarithm of initial metal concentrations in solution. Immobilization effectiveness is expressed as relative dimensionless units (in percentage). When effectiveness values increase, soil stability to heavy metal impact growth.

13.3 Results and Discussion

13.3.1 *Effect of Earthworm Casts on Soil Buffering Capacity Against Different pH Levels*

For Variant No. 1, where loess loam is represented, changes in pH values in the analyzed samples within acid range are shown in Table 13.2. Note that loess loam has conditionally neutral reaction, and casts have neutral one. By adding small amount of acid solution (3 mL), reaction of loess loam was changed from

Table 13.2 Results of buffering capacity measurements in acid and alkaline ranges for Variant No. 1 (Loess Loam, Earthworm Casts) and sand (Reference)

Amount of reagent added, mL	Acidity (pH)		
	Sand	Loess loam	Casts
	Acid range (Reagent 0.1 M HCl)		
0	6.95 ± 0.05	6.60 ± 0.05	7.16 ± 0.06
1.5	3.25 ± 0.10	6.09 ± 0.07	6.78 ± 0.08
3	2.80 ± 0.05	5.07 ± 0.09	6.55 ± 0.05
4.5	2.52 ± 0.04	4.64 ± 0.07	6.42 ± 0.07
6	2.45 ± 0.05	4.43 ± 0.08	6.33 ± 0.11
7.5	2.36 ± 0.06	3.93 ± 0.11	6.24 ± 0.09
9	2.29 ± 0.04	3.73 ± 0.08	6.10 ± 0.08
	Alkaline range (Reagent 0.1 M NaOH)		
0	6.95 ± 0.05	6.60 ± 0.05	7.16 ± 0.06
1.5	11.45 ± 0.05	6.78 ± 0.08	7.43 ± 0.07
3.0	11.65 ± 0.10	7.20 ± 0.10	8.01 ± 0.06
4.5	11.72 ± 0.07	7.72 ± 0.07	8.55 ± 0.05
6.0	11.87 ± 0.06	8.25 ± 0.10	8.87 ± 0.07
7.5	12.05 ± 0.05	8.75 ± 0.05	9.01 ± 0.10
9.0	12.05 ± 0.10	9.35 ± 0.07	9.20 ± 0.07

Here and below, mean and its reference error are given

conditionally neutral (6.60 ± 0.05) to acid one (5.07 ± 0.09), while casts having initially neutral reaction (7.16 ± 0.06) will have conditionally neutral one (6.55 ± 0.05), which can be maintained at adding a larger amount of acid.

By adding a small amount of alkaline solution (1.5 mL), reaction of loess loam remains conventionally neutral (6.78 ± 0.08), while casts get clear slightly alkaline reaction (7.43 ± 0.07) instead original neutral reaction (7.16 ± 0.06) (see Table 13.2). By adding larger amount of alkaline solution (6 mL), reaction of loam and cast becomes alkaline (8.25 ± 0.10 and 8.87 ± 0.07, respectively) and both samples develop strong alkaline reaction by adding maximum amount of alkaline solution (9 mL).

Buffering capacity curves within acid range for Variant No. 1 (loess loam, earthworm casts) and sand (reference) are shown in Fig. 13.2. Buffering capacity area of remediated soil samples (Fig. 13.2a) lies between titration curve of sand and titration curve of bulk soil; buffering capacity area of earthworm casts (Fig. 13.2b) lies between titration curve of sand and titration curve of casts. In acid range, location of titration curves of the studied samples indicates that buffering capacity area of casts is significantly larger than that of loess loam.

Figure 13.3 shows curves of buffering capacity in alkaline range for Variant No. 1 (loess loam, earthworm casts) and sand. Location of titration curves of the studied samples indicates that buffering capacity area of casts is smaller than one of loess loam.

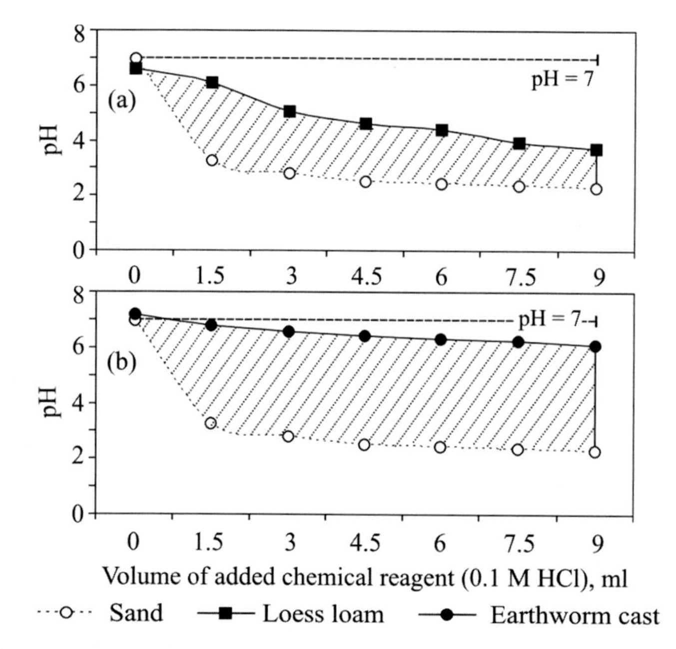

Fig. 13.2 Buffering capacity area of the studied samples in remediated sites within acid exposure range (Variant No. 1): (**a**) loess loam; (**b**) earthworm casts

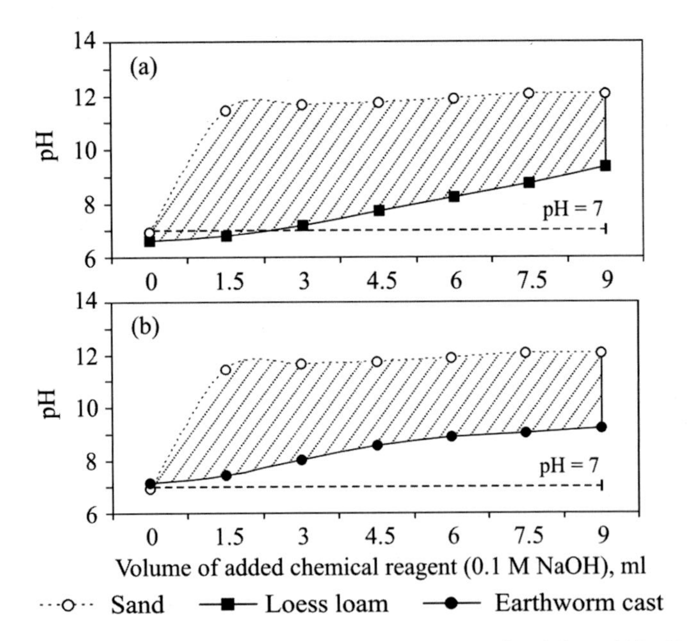

Fig. 13.3 Buffering capacity area of the studied samples in remediated sites within alkaline exposure range (Variant No. 1): (**a**) loess loam; (**b**) earthworm casts

Table 13.3 Estimation of soil buffering capacity indexes in Variant No.1 (loess loam, earthworm casts)

Range	Buffer capacity area, nominal cm^2	
	Loess loam	Casts
Acid	18.1±0.51	32.5±0.52**
Alkaline	33.8±0.43	28.7±0.06*
Acid-Alkaline(Total)	51.9±0.60	61.1±0.53***

Significant difference of mean with level of significance: * – ≤ 0.05, ** – ≤ 0.01, *** – ≤ 0.001

Table 13.3 shows the calculated values of the buffering capacity areas of loess loam and earthworm cast, reflecting buffering capacity of soil of Variant No. 1 and their statistical evaluation. The total area of casts buffering capacity is statistically higher by 9.2 nominal cm^2, i.e. by 17.9% higher than the total area of loess loam buffering capacity. However, their buffering capacity is shown more in the acid range.

For Variant No.2, the upper layer of which consists of humic topsoil filling, changes in pH of the remediated soil samples, earthworm casts and sand (reference) in acid range are shown in Table 13.4. Humic topsoil has conditionally neutral reaction, and casts have neutral one. By adding small amount of acid solution (1.5 mL), reaction of humic topsoil was changed from conditionally neutral in soil solution (6.82±0.12) to subacid reaction (5.86±0.11), while casts having initially neutral reaction (7.16±0.04) get conditionally neutral one (6.33±0.07). By adding large amounts of acid solution (from 4.5 to 9 mL), reaction of humic topsoil solution and casts is the same (acidic).

Table 13.4 Results of buffering capacity measurements within acid and alkaline ranges, Variant No. 2 (humic topsoil, earthworm casts) and sand (reference)

Volume of reagent added, mL	Acidity (pH)		
	Sand	Humic topsoil	Casts
	Acid range (Reagent 0.1 M HCl)		
0	6.95±0.05	6.82±0.12	7.16±0.04
1.5	3.25±0.10	5.86±0.11	6.33±0.07
3	2.80±0.05	5.37±0.10	5.77±0.07
4.5	2.52±0.04	4.86±0.09	5.34±0.08
6	2.45±0.05	4.38±0.08	4.97±0.07
7.5	2.36±0.06	4.12±0.12	4.65±0.10
9	2.29±0.04	4.00±0.10	4.43±0.08
	Alkaline range (Reagent 0.1 M NaOH)		
0	6.95±0.05	6.82±0.12	7.16±0.04
1.5	11.45±0.05	7.68±0.08	7.50±0.06
3.0	11.65±0.10	8.67±0.12	8.23±0.09
4.5	11.72±0.07	9.42±0.09	8.90±0.10
6.0	11.87±0.06	9.94±0.08	9.32±0.07
7.5	12.05±0.05	10.31±0.05	9.56±0.08
9.0	12.05±0.10	10.46±0.13	9.90±0.05

In order to get same changes in pH of the studied samples by means of adding a small amount of alkaline solution (1.5 mL), reaction of humic topsoil and casts was changed to weakly alkaline (7.68±0.08 and 7.50±0.06, respectively) (see Table 13.4). By adding further amounts of alkali, soil and cast developed alkaline and strongly alkaline reaction. Thus, casts in comparison to the humic topsoil represent lower values of pH.

Figure 13.4 shows curves of buffering capacity within acid range for Variant No. 2 (humic topsoil, earthworm casts) and sand (reference). Location of titration curves of the studied samples in this range indicates that area of casts buffering capacity is more than area of humic topsoil buffering capacity.

Figure 13.5 shows curves of buffering capacity within alkaline range for Variant No. 2 (humic topsoil, earthworm cast) and sand. Location of titration curves of the studied samples (curve of cast buffering capacity is lower than one of humic topsoil buffering capacity) indicates that area of casts buffering capacity exceeds area of humic topsoil buffering capacity.

In Variant No.2, values of buffering capacity areas in humic topsoil, earthworm casts and their statistical evaluation are given in Table 13.5. It was found statistically significant difference within each exposure range (see Table 13.5). Thus, casts in this variant of remediation have a large buffering capacity area in the acid and alkaline ranges than that of original humic topsoil. It was found that total area of buffering capacity of casts in Variant No. 2 was significantly larger by 8.5 nominal cm^2 (20.8%) than that of humic topsoil. Both within acid and alkaline ranges, earthworm casts increased values of buffering area.

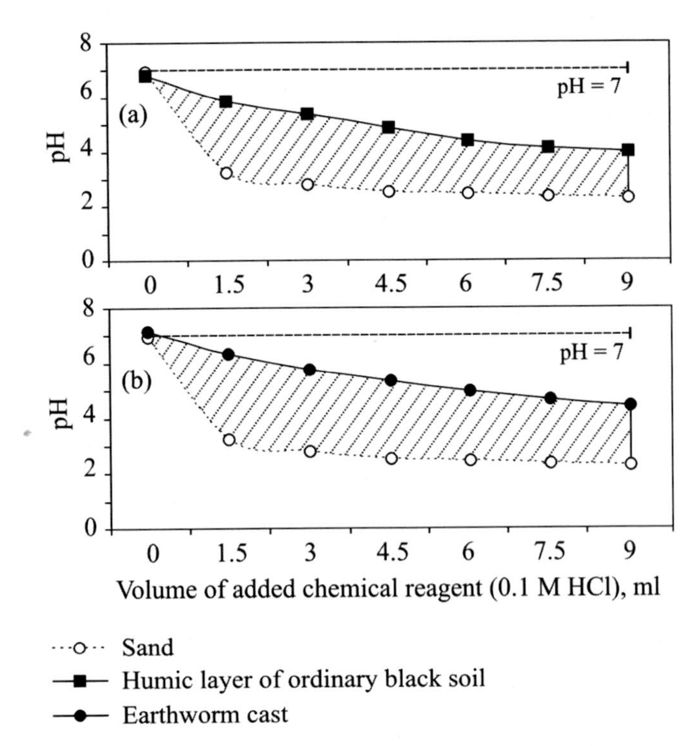

Fig. 13.4 Buffering capacity area of the studied samples in remediated sites within acid exposure range (Variant No. 2): (**a**) humic topsoil; (**b**) earthworm casts

13.3.2 Effect of Earthworm Cast on Soil Buffering Capacity to Different Levels of Copper Contamination

Earthworm casts effect on humus-free loess loam within copper concentration range from 5 to 40 mg Cu/L is less then effect of casts on humus layer of ordinary black soil (197.5 nom. units – humus-free loess loam, 336.1 nom. units – humic topsoil, Table 13.6). Effectiveness of immobilization reflected a degree of resistance to copper contamination was increased from 23.1% to 39.2%, respectively. It is explained by the fact that earthworm casts on humus-free loess loam basically presented as a loam, significantly organic-poor matter, while casts on humus soil are enriched by organic components. Therefore, presence of organic matter in casts is an additional agent of remediated soil resistance to copper toxic concentrations. Areas of toxicant influence ($p \leq 0.01$) between casts of humus-free loess loam and casts of humus layer are different significantly: this difference is more in casts of humus-free loess loam (659.6 nom. units), and smaller in casts of humus soil (426.1 nom. units). Statistical difference between areas of toxicant influence in casts on humus soil (521.0 nom. units) and humus level of soil is missing (534.5 nom. units).

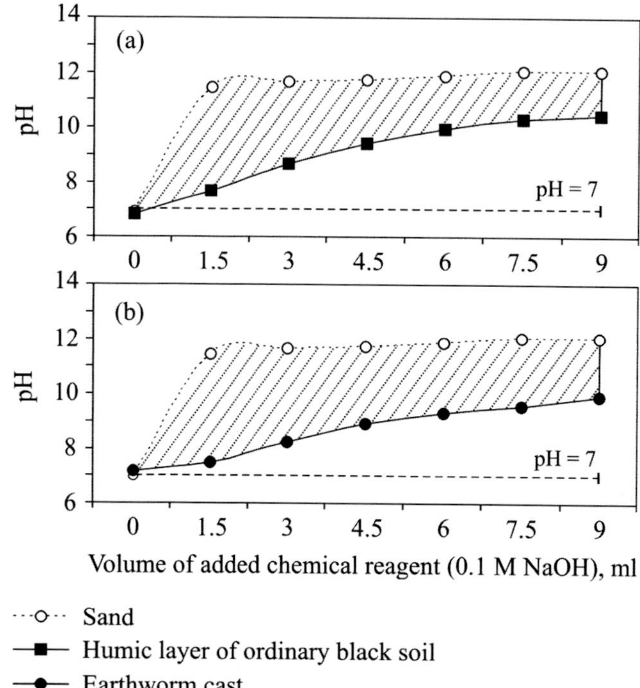

Fig. 13.5 Buffering capacity area of the studied samples in remediated sites within alkaline exposure range (Variant No. 2): (**a**) humic topsoil; (**b**) earthworm casts

Table 13.5 Estimation of buffering capacity values in Variant No. 2 (humic topsoil, earthworm casts)

External exposure range	Buffer capacity area, nominal cm^2	
	Humic topsoil	Casts
Acid	18.9±0.38	23.3±0.52(*)
Alkaline	21.6±0.54	25.7±0.45*
Acid-Alkaline(Total)	40.5±0.57	49.0±0.62***

Significant difference of the mean with significance level: (*) ≤ 0.07, * ≤ 0.05, *** ≤ 0.001

Graphic model of earthworm casts resistance to copper contamination (Variants No. 1, 2) are represented in Figs. 13.6 and 13.7. It indicates higher buffering capacity of casts in humus variant.

Thus, earthworm tropho-metabolic activity within different variants of forest remediation sites affects the soil immobilization capacity maintenance (buffering capacity to heavy metals, including copper). Resistance to elevated copper concentrations increased to casts in range: humus-free loess loam – humic layer of remediated soil.

Table 13.6 Quantitative resistance assessment of earthworm casts and soil to copper contamination

Characteristics	Reference area, nom. units ($S_{reference}$)	Sample area, nom. units (S_{sample})	$\frac{S_{sample}}{S_{reference}} \cdot 100\%$	Effect ($S_{ref} - S_{sam}$) nom. units	Effectiveness of Toxicant Immobilization $\frac{S_{sample}}{S_{reference}} \cdot 100\%$
Earthworm casts on humus-free loess loam (Variant 1)	857.1	659.6 ± 1.55	77.0	197.5	23.0
Earthworm casts on humus layer of ordinary black soil (Variant 2)	857.1	521.0 ± 8.80	60.8	336.1	39.2
Humus layer of ordinary black soil	857.1	534.5 ± 4.23	62.4	322.6	37.6

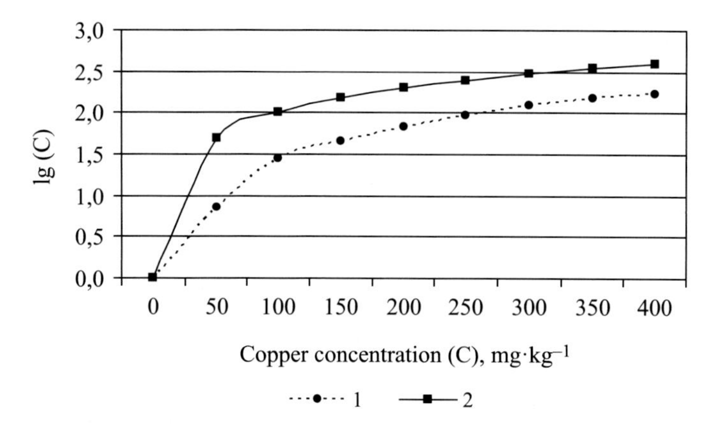

Note. 1 – Earthworm casts (humus-free loess loam, Variant No 1); 2 – Reference.

Fig. 13.6 Graphic model of earthworms casts resistance to copper contamination (Variant No. 1, humus-free)

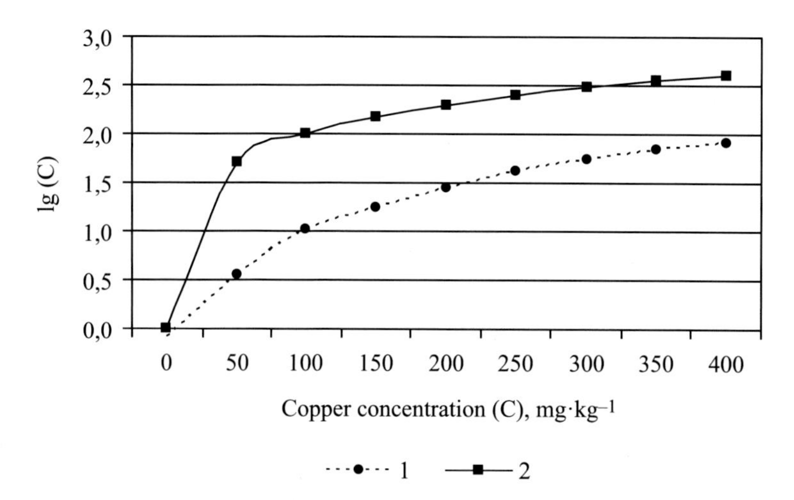

Note. 1 – Earthworm casts (humic layer, Variant No 2); 2 – Reference.

Fig. 13.7 Graphic model of earthworm casts resistance to copper contamination (Variant No. 2, humic layer)

13.4 Conclusion

Ecosystem effectiveness of soil saprophages (earthworms, *Lumbricidae*) was shown to be effected on increase of acid-alkaline (pH-buffering) buffering capacity in remediated soil. The study proves that acid-alkaline buffering capacity of earthworm casts was significantly higher than that of initial remediated soil and subsoil by 17.9% and 20.8%, respectively.

Resistance to copper concentrations increased in casts within follow range: humus-free loess loam – humus layer of remediated soil. Effectiveness of copper immobilization reflecting resistance to copper contamination was increasing from 23.1% to 39.2%, respectively.

That brings positive changes to soil and environmental conditions of remediated soil and naturalization of artificial soil in remediated lands within Steppe zone. Thus, efficiency of remediated land restoration increases with enrichment by earthworm casts; it leads to improvement of ecological quality in remediated soil. Earthworm ecoservice activity had positive changes to environmental conditions of remediated soil and naturalization of artificial edaphotopes within remediated lands in Steppe zone.

References

Aira, M., Monroy, F., & Domínguez, J. (2003). Effects of two species of Earthworms (Allolobophora spp.) on soil systems: A microfaunal and biochemical analysis. *Pedobiologia, 47*(5–6), 877–881.

Aira, M., Lazcano, C., Gómez-Brandón, M., et al. (2010). Ageing effects of casts of Aporrectodea caliginosa on soil microbial community structure and activity. *Applied Soil Ecology, 46*(1), 143–146.

Arranz-González, J. C. (2011). Suelos mineros asociados a la minería de carbón a cielo abierto en España: una revision. *Boletín Geológico y Minero, 122*(2), 171–186.

Atkinson Kendall, E. (1989). *An introduction to numerical analysis*. New York: Wiley.

Behnassi, M., Shahid, S. A., & Gopichandran, R. (2014). Agricultural and food system – Global change nexus: Dynamics and policy implications. In M. Behnassi et al. (Eds.), *Science, policy and politics of modern agricultural system* (pp. 3–13). Dodrecht: Springer Science+Business Media.

Böhm, C., Quinkenstein, A., Freese, D., et al. (2009). Kurzumtriebsplantage auf Niederlausitzer Rekultivierungsflächen: Wachstumsverlauf von vierjährigen Robinien. *AFZDerWald, 10*(64), 532–533.

Böhm, C., Quinkenstein, A., Freese, D., et al. (2011). Assessing the short rotation woody biomass production on marginal post-mining areas. *Journal of Forest Science, 57*(7), 303–311.

Bottinelli, N., Henry-des-Tureaux, T., Hallaire, V., et al. (2010). Earthworms accelerate soil porosity turnover under watering conditions. *Geoderma, 156*(1–2), 43–47.

Brygadyrenko, V. V. (2016). Influence of litter thickness on the structure of litter macrofauna of deciduous forests of Ukraine's steppe zoneю. *Visnyk of Dnipropetrovsk University. Biology, Ecology, 24*(1), 240–248. https://doi.org/10.15421/011630.

Butt, K. R., & Lowe, C. N. (2011). Controlled cultivation of endogeic and anecic Earthworms. In A. Karaca (Ed.), *Biology of Earthworms (Soil Biology 24)* (pp. 107–121). Berlin Heidelberg: Springer.

Chapra, C. S. (2012). *Applied numerical methods with MATLAB® for engineers and scientists*. New York: McGraw-Hill.

Choosai, C., Jouquet, P., Hanboonsong, Y., et al. (2010). Effects of earthworms on soil properties and rice production in the rainfed paddy fields of Northeast Thailand. *Applied Soil Ecology, 3*(45), 298–303.

Cooke, J. A., & Johnson, M. S. (2002). Ecological restoration of land with particular reference to the mining of metals and industrial minerals: A review of theory and practice. *Environmental Reviews, 10*(1), 41–71.

Didur, O., Loza, I., Kul'bachko, Y. (2011). Environmental impact of excretorial activity of earthworms (Lumbricidae) on the buffering capacity of remediated soils. In: Proceeding of NATO ARW "Environmental and Food Security and Safety in Southeast Europe and Ukraine", Dnipropetrovsk, 16–19 May 2011.

Didur, O., Loza, I., Kul'bachko, Y., et al. (2013). Environmental impact of Earthworm (Lumbricidae) excretory activity on pH-buffering capacity of remediated soil. *Visnyk of Dnipropetrovsk University. Biology, Ecology, 62*, 140–145.

Eisenhauer, N. (2010). The action of an animal ecosystem engineer: Identification of the main mechanisms of earthworm impacts on soil microarthropods. *Pedobiologia, 53*(6), 343–352. https://doi.org/10.1016/j.pedobi.2010.04.003.

Gamkalo, Z. G. (2005). Role of organic fertilizer in optimization of the acid-base properties of gray forest soils of Western forest-steppe zone of Ukraine. *Agronomical Chemistry and Soil Science, 66*, 53–58.

Ilyin, V. B. (1995). Estimation of soils buffering capacity to heavy metal contamination. *Agrochemistry, 10*, 109–113.

Jachimko, B. (2012). The influence of lignite mining on water quality. In K. Voudouris & D. Voutsa (Eds.), *Water quality monitoring and assessment* (pp. 373–390). Croatia: Publisher InTech. https://doi.org/10.5772/32897.

Jones, C. G., Lawton, J. H., & Shachak, M. (1994). Organisms as ecosystem engineers. *Oikos, 69*, 373–386.

Kul'bachko, Y., Loza, I., Pakhomov, O., et al. (2014). Tropho-methabolic activity of earthworms (Lumbricidae) as zoogenic factor maintaining the stability of remediated soil against copper contamination. *Visnyk of Dnipropetrovsk University. Biology, Ecology, 22*(2), 104–109.

Kul'bachko, Y., Loza, I., Pakhomov, O., et al. (2011). The zooecological remediation of technogen faulted soil in the industrial region of the Ukraine Steppe Zone. In M. Behnassi, A. S. Shahid, & J. D'Silva (Eds.), *Sustainable agricultural development: Recent approaches in resources management and environmentally-balanced production enhancement* (pp. 115–123). Dordrecht: Springer.

Lavelle, P., Decaëns, T., Aubert, M., et al. (2006). Soil invertebrates and ecosystem services. *European Journal of Soil Biology, 42*(1), 3–15.

Milcu, A., Partsch, S., Langel, R., et al. (2006). The response of decomposers (earthworms, springtails and microorganisms) to variations in species and functional group diversity of plants. *Oikos, 112*(3), 513–524.

Mtui, G. Y. S., Mligo, C., Mutakyahwa, M. K. D., et al. (2006). Vegetation structure and heavy metal uptake by plants in the mining-impacted and non mining-impacted sites of Southern Lake Victoria wetlands. *Tanzania Journal of Science, 32*(2), 39–49.

Novitskiy, M. L. (2011). Granulometric composition of fine soil of sulfide solids and man-made substrates of mine piles. *Bulletin of the Nikitsky Botanical Garden, 103*, 85–87.

Orlov, D. S. (1994). *Soil ecological monitoring*. Moscow: MSU.

Orlov, D. S., Sadovnikova, L. K., & Sukhanova, L. I. (2005). *Soil chemistry*. Moscow: Higher Sch.

Pakhomov, A. E., Kulbachko, Y. L., & Didur, O. A. (2009). Study of ecological interrelations of bigeminate-legged millipeds (Diplopoda) and artificial mixed soils as their habitat in experimental conditions. In I. Apostol, D. L. Barry, W. G. Coldewey, & D. W. G. Reimer (Eds.), *Optimization of disaster forecasting and prevention measures in the context of human and social dynamics* (pp. 163–171). Amsterdam: IOS Press.

Pampura, T. V., Pinsky, D. L., Ostroumov, V. E., et al. (1993). Experimental study of buffering capacity of soil at copper and zinc contamination. *Eurasian Soil Science+, 25*(10), 104–110.

Pecharová, E., & Hrabankova, M. (2006). A concept for reconstructing the post-mining region under the Lisbon strategy. *Ekológia, 25*(3), 194–205.

Pecharova, E., Hezina, T., Prochazka, J., et al. (2001). Restoration of spoil heaps in Northwestern Bohemia using wetlands. In J. Vymazal (Ed.), *Transformations of nutrients in natural and constructed wetlands* (pp. 129–142). Leiden: Backhuys Publishers.

Pecharová, E., Martis, M., Kašparová, I., et al. (2011). Environmental approach to methods of regeneration of disturbed landscapes. *Journal of Landscape Studies, 4*(2), 71–80.

Pokarzhevsky, A. D. (1985). *Geochemical ecology of terrestrial animals*. Moscow: Nauka.
Ripl, W., Pokorny, J., Eiseltova, M., et al. (1994). Holistic approach to structure the function of wetlands and their degradation. In M. Eiseltova (Ed.), *Restoration of lake ecosystems – A holistic approach* (pp. 16–35). Oxford: IWRB Publ.
Safonov, A. I. (2005). Phytogeochemistry of copper in man-made environment. *Problems of Ecology and Nature Protection of Technogenic Region, 5*, 68–74.
Singh, A. N., & Singh, J. S. (2006). Experiments on ecological restoration of coal mine spoil using native trees in a dry tropical environment, India: A synthesis. *New Forests, 31*(1), 25–39. https://doi.org/10.1007/s11056-004-6795-4.
Sklenicka, P., Prikryl, I., & Svoboda, I. (2004). Non-productive principles of landscape rehabilitation after long-term opencast mining in north-west Bohemia. *Journal-South African Institute of Mining and Metallurgy, 104*(2), 83–88.
Skousen, J., Sencindiver, J., Owens, K., et al. (1998). Physical properties of minesoils in West Virginia and their influence on wastewater treatment. *Journal of Environmental Quality, 27*(3), 633–639.
Striganova, B. R. (1980). *Feeding of soil saprophages*. Moscow: Nauka.
Strzyszcz, Z. (1996). Recultivation and landscaping in areas after brown-coal mining in Middle-East European countries. *Water Air and Soil Pollution, 91*, 145–157.
Thassitou, P. K., & Arvanitoyannis, I. S. (2001). Bioremediation: a novel approach to food waste management. *Trends in Food Science & Technology, 12*(5–6), 185–196.
Truskavetskiy, R. S. (2003). *Buffering capacity of soils and their main functions*. Kharkiv: New Word.
van Emden, H. (2008). *Statistics for terrified biologists*. Oxford: Blackwell Publishing.
Wang, Y., Dawson, R., Han, D., et al. (2001). Landscape ecological planning and design of degraded mining land. *Land Degradation and Development, 12*(5), 449–459.
Weiss, N. A. (2012). *Introductory statistics*. Boston: Addison-Wesley.
Zar, J. H. (2010). *Biostatistical analysis*. Upper Saddle River: Pearson Prentice-Hall.

Sergiy Nazimov is a junior researcher at the Laboratory of Biological Monitoring, Research Institute of Biology, Oles Honchar Dnipropetrovsk National University, Dnipropetrovsk, Ukraine.

Dr. Iryna Loza is a Senior Researcher at the Laboratory of Biological Monitoring, Research Institute of Biology, Oles Honchar Dnipropetrovsk National University, Dnipropetrovsk, Ukraine.

Yurii Kul'bachko is Doctor of Biological Sciences and Professor at the Faculty of Biology, Ecology and Medicine, Oles Honchar Dnipropetrovsk National University, Ukraine.

Oleg Didur PhD, Senior Researcher, Laboratory of Biological Monitoring, Research Institute of Biology, Oles Honchar Dnipropetrovsk National University, Dnipropetrovsk (Ukraine).

Dr. Oleksandr Pakhomov is Professor of Biological Sciences at the Faculty of Biology, Ecology and Medicine, Oles Honchar Dnipropetrovsk National University, Ukraine.

Angelina Kryuchkova is a Junior Researcher at the Laboratory of Biological Monitoring, Research Institute of Biology, Oles Honchar Dnipropetrovsk National University, Dnipropetrovsk, Ukraine.

Maria Shulman is a junior researcher at the Laboratory of Biological Monitoring, Research Institute of Biology, Oles Honchar Dnipropetrovsk National University, Dnipropetrovsk, Ukraine.

Tatiana Zamesova is a junior researcher at the Laboratory of Biological Monitoring, Research Institute of Biology, Oles Honchar Dnipropetrovsk National University, Dnipropetrovsk, Ukraine.

Part III
Forest Management from a Climate Change and Sustainability Perspective

Chapter 14
Understanding Stakeholders' Perspective on REDD+ Implementation as a Multi-Sectoral Approach

Himangana Gupta

14.1 Introduction

Until the Paris Agreement on Climate Change, which accorded equal importance for mitigation and adaptation, the main focus of the Kyoto Protocol to the UNFCCC remained on mitigation by reducing the emissions of carbon dioxide (CO_2), a major greenhouse gas (GHG), and its removal from the atmosphere through sequestration. The sequestration part was handled by Land Use, Land Use Change and Forestry (LULUCF). The only Kyoto mechanism dealing with this was the Clean Development Mechanism (CDM), which under the afforestation/reforestation scheme, allowed carbon credits to be earned for planting forests. Later, it came under criticism for discouraging the prevention of deforestation, while unwittingly encouraging actions that could result in the destruction of ecosystems and their associated biodiversity (Totten et al. 2003).

The 11th Conference of Parties (COP) to the UNFCCC at Montreal in 2005 considered a proposal by Papua New Guinea and Costa Rica on behalf of the Coalition of Rainforest Nations to compensate the developing countries for not cutting their forests and keeping them in pristine form. The proposal was added as an agenda item on Reducing Emissions from Deforestation (RED) in developing countries in order to enhance forest carbon stocks in developing countries (ENB 2005). The acronym became REDD when words forest degradation were added at COP-13 in Bali (COP-13 2008). Later it became REDD+ to account for conservation and sustainable management of forests.

Angelsen (2008) argues that REDD could help in making the overall GHG emission targets more ambitious without raising overall costs. According to the Secretariat of the Convention on Biological Diversity (CBD) (2009), it has the potential to deliver significant co-benefits for forest biodiversity if mechanisms are

H. Gupta (✉)
National Communication Cell, Ministry of Environment, Forest and Climate Change, Government of India, New Delhi, Delhi, India

M. Behnassi et al. (eds.), *Climate Change, Food Security and Natural Resource Management*, https://doi.org/10.1007/978-3-319-97091-2_14

designed appropriately. It should recognize the contribution of diverse forests, in particular primary forests, to long-term carbon sequestration/storage; consider the rights of indigenous and local communities; and address important forest governance issues, such as illegal logging.

Major efforts of the initiative are focused on the old growth forest ecosystems which are dominated by large-sized, slow-growing species and large, slow-moving carbon stocks (Díaz et al. 2009). Apart from storing carbon, trees and shrubs provide food, fuel, fodder, timber, and medicines for use and sale. They can restore degraded lands, protect watersheds, and provide green spaces for relaxation and culture (Kiarago 2014).

A core idea behind REDD+ is to make performance-based payments, that is, to pay forest owners and users to reduce emissions which can provide strong incentives directly to forest owners and users to manage forests better and clear less forestland (Angelsen et al. 2009). But, reducing deforestation is socially and politically costly (Karsenty 2008).

14.2 REDD+ Governance

REDD+ exemplifies how a scientifically informed policy idea permeates through multiple spheres of decision-making and organization, creates contested interests and claims, and translates into multiple implementation actions running ahead of policy processes and state-driven decisions (Corbera and Schroeder 2011). However, linking co-benefits to carbon benefits remains a significant methodological challenge, and prioritizing both in evaluation design is difficult (Caplow et al. 2011).

The shift towards REDD, and the increasing global regulation of timber trade and products, increasingly re-orients forest policies and regimes towards a more 'systemic' basis, where deforestation in various locations can be addressed by centralized and more uniform policy approaches (Forsyth and Sikor 2013).

One of the ways to enhance REDD+ outcomes is to select existing and new community forest management sites with user group and contextual characteristics associated with successful forest outcomes, which include a stable policy environment, low levels of intergroup conflict, and forest-dependent user groups that have management experience (Agrawal and Angelsen 2009).

Though the international architecture will set the framework for REDD implementation, the realization of co-benefits, particularly for poverty and equity will largely depend on the ways in which REDD incentive payments are translated into strategies for emissions reductions at the national level (Brown et al. 2008). Further, REDD effectiveness will be subject to participant countries' ability to address the underlying drivers of deforestation and liaise with the agents involved in land-use change (Corbera et al. 2010).

Therefore, for a successful market-based REDD mechanism, we need a collective capacity to agree upon a baseline which would either take the form of a reference period in the past or a scenario which could be used as a convincing projection of the future trends of deforestation (Karsenty 2008).

14.3 Possible Negative Impacts of REDD+ and Their Solution

Several negative aspects of REDD+ have been listed by the Secretariat of CBD (2009), which include: (i) Methodologies based on assessments of only net deforestation rates could have negative impacts on biodiversity; (ii) the use of net, rather than gross, deforestation rates could obscure the loss of mature forests by their replacement in situ or elsewhere with areas of new forest growth; (iii) significant losses of biodiversity could be accompanied by unrecorded emissions; and (iv) some high-biodiversity regions would not benefit from carbon-focused conservation, and could come under increased pressure if REDD is implemented.

Additional gains for biodiversity conservation are possible without compromising the effectiveness for climate change mitigation if REDD takes biodiversity distribution into account (Strassburg et al. 2010). The policy could leave out forest communities or harm them by undermining tenure rights, disempowering local decision making, limiting local livelihoods in the name of conservation, and promoting elite capture of lands and carbon payments (Larson 2011). Therefore, binding agreements to protect local rights may be needed.

Inclusion of REDD+ in a global compliance system will necessitate clearly-defined and allocated forest carbon property rights in the form of carbon credits with liability assigned for possible future carbon release into the atmosphere (Palmer 2011). The significance of command over complementary productive resources can undermine the ability of community-based carbon forestry projects to include marginalized stakeholders despite their inclusive aims. In the worst case, local elites may be able to capture financial and other benefits available through REDD+ actions (Forsyth and Sikor 2013).

Effective institutional arrangements to ensure continued equitable benefit distribution and prevent elite capture of community forestry resources become more important. Otherwise, the sustainability of carbon stored in community forests will be threatened by those who do not receive benefits or if local communities and forest-dependent poor users are excluded from REDD+ projects (Agrawal and Angelsen 2009).

Many REDD+ interventions are likely to be affected by poor governance and corruption, but Measuring, Reporting and Verification (MRV) mechanisms – both for carbon and financial flows – can also contribute to reducing corruption (Angelsen et al. 2009). As an example, in Papua New Guinea too much initial emphasis was placed on carbon accounting and valuation at the expense of community engagement which derailed REDD efforts. Also, the rush by some businesses, NGOs, researchers, and various political interests to establish pilot REDD projects and develop carbon markets has occurred with virtually no involvement or understanding of most of the forest communities (Melick 2010).

In India, a phased approach for REDD+ implementation, having safeguards for local communities and biodiversity along with a system of their reporting and capacity building, has to be developed. Success of such projects will depend on a rigid, scalable, and reliable finance mechanism, technological assistance, and effective forest-related legislation along with transparent and equitable political momentum with the support of core stakeholder groups (Sharma and Chaudhry 2013).

14.4 Forests in Climate Change Debate

From 1750 to 2011, CO_2 fossil fuel combustion and cement production released 375 GtC of GHGs to the atmosphere, while deforestation and other land use changes are estimated to have released 180 GtC (IPCC 2013). Land use change is responsible for annual emissions of only 1.0 ± 0.5 GtC, whereas the fossil fuel combustion and cement production emit 8.3 ± 0.4 GtC year^{-2} (Le Quéré et al. 2013).

The countries responsible for higher CO_2 in the atmosphere were reluctant to take the entire burden of reducing emissions, and wanted the developing countries not responsible for historical emissions to share the burden by bringing forestry initiatives into the equation. This has brought forests to center stage of global climate change debates.

Apart from delay in reduction of industrial emissions, a more positive reason for bringing forests into climate change debate is their inextricable linkage to human survival. Biodiversity in the forests can not only keep carbon reserves safe, but also provide adaptation mechanisms in future. Forests have the unique ability to reduce emissions by capturing and storing carbon, and at the same time reduce the climate vulnerability of people and ecosystems (Pandey et al. 2012; Pandey and Jha 2012). This distinctive character of forests to address both the mitigation and adaptation has focused the attention of scientists and policy makers on forests.

There are potential risks for biodiversity if REDD+ efforts are poorly designed. Some of them are: (i) the conversion of natural forests to plantations and other land uses of low biodiversity value; (ii) the introduction of growing of biofuel crops; (iii) the displacement of deforestation and forest degradation to areas of lower carbon value and high biodiversity value; and (iv) the afforestation in areas of high biodiversity value (Christophersen et al. 2011).

14.5 REDD Negotiations

The REDD concept derived from the discussion in the UNFCCC side events since COP-9 in 2003 under the labels of 'Avoided Deforestation', 'Compensated Reduction', 'Reducing Emission form Deforestation (RED)', etc. was brought on the agenda of COP-11 at Montreal in 2005 by Papua New Guinea and Costa Rica as on Reducing Emissions from Deforestation in Developing countries. Soon, the agenda came under strong criticism for its narrow and skewed focus.

India led the charge to correct the skewed focus with strong support from China. Other developing countries joined India on and off but also disembarked unannounced. Brazil was consistent and vehement in using every opportunity to stop an agreement. Coalition for Rainforest Nations (CFRN) led by Papua New Guinea played a positive role but diverse interests within its members made the process difficult. However, the coalition was able to hold together its members (Rawat 2011).

At COP-12 in Nairobi, India proposed that countries which have applied strong conservation measures be adequately compensated under the instrument of REDD. The policy approach presented by India was named *Compensated Conservation.* The proposal was intended to compensate the countries for maintaining and increasing their forests carbon pools backed by a verifiable monitoring system.

In the UNFCCC debates, REDD+ refers to 'Reducing Emissions from Deforestation and forest Degradation, and the role of conservation, sustainable management of forests and enhancement of forest carbon stocks in developing countries'. REDD+ was born at COP-13 in Bali as an expanded concept of REDD with the addition of words 'the role of conservation, sustainable management of forests and enhancement of forest carbon stocks' and was permanently embedded in COP decision 1/CP.13 also known as 'Bali Action Plan' (UNFCCC 2007). The achievement did not come without hard bargaining and haggling over intricacies of language.

India's contention throughout the negotiation process was that the 'conservation, sustainable management of forests and enhancement of forest carbon stocks' should be accorded the same level of priority as deforestation and forest degradation. This was achieved by replacing a semicolon (;) with a comma (,) in the original text. It may appear to be a trivial matter, but for the REDD+ proponents and close watchers of the debate, "replacement of ';' with a ',' was a watershed change" (The Economist 2008). COP-16 at Cancun in 2010 put a final seal of approval when five activities comprising REDD+ were explicitly defined in Cancun Agreements. The agreements asked the Parties to guard against the conversion of natural forests to forest plantations (UNFCCC 2011). The rights of indigenous peoples and local communities and the extent of their engagement has become the part of the text with the introduction of the concept of equitable distribution of incentives. Consensus for financing mechanism has moved towards acceptance of a hybrid model with market, non-market, and fund-based approaches embedded in it.

Warsaw framework on REDD+ was agreed during the COP-19 in 2013. It noted that livelihoods may be dependent on activities related to drivers of deforestation and forest degradation, and that addressing these drivers may have an economic cost and implications for domestic resources (UNFCCC 2013).

Subsequently, the COP-21 in Paris recognized the key role that resilient forests and landscapes played for both climate change and development. Germany, Norway and the United Kingdom promised annual support for REDD+ for countries that come forward with ambitious and high quality proposals, with an aim to provide US$1 billion per year by 2020, or over US$5 billion in the period 2015–2020 (World Bank 2015). Article 5 of the Paris Agreement encourages the Parties to take action to implement and support, including through results-based payments, the existing framework as set out in related guidance and decisions already agreed under the Convention for policy approaches and positive incentives for activities relating to reducing emissions from deforestation and forest degradation, and the role of conservation, sustainable management of forests and enhancement of forest carbon stocks in developing countries; and alternative policy approaches, such as joint mitigation and adaptation approaches for the integral and sustainable management of forests, while reaffirming the importance of incentivizing, as appropriate, non-carbon benefits associated with such approaches (UNFCCC 2015).

The CBD has recognized that if well designed and implemented, REDD+ can provide considerable benefits for biodiversity. COP-9 to the CBD invited Parties, Governments, and relevant international and other organizations to ensure that the actions under the initiative do not run counter to the objectives of the CBD (Decision IX/5), but support the programme of work on forest biodiversity, and provide biodiversity benefits to indigenous and local communities (Christophersen et al. 2011).

14.6 Drivers of Deforestation and Forest Degradation

Drivers of deforestation and forest degradation fall into two categories in the Indian perspective. The first category is planned deforestation for road and railway construction, mining, hydro-electric projects, industrial requirements, and expansion of cities. The government is in control of related decisions and actions since they are projected in accordance with policies, legal framework and management plans. Under the second category, the deforestation is unplanned and spontaneous, and is beyond government control (India 2011). Both planned (controlled) and unplanned (uncontrolled) withdrawals from forests affect the forest carbon stocks, besides compromising other services flowing from forest ecosystems. Unauthorized activities and natural causes – such as encroachment of forest land, uncontrolled felling, fuelwood, timber and NTFP extraction; unregulated livestock grazing; fodder collection; forest fires, insect attack, disease outbreak, and illegal mining operations – are unplanned drivers.

Minimizing the impacts of planned or controlled drivers is possible by introducing appropriate policy instruments and management options including effective legal framework and site-specific mitigation measures. Challenge lies in addressing and managing the unplanned or uncontrolled drivers and activities which are mainly a direct outcome of local people's dependence on the adjoining forest areas to meet their livelihood needs of fuelwood, grazing, fodder, and food supplements, etc., and to a small extent illegal mining activities within the forest.

According to India's submission to Subsidiary Body for Scientific and Technological Advice (SBSTA) of the UNFCCC (India 2011), deforestation and forest degradation can be assessed based on two types of data which may be available from the Forest Survey of India (FSI) report, the state forest departments, as well as from published research papers. To assess the total loss of forest carbon stocks in a given period and area, two components have to be considered:

- The carbon stock loss in areas that changed from forest land to other land uses in the respective period; and
- The reduction of average carbon stock in areas that remain forest land.

14.7 Methodology

Top scientists, NGOs, administrators, policy makers, and stakeholders were interviewed to critically analyze the promises and potential of the REDD+ initiative. The interviews were semi-structured with open-ended questions providing lengthy discussions.

The questions asked from experts were: do you think REDD+ is a promising mechanism? Will it really lead to a reduction in emissions? And what is its potential to deal with the issues of biodiversity and local communities' adaptation to climate change?

From the responses, five frames were identified. REDD+ is framed as: multi-benefit approach; biodiversity conservation; a pure carbon storage initiative; an unreal solution and a diversionary tactic; and corporate bonanza.

Following the assessment based on expert interviews, Rampur tehsil (an administrative sub-unit of a district) in Shimla district of the northern Indian state of Himachal Pradesh was selected for field visits and group discussions with the local people in the area. The USAID-funded pilot project, named USAID-India Forest Plus Project, is implemented at six locations in the Kotgarh, Ani, and Rampur forest divisions of Satluj landscape with the participation of Himachal Pradesh Forest Department and USAID (MoEF 2014). The Rampur tehsil has protected zones where tourism is restricted and cannot provide livelihood opportunities. Tree felling is prohibited and, therefore, people carry out agriculture in areas that are covered by grasses in the rainy season. People in Rampur did not have complete awareness of the project.

Six group discussions were conducted with six to seven people in each group. The area is rich in forests and maximum people are dependent on agriculture for their livelihoods. The assessment reveals that people are not very likely to understand REDD+ as a concept and do not expect anything from its implementation. The main questions put to them were:

- What problems do you anticipate in the working of the program?
- Would you participate in the project if it is implemented in your area?

These, and other supplementary questions that came up during the discussion, were designed to indirectly answer the following questions:

- Do the forest resources and services impact the social and economic capital of the rural poor?
- Does the delivery of forest services to rural community impact the conservation aspects of the forest?

The discussion with the people was open, focusing on their life, what they consider as important, and what the government needs to do for their needs. Two group discussions were carried out with the youngsters who are pursuing higher education. Two discussions were carried out in the city area of Rampur and the rest four were with the villagers and pastoralists of the region. The discussions linked the forest-biodiversity aspects, the protection of which is inherent among the people of the area, to their social and economic conditions. The main point noted during the discussions was that the people would participate in carbon mitigation and adaptation strategies only if they were benefitted directly or indirectly.

14.8 Expert Views and Assessment

Out of 24 experts approached for their views, only nine interviewees answered questions on REDD+. But they are some of the very few experts in the field and have enriched the discourse in this section. None framed it as 'corporate bonanza' because the corporate sector has not yet been involved in the REDD+ projects in India, unlike in many other countries and UN-REDD guidance on such projects. Therefore, the responses were grouped into four frames. The subscription to frames was not mutually exclusive: for instance, an expert who feels that REDD+ is a multiple benefit approach may also draw attention to possible negative impacts on local communities through maladaptation due to the projects.

It is clear from Table 14.1 that more scientists/experts consider REDD+ as an unreal solution and a diversionary tactic in order to delay taking emission reduction commitments. Only one considered it as a biodiversity conservation proposal. Two considered it as a multi-benefit approach and two considered it a purely carbon storage mechanism. In view of expert assessments, it can be inferred that REDD+ may not turn out to be a good mechanism achieving all the goals – carbon stock preservation, biodiversity conservation and community benefits. This is mainly due to lack of an efficient mechanism to incentivize multiple benefits. However, some experts believe that the initiative has a good future potential and can help in achieving objectives meant for long term sustainability.

Any positive mechanism proactively put in place to deal with reducing forest degradation and destruction would definitely yield results. But an overnight result should not be expected because it requires long-term commitment and investments to go in. There are a range of options for adaptation. But, if they do not connect with each other, it may become an agenda of climate change mitigation only and not the agenda of biodiversity.[1]

However, both REDD+ and CDM have been criticized by civil society organizations across the globe. "The developed countries have committed environmental sins and they want to minimize their effects by acquiring rights over the good work of the communities in the developing countries. There is hardly any sincere commitment for environmental protection or for reducing carbon footprint".[2]

Table 14.1 Number of experts subscribing to different frames

Frame	Number of responders
Multiple benefit approach	2
Unreal solution	4
Purely carbon storage	2
Biodiversity conservation	1

[1] Interview with Dr. Balakrishna Pisupati, former chairman of National Biodiversity Authority of India.

[2] Interview with Bikash Rath, Regional Centre for Development Cooperation (RCDC), Odisha, India.

"Both A/R CDM and REDD+ are unreal solutions to a real problem. Instead of supporting the natural forests, they help the polluters to get away with their green crimes in the guise of a very marginal green action".[3] Such projects are considered as gimmicks of rich countries: "They are aimed at serving the interests of multinationals as well as local industrial houses. Hence, they are not helpful for the local people".[4]

A top scientist and proponent of unreal and diversionary frame says that fossil fuels account for 83% of the emissions and only 17% are from the land use change. He questions the focus on deforestation with its low contribution. "These are delaying and diversionary tactics. The calculation, monitoring and verification of carbon sequestration projects is so complicated that at the end of the day we ourselves don't know whether there was carbon sequestration or not".[5]

Another scientist and India's lead negotiator on REDD+ thinks that the initiative is not being understood in the correct perspective. The new comprehensive approach is more appropriate than only focusing on deforestation and forest degradation. It can play a significant role in lowering the emissions and protecting biodiversity. "The countries which have controlled their deforestation and have started growing their forest resource are sequestering more carbon. If these countries were not doing this, then the emissions would have been much more in forestry sector than they are at present. So, those countries, which are adding to the forest cover, are sequestering more CO_2 from the atmosphere could be given carbon credits and be taken as a part of REDD strategy".[6]

An IPCC lead author for AR5 sees no clear guidelines on essential issues in the Warsaw framework:

> The biodiversity safeguards are mentioned but it does not say how they will be assessed and enforced. If you follow back to what is agreed, they have agreed respect for the knowledge and rights of indigenous people and members of local communities, consistency with the conservation of natural forests and biological diversity ensuring that the actions that support REDD are not used for the conversion of natural forests, but are instead used to incentivize the protection and conservation of natural forests and their ecosystem services, which sounds good but they do not mention a mechanism. There is no agreed definition of forests. If one country decides that eucalyptus plantations are fine, then that seems to be okay. It is not clear where the money is going to come from. If it is going to come from the carbon markets, there will be a huge amount of money, but nobody is interested in biodiversity. If it is just going to come from voluntary payments from governments, there isn't going to be very much money. So, neither is very clear.[7]

---

[3] Interview with Ranjan Panda, Water Initiatives Odisha, India.

[4] Interview with Prof Binayak Rath, Professor, National Institute of Scientific Education and Research (NISER), Bhubaneshwar, Odisha, India.

[5] Interview with Dr. Govindaswami Bala, Professor, Centre for Atmospheric and Oceanic Sciences, Indian Institute of Science, Bangalore, India. Lead author, Carbon Cycle, IPCC WGI, AR5.

[6] Interview with Dr. Jagsish Kishwan, Chief Advisor-Policy, Wildlife Trust of India, Uttar Pradesh, India. India's lead negotiator on REDD+.

[7] Interview with Prof. Richard Corlett, Director, Center for Integrative Conservation, Xishuangbanna Tropical Botanical Garden, Yunnan, China. Lead author, Asia, in IPCC WGII, AR 5.

14.9 Project Area

The Rampur circle of district Shimla has five forest divisions – Rampur, Kotgarh, Kinnour, CAT Plan, and Ani. Maximum people in the villages are associated with agriculture related activities. In spite of this, villagers do not cut natural forests illegally for their agricultural needs. Since the CBD believes that involving indigenous peoples and local communities is key to the success of REDD+ (Christophersen et al. 2011), the views of stakeholders were collected through group discussions to bring to fore issues considered important by them and the possibilities of REDD implementation and the potential of such a regime in these areas.

14.9.1 General Observations

Area gets good rainfall, so there are grass-covered areas where the local people replace those grasses with horticultural crops for their livelihoods. The vegetables grown there are sold by farmers in the Rampur city market. However, they do not touch the forests for agricultural needs mainly because forests fall under the protected zone. Another source of livelihood for the people is livestock and pastoralism.

The people of the area are in tune with the ecosystem and are actually acting as guardians of forests by living on meager amounts of vegetation produce and some Non-Timber Forest Produce (NTFP), unlike other Indian states such as Punjab, Haryana and Uttar Pradesh, where natural vegetation has been removed to accommodate agricultural needs. Urbanization has increased in the past few years which has improved the living standards of the people. The area has started to attract people from Kinnaur and Shimla regions to settle there as there are more schools, colleges and hospitals in the Rampur city. Roads and other infrastructure are well built and transport facilities are good. Taken together, these infrastructures increase the socio-economic status of the people. The city is situated in the valley, which is a blessing for people who live in higher reaches as they can travel to Rampur to avoid harsh winters.

However, it lags behind other neighboring regions in agriculture. For instance, Shimla and Kinnaur have many apple orchards and the sale of the harvest is enough for the complete year and for extra activities even if there is no other source of income. However, this forest-rich area cannot support apple trees due to higher temperatures. However, in some areas, plums and pears are grown. In the interior villages, people depend on growing and selling vegetables. There are forest dwelling communities also. The people said that there was less money or no money at all from tourism in the area as people usually visit higher reaches like Shimla and Kinnaur rather than Rampur. However, a temple of *Bhim Kali* in the region is very famous and is the only source of tourism in the area.

14.9.2 Stakeholder Views

The group discussions with stakeholders threw up mixed responses to the REDD+ regime. Some people said they would be happy to be involved with such a program, and others said it was difficult to understand the scheme and it may not bring benefits at the end. They collectively believed that any project should not displace people and threaten their livelihoods or infringe on their rights in any way.

Another outlook was that if some useful native plants are grown in degraded areas adjoining the forests, they could receive benefits of NTFP and it would be of immense use if they have medicinal value. If the plants are not useful, they may not take part in such initiatives. Their main fear is that in the hope of getting something at a later stage, they may lose what they already have. The REDD+ concept is unclear to them, especially its role in village communities. A well-educated interviewee from the area said that a lot of time and energy would be required to explain the complexities of the projects to the village people. "We have all that is needed for survival and people from outside may destroy our present livelihood options". Some people of the area, therefore, were reluctant to take up such initiatives as the forests are already protected and they could not receive benefits through such projects. Rather, they feared infringement of their present rights.

The people see no threat to the forests from anthropogenic activities, but the region is prone to forest fires. Fires are sometimes natural but burning of old grasses to clear up some area or for growth of new grasses spreads too far and causes forest fires. In fact, some areas had dry grasses only with sparse vegetation.

The potential for growing new and useful native species in the area is immense. However, forest fires need to be tackled quickly to avoid direct carbon emissions and destruction of habitat.

14.9.3 Loss of Trust

Industries are being quickly set up in the region, which are using the natural resources of the mountain area without any benefit for the local people. In a nearby district, Kinnaur, people allege that a prominent industrial house 'Jaypee Group', engaged in river valley and hydro-power projects, has cut protected forests. They say that the industrialists are pumping money to government officials for setting up such projects, which has led to increased corruption. No new forests are grown in return for so much of deforestation. Only 50–60 Rubinia trees have been planted in lower regions under the compensatory reforestation scheme of the government of India. This is one of the reasons why people have lost trust on both private and public initiatives. People have learned from the examples of bad incidents happening to their near and dear ones in the neighboring districts which have created an atmosphere of fear and loss of trust among the locals.

There have been several agitations against such projects in the Kinnaur region and people have been imprisoned, but the government takes no action. Too many hydel projects have destroyed the biodiversity of the mountains. A new hydel project recently approved by the central government is about to be set up. However, the locals of *Poo* village in the Kinnaur district near the Indo-Tibet border have protested against this. The local people are being promised money and land in lieu of setting up this project. People have refused to take anything in return of their own land where they have grown apple trees which is a constant source of income. People have lived in harmony with nature but private corporate interests are destroying their natural survival mechanisms.

Outsiders, who set up projects in the area, are not aware of local conditions, and hence destroy the environment. "We are the ones which are well aware of the region's plant diversity and its requirements. Outsiders just see their own benefit and leave us out", said one local. The people of *Kilba* village in Kinnaur district complain that some industrial projects have destroyed their lands and their apple trees do not flower properly because of too much dust in the atmosphere.

Heavy drilling by Jaypee has increased incidences of landslides in the region. The agitations have not moved the government. "The hydel projects have been set up to give electricity to the big cities and we are left out. Our lands are taken but we get nothing in return", a villager complained. The result is that local communities' adaptation mechanisms are losing efficacy and their livelihoods have been challenged. Everyday, there is a danger of losing something that was their own.

Stakeholders' responses show that they feel threatened by some ongoing projects that are not linked to REDD+ in any way, but makes them hesitant towards any new initiative which they think may take that shape. They have been protecting forests and benefitting from them for a long time and they would not like to participate in such programs. Under such circumstances, the potential of REDD+ will remain minimum until these people are brought into the picture. They are not enthused by the potential carbon credits for such activities.

14.10 Conclusion

The analysis of REDD+ negotiations and debates brings out two dimensions: the policy perspective and the technological perspective. The policy perspective will help in further fine-tuning the policy and its implementation. The technological perspective will be helpful in measurement activities such as monitoring, reporting and verifiable (MRV) activities, including the setting up of reference level and development of safeguards that fulfill the intended objectives of the REDD+ activities.

A detailed perspective and views of the top scientists and experts point towards the fact that a lot still needs to be done in incentivizing implementation, enhancing safeguards, bringing out multiple benefits, and setting up social mechanisms. Some deem it as a good policy for both climate change and biodiversity but fear that climate change issues may hijack the agenda rather than biodiversity issues. Others consider it as a diversionary tactic. And since there is a lack of mechanism for finance and safeguards, this policy may take time to pick up.

Discussions with stakeholders in a forest-rich Rampur region of Himachal Pradesh show that people may be unwilling to accept such a mechanism doubting the forest conservation practices, especially by private companies. Under such a situation, it is extremely difficult to create trust among people that the projects will be helpful to them. However, pilot projects, if implemented in consonance with the local people, assuring good benefits and increased resilience of their forests, and hence their livelihood, may win their trust. Such model projects can be used as an example elsewhere.

Acknowledgements This work was carried out as a part of my PhD research in Panjab University, Chandigarh, India. I am, therefore, thankful to the University Grants Commission (UGC), New Delhi, India for providing funding support in the form of Senior Research Fellowship at the time of PhD. I also thank Mr. Raj Kumar Gupta (Environmental and Social Policy Analyst), Dr. Rajiv Pandey (Scientist-E, Indian Council of Forestry Research and Education), and the interviewees for their valuable inputs.

Disclaimer The views expressed in this chapter are those of the author and do not reflect the official policy or position of any department of the Government of India.

References

Agrawal, A., & Angelsen, A. (2009). *Using community forest management to achieve REDD+ goals. Realising REDD Natl. Strategy Policy Options* (pp. 201–212). Bogor: Center for International Forestry Research (CIFOR).

Angelsen, A. (2008). *Moving ahead with REDD: issues, options and implications*. Bogor: Center for International Forestry Research.

Angelsen, A., Brockhaus, M., Sills, E., et al. (2009). *Realising REDD+: National strategy and policy options*. Bogor: Center for International Forestry Research.

Brown, D., Seymour, F., & Peskett, L. (2008). *How do we achieve REDD co-benefits and avoid doing harm? Mov. Ahead REDD Issues Options Implic* (pp. 107–156). Bogor: Center for International Forestry Research.

Caplow, S., Jagger, P., Lawlor, K., & Sills, E. (2011). Evaluating land use and livelihood impacts of early forest carbon projects: Lessons for learning about REDD+. *Environmental Science & Policy, 14*, 152–167. https://doi.org/10.1016/j.envsci.2010.10.003.

Christophersen, T., Stahl, J., & Secretariat of CBD. (2011). *REDD-plus and biodiversity. Secretariat of the Convention on Biological Diversity*. Quebec: Montreal.

COP-13. (2008). *Report of the Conference of the Parties on its thirteenth session*, Held in Bali from 3 to 15 December 2007. UNFCCC.

Corbera, E., & Schroeder, H. (2011). Governing and implementing REDD+. *Environmental Science & Policy, 14*, 89–99. https://doi.org/10.1016/j.envsci.2010.11.002.
Corbera, E., Estrada, M., & Brown, K. (2010). Reducing greenhouse gas emissions from deforestation and forest degradation in developing countries: Revisiting the assumptions. *Climatic Change, 100*, 355–388. https://doi.org/10.1007/s10584-009-9773-1.
Díaz, S., Hector, A., & Wardle, D. A. (2009). Biodiversity in forest carbon sequestration initiatives: not just a side benefit. *Current Opinion in Environment Sustainability, 1*, 55–60. https://doi.org/10.1016/j.cosust.2009.08.001.
ENB. (2005). *Summary of the eleventh conference of the parties to the UN framework convention on climate change and first meeting of the parties to the Kyoto Protocol*. 28 November–10 December 2005. International Institute for Sustainable Development (IISD).
Forsyth, T., & Sikor, T. (2013). Forests, development and the globalisation of justice. *The Geographical Journal, 179*, 114–121. https://doi.org/10.1111/geoj.12006.
India. (2011). *Submission by India to SBSTA, UNFCCC for SBSTA Agenda item 4*. Government of India: Ministry of Environment and Forests.
IPCC. (2013). In T. F. Stocker, D. Qin, G.-K. Plattner, M. Tignor, S. K. Allen, J. Boschung, A. Nauels, Y. Xia, V. Bex, & P. M. Midgley (Eds.), *Climate Change 2013: The physical science basis. Contribution of working group I to the fifth assessment report of the intergovernmental panel on climate change*. Cambridge: Cambridge University Press.
Karsenty, A. (2008). The architecture of proposed REDD schemes after Bali: facing critical choices. *International Forestry Review, 10*(3), 443–457.
Kiarago, H. (2014). *The value of trees in landscapes goes way beyond carbon*. Transformations.
Larson, A. M. (2011). Forest tenure reform in the age of climate change: Lessons for REDD+. *Global Environmental Change, 21*, 540–549. https://doi.org/10.1016/j.gloenvcha.2010.11.008.
Le Quéré, C., Andres, R. J., Boden, T., et al. (2013). The global carbon budget 1959–2011. *Earth System Science Data, 5*, 165–185. https://doi.org/10.5194/essd-5-165-2013.
Melick, D. (2010). Credibility of REDD and experiences from Papua New Guinea. *Conservation Biology, 24*, 359–361. https://doi.org/10.1111/j.1523-1739.2010.01471.x.
MoEF. (2014). *Sustainable management of forests as per agreed methodology for REDD+*. Government of India: Ministry of Environment and Forests.
Palmer, C. (2011). Property rights and liability for deforestation under REDD+: Implications for "permanence" in policy design. *Ecological Economics, 70*, 571–576. https://doi.org/10.1016/j.ecolecon.2010.10.011.
Pandey, R., & Jha, S. (2012). Climate vulnerability index – measure of climate change vulnerability to communities: A case of rural Lower Himalaya, India. *Mitigation and Adaptation Strategies for Global Change, 17*, 487–506. https://doi.org/10.1007/s11027-011-9338-2.
Pandey, R., Rawat, G. S., & Kishwan, J. (2012). Carbon balance assessment in Indian Himalayan managed forests: Analysis of anthropogenic extractions for domestic combustion vis-a-vis accretions of forest biomass. *SAARC Forestry, 1*, 132–153.
Rawat, V. R. S. (2011). *REDD Plus in India: From negotiations to implementation*. Pre-Congress workshop of 1st Indian forests congress for Theme: Forest and Climate Change, Shimla.
Secretariat of CBD. (2009). *Biodiversity and climate change action*. Secretariat of the Convention on Biological Diversity.
Sharma, V., & Chaudhry, S. (2013). An overview of Indian forestry sector with REDD. *International School of Research Notice, 2013*, e298735. https://doi.org/10.1155/2013/298735.
Strassburg, B. B. N., Kelly, A., Balmford, A., et al. (2010). Global congruence of carbon storage and biodiversity in terrestrial ecosystems. *Conservation Letters, 3*, 98–105. https://doi.org/10.1111/j.1755-263X.2009.00092.x.
The Economist. (2008). Fiddling with words as the world melts. *The Economist*.
Totten, M., Pandya, S. I., & Janson-Smith, T. (2003). Biodiversity, climate, and the Kyoto Protocol: risks and opportunities. *Frontiers in Ecology and the Environment, 1*, 262–270. https://doi.org/10.1890/1540-9295(2003)001[0262:BCATKP]2.0.CO;2.
UNFCCC. (2007). *Bali Action Plan*. Bonn: UNFCCC.

UNFCCC. (2011). *Report of the conference of the parties on its sixteenth session*, Held in Cancun from 29 November to 10 December 2010.

UNFCCC. (2013). *Addressing the drivers of deforestation and forest degradation.*

UNFCCC. (2015). *Report of the Conference of Parties on its twenty-first session*, Held in Paris from 30 November to 13 December 2015. United Nations, Paris.

World Bank. (2015). Outcomes from COP21: Forests as a key climate and development solution. *World Bank*. http://www.worldbank.org/en/news/feature/2015/12/18/outcomes-from-cop21-forests-as-a-key-climate-and-development-solution. Accessed 4 Dec 2016.

Himangana Gupta, Ph.D., is an expert in climate change and biodiversity policy and diplomacy. In her present job, she is coordinating with scientists and climate experts to compile the Biennial Update Reports and Third National Communication to the UNFCCC. She is a doctorate in environment science with specialization in climate change and biodiversity policy and was a University Gold medallist in masters. She has written research papers in reputed international and national journals on current state of climate negotiations, forestry, industrial efficiency, rural livelihoods and women in climate change mitigation and adaptation.

Chapter 15
Conserving Carbon and Biodiversity Through REDD+ Implementation in Tropical Countries

Lokesh Chandra Dube

15.1 Introduction

Agriculture, Forestry and Other Land Use (AFOLU) sector contributes to around 24% of total global greenhouse gas (GHG) emissions from anthropogenic sources, dominated by tropical deforestation. Between the era from1750 to 2011, about 180 PgC was released to the atmosphere due to land use change, mainly deforestation (IPCC 2014). Table 15.1 shows rate of net emissions from forests (gross deforestation adjusted for forest regrowth) for different time periods.

The world has experienced about 0.2% decrease every year in forest areas between 1990 and 2005, corresponding to 13 million ha per year (FAO 2006). Carbon stocks in forest biomass decreased by about 5.5% at the global levels during this period (FAO 2007). According to FAO (2015), the carbon stocks in forest biomass decreased by almost 17.4 Gt, equivalent to a reduction of 697 million tonnes per year or about 2.5 Gt of carbon dioxide (CO_2) between 1990 and 2015. According to Kindermann et al. (2006), 200 Mha or around 5% of forest area will be lost between 2006 and 2025, resulting in a release of additional 17.5 GtC. Forest cover will shrink by around 500 million hectares, which is 1/8 of the current forest cover, within the next 100 years. The accumulated carbon release during the next 100 years would amount to 45 GtC, which is 15% of the total carbon stored in forests.

Tropical rainforests are spread on about 7% of earth's land area but around half of the world's biodiversity is present in these forests (Corlett and Primack 2011). Tropical ecosystems have the highest species richness. Deforestation and forest degradation are leading the loss of biodiversity in tropical regions. Anthropogenic disturbance in the form of deforestation in the tropical forests can double biodiversity

L. C. Dube (✉)
NATCOM Project Management Unit, Ministry of Environment, Forest and Climate Change, New Delhi, India

TERI University, New Delhi, India

M. Behnassi et al. (eds.), *Climate Change, Food Security and Natural Resource Management*, https://doi.org/10.1007/978-3-319-97091-2_15

Table 15.1 Rate of carbon loss from forests

Period	Rate of carbon loss from forest (PgC per year)	Region	Source
1980–1990	~0.9	Tropical Asia	Houghton (2003)
1990–2000	1.1 ± 0.5	Tropical Asia	
1990–2000	2.2 ± 0.8	Global	
1980–1990	~1.4	Global	Denman et al. (2007)
1990–2000	~1.6	Global	
2000–2005	~0.81	Tropical forests	Harris et al. (2012)
2000–2010	~1.0	Tropical forests	Baccini et al. (2012)
2002–2011	~0.9	Tropical forests	Stocker et al. (2013)

loss (Barlow et al. 2016). However, the biodiversity loss can be substantially reduced by prompt and stringent mitigation actions (Warren et al. 2013).

In this chapter, an assessment of carbon market and biodiversity market has been made to find out the possibilities of integrating climate change mitigation and conserving biodiversity through the group of activities collectively known as 'reducing emissions from deforestation and forest degradation, conservation of forest carbon stocks, sustainable management of forest and enhancement of forest carbon stocks (REDD+)'. The objectives of this assessment include:

- Examining Biodiversity – REDD+ interrelations and present an overview of potential biodiversity markets on parallels of carbon markets
- Presenting a snapshot of financing REDD+ implementation in tropical countries
- Estimating emission reduction potential for selected tropical countries

15.2 Materials and Methods

Different proposals and approaches pertaining to REDD+ implementation in tropical forests, as available in literature were reviewed and evaluated. Status of available Forest Reference Levels (FRL)/Forest Reference Emission Levels (FREL) was compiled. On the basis of evaluation and analyses, conclusions were drawn and policy oriented synthesis approach has been proposed in the discussion section. No detailed analysis of 'additionality' aspects has been carried out. Standards and/or approaches were reviewed including UN REDD+, Aichi Targets under CBD, New York Declaration on Forests, REDD under Verified Carbon Standards (VCS), Climate Community Biodiversity Alliance (CCBA) standards, and the Gold Standard.

15.3 Convergence in Financing Mechanisms for REDD+

15.3.1 Climate Change and Biodiversity: Science and Policy Interrelations

Forests act as sink for CO_2 and also as natural habitat for plants and animal species. Climate change and biodiversity loss are the two major threats closely linked with forests. Forests are influenced by these threats, but at the same time, also have potential to influence climate and biodiversity. Recognizing the three aspects – Forests, Climate and Biodiversity, United Nations Conference on Environment and Development (UNCED), in 1992, adopted the Statement of Forest Principles, the United Nations Framework Convention on Climate Change (UNFCCC), the United Nations Convention on Biological Diversity (CBD) and the United Nations Convention to Combat Desertification (UNCCD). Linkages between these agreements were not specified. Later, in 2001, a Joint Liaison Group (JLG) was formed including Secretariats of UNFCCC, CBD and UNCCD to work on interrelations and linkages. So far 14 meetings of JLG have taken place, latest one being in August, 2016. Current issues under discussion in JLG include contribution of the Rio Conventions towards achieving the 2030 Agenda for Sustainable Development, Ecosystem-based approaches (including ecosystem restoration) in Land Degradation Neutrality and Nationally Determined Contributions along with Common indicators among the Rio Conventions and the SDG indicator framework (UNFCCC et al. 2016). The recent activities of JLG indicate that it can play an important role in linking the climate and biodiversity mechanisms for synergistic impact.

World's forests and wetlands roughly contain 50% of global carbon. Deforestation and degradation of natural forests is a major cause of human induced greenhouse gas emissions and loss of biodiversity. Climate influences species distribution pattern and stability of climatic patterns promotes biodiversity (Robertson and Chan 2011). With changing climate, the biological diversity of tropical forests is facing threats of depletion. According to CBD (2009), for every 1 °C increase in mean global temperature, about 10% of species will be at a greater risk of extinction. Since the problems of deforestation and biodiversity depletion are interconnected in their cause and effect, the solution should also be obtained in an integrated manner. Jantz et al. (2014) found that protected area corridors not only facilitate habitat connectivity, but also help realizing the benefits of climate change mitigation in tropics.

Natural forests, when overexploited, create a potential threat of climate change and a real loss to biodiversity. Climate change further accelerates depletion of biodiversity which in turn influences climatic changes. Hence, conservation of forest

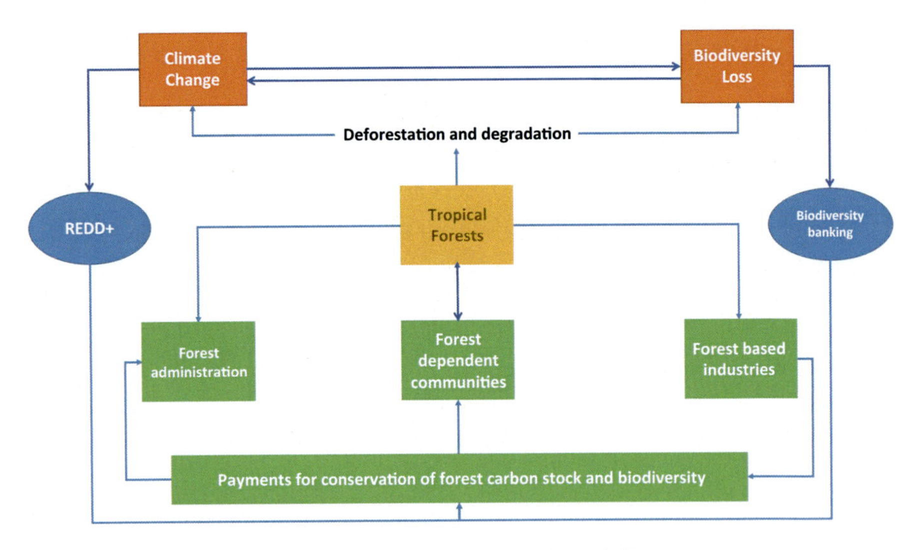

Fig. 15.1 Deforestation, climate change and biodiversity interrelations

carbon stock and sustainable management of forest habitats are the key factors that can be effectively used to minimize the adverse impacts of global climate change. REDD+ and biodiversity banking are two important tools to address the twin challenges. Stakeholders in operationalizing payments for conservation of forest carbon stock and biodiversity include forest administration, forest dependent communities and forest based industries. Forest based industries will compensate forest dependent communities and forest administration for conserving carbon stock and biodiversity. An indicative framework of interactions is provided in Fig. 15.1.

Tropical forests and certain other ecosystems such as wetlands and grasslands are not only major stores of carbon but also serve as the areas of rich biodiversity. Thus, conserving these areas would deliver both, carbon as well as biodiversity benefits (Koziell and Swingland 2002). In the past attempts have been made to integrate the carbon and biodiversity components in forest conservation, management and enhancement projects. Under UNFCCC, prevention of potential harm to biodiversity is included as one of the safeguards related to REDD+ implementation.

15.3.2 International Mechanisms to Finance REDD+

Stern (2007) identified curbing deforestation as a highly cost-effective way of reducing GHG emissions and having the potential to offer significant reductions fairly quickly. It was also identified that avoiding deforestation helps preserve biodiversity and protect soil and water quality. The review made by Stern serves as a blueprint for the actions to avoid deforestation aiming at climate change mitigation. Kindermann et al. (2008) described avoidance of deforestation as competitive and

Table 15.2 Status of REDD+ finance

Type of funding/donor	Total financial pledge/investment reported in millions USD
Bilateral	4981
Multilateral	3227
Other multilateral (multiple channels)	53
Unknown	463
Private foundations	101
Private sector	1000
Total	9825

Source: Norman and Nakhooda (2015)

low cost abatement option. According to Strassburg et al. (2009), 80% of avoided deforestation costs less than USD5.00 per ton of carbon dioxide (CO_2).

By establishing the Global Forest Fund in March 2007, the Australian Government has devoted USD 160 Millions to help reduce deforestation in the Asia-Pacific region (Neeff et al. 2007). The Forest Carbon Partnership Facility (FCPF) was launched by the World Bank in 2007. FCPF has created two distinct funds- the Readiness Fund and the Carbon Fund to promote results based payments for scalable REDD+ activities in developing countries. The FCPF influences private players to facilitate countries build up approach for avoiding forest degradation through tradable carbon credits. The Carbon Fund is operational since 2011.

Currently, REDD+ finance is concentrated on emissions and forest loss suggesting that availability of finance for conservation of forest carbon stock is limited (Wolosin et al. 2016). Broadly three kinds of financial support is available for REDD+: readiness, implementation, and results-based payments (Lee and Pistorius 2015).

A total of global REDD+ financing is of the order of USD 9.8 Billion. Status of REDD+ finance as of 2014/2015 is given in Table 15.2.

According to Parker (2014), scale of REDD+ finance was of the order of USD 15 billion, 90% of which came from public finance. Parker (2014) also indicated that strategies to provide long term finance support for REDD+ are inadequate.

Streak (2016) commented that "advanced economies so far failed to create 'adequate and predictable' demand or finance for verified emission reductions from REDD+" and suggested three options for international mitigation actions namely: General Mitigation Target, International Mitigation Target (as part of a Dual Target), and REDD+ Target (Dual Target limited to REDD+).

Market based mechanisms of conserving forest carbon stocks and protecting biodiversity are in different stages of development. Conservation of forest carbon stocks are eligible for results based payments towards REDD+ under the UNFCCC. Biodiversity is identified as a major non-carbon benefit of REDD+. So far the emphasis has been on results based payments to support REDD+; alternative approaches such as non-market approaches are also under development. Mechanisms on conservation banking and biodiversity crediting have also been proposed and implemented in some parts of the world. Historical trends suggest that Voluntary Carbon Market has offered premium to the credits generated for carbon forestry projects with biodiversity co-benefits, but looking at the interrelated nature and need of integrated solution, these measures are not sufficient.

15.3.3 REDD+ in International Carbon Market

International carbon market can be categorized into two structures. First is the compliance market, also known as regulatory market, governed by international, national or sub-national regulations such as Kyoto Protocol, European Union Emissions Trading Scheme, Regional Greenhouse Gas Initiatives in USA and New South Wales GHG Abatement Scheme in Australia. Second structure of the carbon market is voluntary market; segments of voluntary carbon market are government purchasing programs like Australian Government's Greenhouse Challenge Plus program and the retail market that includes the activities of companies and individuals who wish to offset their GHG emissions arising from activities, products or services. Retail entities or individuals often purchase Voluntary Emission Reductions (VERs) in small quantities. These VERs are usually not intended for compliance. Deforestation avoidance is not an eligible activity to earn carbon credits under Clean Development Mechanism (CDM) of Kyoto Protocol and most of the Non- Kyoto type compliance markets. Voluntary market players like Verified Carbon Standard and Social Carbon allow avoided deforestation as eligible activity for voluntary carbon credit generation. It should be noted that some of the first deals in the voluntary carbon market were for avoided deforestation projects (Hamilton et al. 2007).

The Chicago Climate Exchange was a voluntary greenhouse gas emission cap and trade scheme (discontinued in 2010) located in North America. Chicago Climate Exchange's Forestry Offset Projects for the member parties were under three categories: forestation and forest enrichment, combined forestation and forest conservation and urban tree planting. Forest conservation credits were provided for combined forestation and forest conservation projects in specified locations if the two activities occurred on contiguous sites. Crediting was quantified on the basis of avoided deforestation rates specified for eligible geographic regions.

Observations from voluntary carbon market (Hamrick and Goldstein 2015) suggest that Land Use, Land Use Change and Forestry (LULUCF) projects have accounted for more than 50% share in these markets by volume of trade in 2014. In the voluntary offset market trade, avoided deforestation projects and Afforestation/Reforestation projects fetched average prices of USD 5.2/tCO_{2e} and USD 7.7/tCO_{2e} respectively, between 2007 and 2014.

15.3.4 Status of REDD+ Mechanism Under UNFCCC

'Reducing emissions from deforestation and forest degradation, conservation of forest carbon stocks, sustainable management of forest and enhancement of forest carbon stocks' are collectively termed as REDD+. The REDD+ originated as concept of avoided deforestation and was named as reducing emissions from deforestation in developing countries (REDD). It was recognized that the developing countries face problem of forest degradation which is also a contributor to greenhouse gas emissions, hence 'reducing emission from forest degradation' was also added to REDD.

Santilli et al. (2003) presented a proposal called 'compensated reduction' to include deforestation avoidance in tropical countries under the Kyoto Protocol. The proposal includes a voluntary, national deforestation stabilization and reduction target for non-Annex I countries as its main aspect. Forneri et al. (2006) argued that a separate protocol for forest based actions may be the most viable option, as it could offer the necessary flexibility and avoid some technical and political pitfalls that would be likely to beset new efforts under the Kyoto Protocol. Kurg (2007) proposed incentives for emission reduction from base emissions and gave detailed method and steps for the same. He advocated completeness to be taken into account that means gross emissions from deforestation to be calculated rather that net emissions (avoidance of deforestation minus sequestration from afforestation/reforestation). India, as a Party to UNFCCC proposed the concept of 'compensated conservation' of forest carbon stocks.

Years of intense negotiations on the possible pathways to reduce emissions culminated as the inclusion of ".....role of conservation, sustainable management of forests and enhancement of forest carbon stock...." to REDD in Bali Action Plan during COP 13 in 2007. The overall approach is now referred to as REDD-plus (also written as REDD+). Thus, REDD+ in present from, as it exists has resulted from a synthesis of the two approaches of 'compensated conservation' and 'compensated reduction'.

The 'Bali Action Plan' of the UNFCCC mandated Parties to negotiate a post-2012 instrument to provide financial incentives for the mitigation of climate change from forest actions in developing countries. Subsequently, discussions and deliberations among country governments took place to further elaborate and take necessary actions in this regard. This debate finally led to the acceptance of the full range of REDD+ issues in COP-15 at Copenhagen in 2009. A considerable progress has been made since then to include issues of financing for REDD+ in international negotiations, most recent being the adoption of the Paris Agreement under UNFCCC which includes positive incentives and results-based payments for REDD+ activities under its article 5.1 and 5.2 (UNFCCC 2015). The Agreement, through its Article 6.2 and 6.4, also defines a mechanism of internationally transferable mitigation outcomes towards nationally determined contributions. The mechanism still needs details on operationalization.

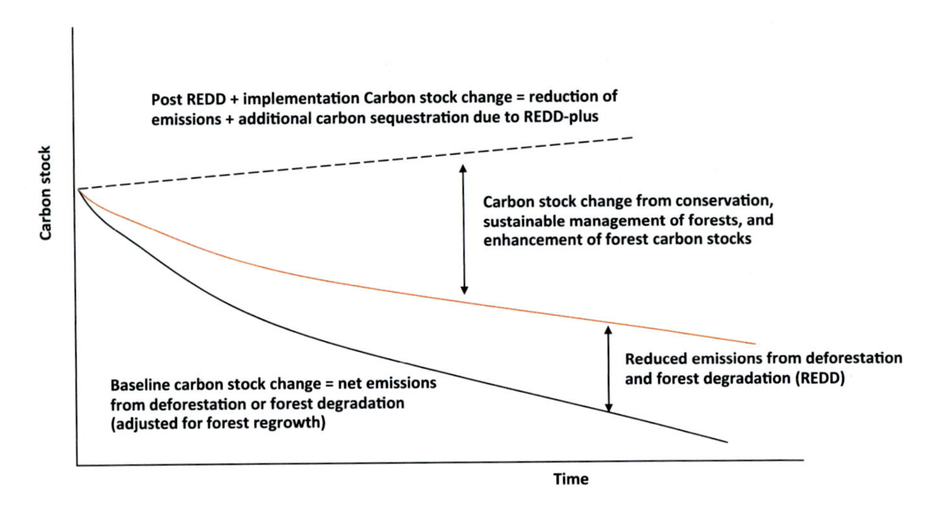

Fig. 15.2 Conceptual representation of REDD+ implementation

Figure 15.2 depicts conceptual representation of carbon stock change due to REDD+ implementation. The figure shows that the baseline carbon stock decreases over time due to net emissions from deforestation and forest degradation. Some amount of emission is avoided by activities that reduce deforestation and degradation. A net positive change is seen with increase in carbon stock through interventions that tend to conserve and sustainably manage forests and add to the sequestration levels by means of afforestation, reforestation and revegetation.

To access the benefits of REDD+, a developing country needs to specify its Forest Reference Level (FRL) and/or Forest Reference Emission Level (FREL). FREL is understood as gross carbon emissions from deforestation and forest degradation while FRL includes net removals from forest regrowth (adjusted to emissions). Having adopted the modalities to develop FRELs and FRLs through its decision 12/CP.17, the COP invited developing countries to propose their FREL and/or FRL, on a voluntary basis. Countries agreed that reference levels should be determined at national levels considering country-specific circumstances, and yet reviewed by UNFCCC. Conceptually, FRELs/FRLs are performance benchmarks against which emission reductions and removals will be measured, reported and verified (Kissinger et al. 2012). These proposed FRELs and FRLs undergo a technical assessment process which is conducted by a pool of experts appointed by UNFCCC. Table 15.3 provides summary of key features, assumptions and methodology of the FRL/FREL proposed as part of their respective Biennial Update Reports (BURs). It may be noted that all these countries (except Chile) fall in tropical region.

Table 15.3 Summary of the main features of FRL/FREL

Year of submission	Countries proposing FREL/FRL	Type and duration of FREL/FRL	FREL/FRL $MtCO_2eq$/year	Activities	Pools	National/ sub-national
2014	Brazil	FREL for 1996–2005 and 1996–2010	1106.03 (2006–2010)	Reducing emissions from deforestation	AB, BB and L	Sub-national
			907.96 (2011–2015)			
2015	Colombia	FREL based on average historical emissions for 2000–2012	51.59	Reducing emissions from deforestation	AB and BB	Sub-national
	Ecuador	FREL = historical emissions for 2000–2008	43.42	Reducing emissions from deforestation	AB, BB, DW and L	National
	Guyana	Combined reference level approach	46.30	Reducing emissions from deforestation	AB, BB and DW	National
				Reducing emissions from forest degradation		
	Malaysia	FRL based on historical emissions for 1992–2005 and 1997–2010	−183.55 (2006–2010)	Sustainable management of forests	AB and BB	National covering permanent reserved forests only
			−197.83 (2011–2015)			
	Mexico	FREL = average annual CO_2 emissions for 2000–2010	44.39	Reducing emissions from deforestation	AB and BB	National
2016	Chile, Congo, Costa Rica, Ethiopia, Indonesia, Paraguay, Peru, Vietnam, Zambia	Technical assessment of Biennial Update Reports in progress.				

Source: (UNFCCC 2016)
AB Aboveground biomass, *BB* Belowground biomass, *L* Litter, *DW* Deadwood

15.3.5 REDD+ in International Biodiversity Market

The carbon market is already in operation with international trading of offsets while markets for biodiversity conservation are emerging in a wide variety of forms around the globe. Legally mandated biodiversity offsets have been initiated in Australia, Brazil, Canada, EU and the USA. Potential of voluntary offsets is accelerated by some companies which have made public commitments to implement biodiversity offsets linked to their 'footprints'. Biodiversity offsets are conservation activities intended to compensate for the residual, unavoidable harm to biodiversity caused by economic development projects.

Conservation banking (CB) in the USA is perhaps the most developed example of a market for biodiversity mitigation (Koziell and Swingland 2002). According to US Fish and Wildlife Service, conservation banks are permanently protected privately or publicly owned lands that are managed for endangered, threatened, and other at-risk species. A conservation bank is a storehouse of biological diversity and resources. Instead of currency and financial assets, the bank-owner has habitat or species credits to trade. Creation of biodiversity banks involves acquiring existing habitat and protecting it through conservation easements (a voluntary legal agreement between a landowner and an agency for facilitating conservation on the land), restoration or enhancements of disturbed habitat, creation of new habitat in some situations and prescriptive management of habitats for specified biological characteristics. Under the U.S. Fish and Wildlife Service Endangered Species Program,[1] each biodiversity credit may be equivalent to an appropriate unit of indicators of conservation of species at risk. Few examples include, unit area of natural habitat for a particular species; the habitat required to support a breeding pair; and a wetland unit along with its supporting uplands. Methods of determining available credits may rely on ranking or weighting of habitats based on habitat condition, size of the parcel, or other factors. A conservation bank may have more than one type of credit if more than one listed species or habitat type occurs at the bank.

According to Madsen et al. (2011), the global biodiversity market size is at least USD 2.4–4.0 billion annually, most of which is concentrated in North America, more precisely in USA. With support from biodiversity market, at least 187,000 ha of land is brought annually under some kind of conservation management or permanent legal protection.

Parties to the CBD, at COP-10 in 2010, adopted vide decision X/2, a revised and updated Strategic Plan for Biodiversity 2011–2020. The plan included the Aichi Biodiversity Targets for the 2011–2020 period, comprising a set of five strategic goals, each including several targets aiming at protecting and enhancing forest biodiversity. Some of these targets can be directly linked to REDD+.

[1] https://www.fws.gov/ventura/docs/hc/conservationbanks.pdf

15.3.6 Convergence of Carbon and Biodiversity Markets

There are many possibilities of conserving biodiversity and sustainably using biological resources through investing in carbon mitigation or offsets. One increasingly attractive option is to use carbon finance to tackle deforestation in the tropics. Another possibility is creating carbon offsets in biodiversity-friendly agriculture through no-till cultivation. Ideas of congregating international carbon and biodiversity policies and markets through interconnected activities such as REDD+ have been contemplated by many authors such as Bekessy and Wintle (2008), Swingland et al. (2003), Dube and Sen (2009), Phelps et al. (2012a), (b).

A research by Engel (2014) suggests that successful implementation of REDD+ could have synergistic effects on successfully reaching the Aichi Biodiversity Targets. Similarly, implementation of Aichi Biodiversity targets, could support successful implementation of REDD+ activities. Panfil and Harvey (2015) studied 80 REDD+ projects under implementation and found that all 80 REDD+ projects described biodiversity conservation goals. They opine that measurement of biodiversity impacts will be limited to the extent goals are clearly specified and logical links exist between these goals, project interventions and monitoring arrangements. Similar views were expressed by Milesa et al. (undated) in UN-REDD Policy Brief Issue #05. The brief suggests that “joint planning for REDD+ implementation and achievement of the CBD Aichi Targets could help countries to develop cost-effective and complementary approaches to climate change mitigation and biodiversity conservation”.

UNFCCC through its decision 18/CP.21 clarified that methodological issues related to non-carbon benefits resulting from the implementation of the REDD+ activities do not form pre-requisite for developing countries seeking to receive support or results-based payments, however, countries that want to integrate these benefits into REDD+ are encouraged to share the relevant information with UNFCCC.

EBEX21 (Emissions-Biodiversity Exchange for the twenty-first century) is a scheme in New Zealand created by Crown Research Institute to help initiating New Zealand carbon farmers create high quality carbon credits. Under EBEX21, eligible land is set to revert back to native forest through a change in land use – there are no soil carbon emissions associated with the creation of EBEX credits. Harvesting is prohibited in these forests by covenanting which makes the status of forest permanent on these lands.

One major challenge in convergence will be to strengthen the links between biodiversity performance indicators and carbon metrics and standards. The Climate, Community and Biodiversity Alliance (CCBA) is tackling this challenge by developing standards for evaluating land-based carbon projects. These standards aim to identify land-based climate change mitigation projects that simultaneously generate climate, biodiversity and sustainable development benefits. The CCB Project Design Standards comprises the required criteria and optional ‘point-scoring’ criteria for carbon sequestration projects. Once a project has been designed, a third-party evaluator uses standard indicators to determine which criteria are satisfied. Projects that use best practices and deliver significant climate, community and biodiversity benefits earn CCB approval. Gold status is awarded by CCBS to exceptionally designed projects that go beyond the basic requirements (e.g. projects that use primarily native species, enhance water and soil resources, build community capacity, and adapt to climate change and climate variability or deliver net positive biodiversity impacts). A separate Social and Biodiversity Impact Assessment Manual is in place for REDD+ Projects under CCB standards.

15.4 REDD+ Beyond UNFCCC and CBD: Emission Reduction Potential

Through the New York Declaration on Forests (2014), a non-legally binding political agreement under the Aegis of United Nations, a coalition of governments, businesses, civil society and indigenous leaders pledged to halve deforestation by 2020 and strive to end it completely by 2030. The declaration states that achieving these outcomes could reduce emissions by 4.5–8.8 billion tonnes per year by 2030. Further, it recognizes the role of scaling up of payments for Verified Emission Reductions (UN 2014).

Tropical developing countries in relation to New York Declaration on Forests (NYDF) may be categorized as:

1. Countries whose national governments have signed the NYDF: A total of 36 national governments have signed the Declaration. These nations include both, developed and developing economies, tropical and non-tropical climates. Indonesia as a developing country in tropical region has been chosen for estimation.
2. Countries whose sub-national governments have signed the NYDF: A total of 20 sub-national governments endorsed the Declaration. Brazil and Nigeria have been chosen for analysis as these represent tropical developing countries whose national governments have not signed the Declaration, but, some of their sub-national governments have endorsed it. Data of these countries were subjected to analysis to examine country-wide potential of emission reduction, assuming NYDF pledge is expanded to the forests in entire jurisdiction of the country.
3. Countries which are not part of the NYDF: Considering data availability, Cambodia, Myanmar and Zimbabwe were chosen for estimating the emission reduction potential that lies outside the NYDF. It was hypothetically assumed as if the commitments of Declaration were applicable to these countries.

All six selected countries are facing carbon loss (i.e. negative rate of change of carbon stock in living biomass) from deforestation and forest degradation. For estimating emission reduction potential, targets for deforestation reduction in these countries have been assumed based on their historic deforestation rates. These rates are available from local/national/global assessments. FAO (2015) has given rate of

Table 15.4 Carbon stock change in selected tropical developing countries

Country	Carbon stock change in living forest biomass (1990–2015) (kilotonne/year)	Annual change in carbon stock (%)	Carbon Stock in 2015 (million tonne)
Brazil	−199,620	−0.3	59,222
Cambodia	−7040	−1.4	433
Indonesia	−195,640	−1.3	12,488
Myanmar	−19,549	−1.1	1551
Nigeria	−47,240	−3.5	835
Zimbabwe	−10,195	−1.8	442

change of carbon stock in living forest biomass for the period 1990–2015 for countries of the world. Summarized in Table 15.4 are the rates of carbon stock change for the selected tropical countries.

Following two scenarios have been developed:

1. Business as usual (BAU) scenario: Future carbon stocks have been projected at the current rate of carbon stock change, for the period between 2015 and 2030. Rate of carbon stock change is a function of deforestation and forest degradation rate. Annual change in carbon stock (%) applied in this scenario are same as shown in Table 15.4 and remain static throughout the period of analysis (2015–2030). This scenario represents proxy for baseline.
2. End of Deforestation (EOD) scenario: In line with New York declaration, emission reduction target of 50% by the year 2020 was assumed. This translates to average annual reduction of 87% in the rate of carbon stock change for the period of 2015–2020. Further emission reduction target of 100% by the year 2030 (ending deforestation by 2030) translates to average annual reduction of 57% in the rate of carbon stock change for the period between 2020 and 2030. The average annualized reduction rates were derived so that the desired levels of reductions are met for years 2020 and 2030. It may be noted that negative values and rates of carbon stock change represent net emissions.

Difference between the carbon stock estimates of two scenarios is emission reduction potential. Country specific results are shown in Fig. 15.3.

The country-wise graphs show the carbon stock saved (or emissions reduced) from slowing down and stopping deforestation in six tropical countries for the period 2015–2030. It may be noted that even if deforestation is completely stopped by 2030, the carbon stock will still be lower as compared to 2015 levels. Forest regrowth from sustainable management of forests and other means would help recover the potential carbon losses. Reaching these targets would need international support on finance and technology. This international support can come from various avenues of REDD+ finance including market based mechanisms such as sale of voluntary emission reductions in international carbon market.

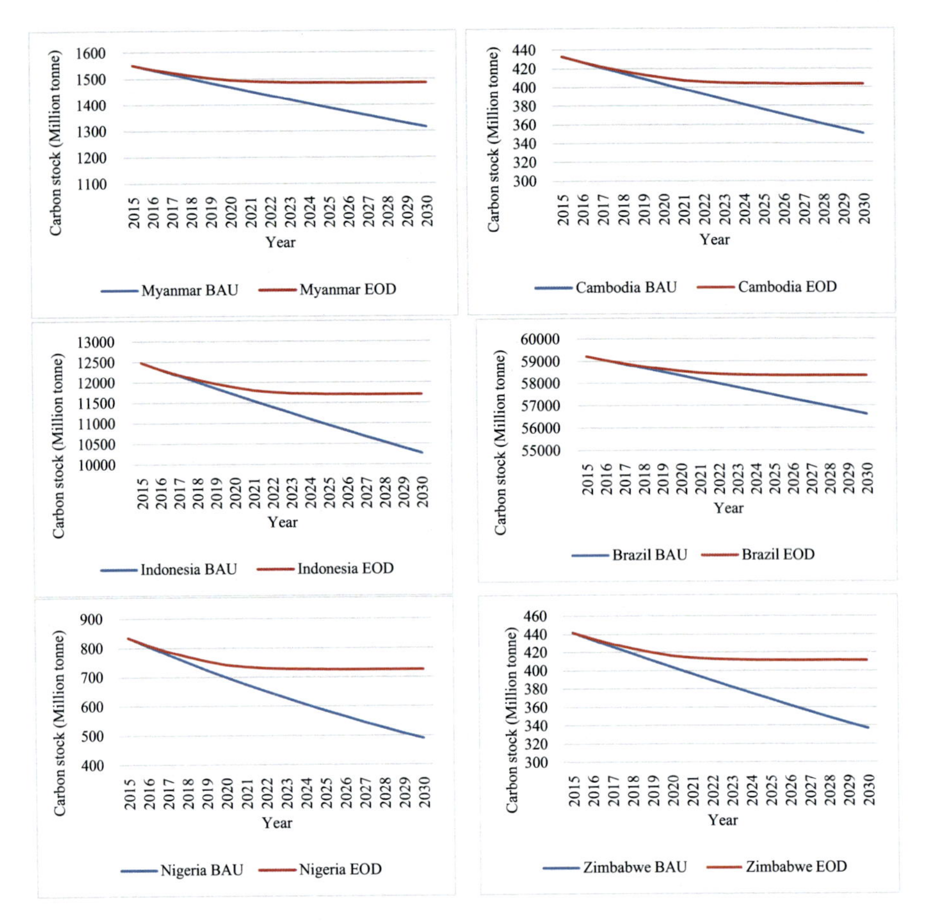

Fig. 15.3 Carbon stock change for selected tropical countries in BAU and EOD scenarios

15.5 Discussion and Conclusions

From the above review of literature and analysis of data, it is concluded that an integrated approach would be required to tackle the problem of deforestation and biodiversity loss. This would require a composite of proactive approach in the form of forest conservation, a curative approach in form of enhancement and sustainable management of forests including afforestation, reforestation, assisted natural regeneration and a preventive approach in the form of activities that reduce deforestation and forest degradation.

Under the New York Declaration, countries have taken collective voluntary deforestation reduction targets. Individual country level targets can be decided from business as usual rate of deforestation in a base year based on historical emission rates. These targets need to be chosen by the country itself so as to become self-committed objective. Deforestation reduction measures may be financed at the planning stage itself by means of forward sale of VERs (or Verified Deforestation Reductions- VDRs) or afterwards by getting the VERs verified and then selling those in voluntary/offset market. The reduction units (VERs/VDRs) may be linked to carbon stock of forest/biomass content of forest or a combined unit indexed on the basis of carbon-biodiversity-livelihood and other services. Countries may also be interested in enacting a policy decision/legislation under the framework of voluntary deforestation reduction target. This policy/legislation can be referred to as F+ policy i.e. those intend to increase forest sinks and reduce deforestation. On the basis of their impact on climate, these F+ policies, in principle, are analogous to E- policies referred in the CDM terminology i.e. those intend to reduce emissions. Implications of these policy decisions may be difficult to monitor as such, while it will be easier to monitor the individual project/activity.

If strong international support is demonstrated for REDD+ projects generating high levels of carbon and biodiversity benefits, developing country Parties to the Paris Agreement will be motivated to explore additional opportunities of including forest related voluntary targets in their updated Nationally Determined Contributions (NDCs).

Carbon has been evolved as a global environmental commodity whereas owing to inherent complexities of biological systems, it will not be easy to develop a biodiversity trading system. A full-fledged market system for international trade in biodiversity credits comparable on scales of carbon markets are still remote, but an opportunity of developing biodiversity offsets as a new business sector at national, sub-national and corporate levels exists in developing countries of tropical region. Integration of elements from instruments available within UN system (such as UNFCCC, UNCCD and CBD) and initiatives outside UN system (such as voluntary carbon markets and conservation banking) can create a hybrid system of carbon-biodiversity credit trading and other financial mechanisms to achieve conservation of carbon and biodiversity through REDD+.

Disclaimer Views presented in this article are author's personal views and do not represent the position of NATCOM project or other departments of the Government of India in whatsoever manner.

References

Baccini, A., Goetz, S. J., Walker, W. S., et al. (2012). Estimated carbon dioxide emissions from tropical deforestation improved by carbon-density maps. *Nature Climate Change, 2*, 182–185. https://doi.org/10.1038/nclimate1354.

Barlow, J., Lennox, G. D., Ferreira, J., et al. (2016). Anthropogenic disturbance in tropical forests can double biodiversity loss from deforestation. *Nature, 535*, 144–147.

Bekessy, S. A., & Wintle, B. A. (2008). Using Carbon Investment to Grow the Biodiversity Bank. *Conservation Biology, 22*, 510–513. https://doi.org/10.1111/j.1523-1739.2008.00943.x.

CBD. (2009). *Biodiversity and climate action.*

Corlett, R. T., & Primack, R. B. (2011). Many tropical rain forests. In: *Tropical rain forests* (pp. 1–31). Wiley.

Denman, K. L., Brasseur, G., Chidthaisong, A., et al. (2007). *Couplings between changes in the climate system and biogeochemistry*. New York: Cambridge University Press.

Dube, L. C., & Sen, A. (2009). *Avoided deforestation coupled with biodiversity banking* (p. 118). Raipur: VRM Foundation.
Engel, A. (2014). Forest interactions between the CBD and UNFCCC An analysis of forest-related institutional interactions and proactive interaction management between the biodiversity and climate change regimes. MSc Thesis, Wageningen University.
FAO. (2006). *Global forest resources assessment 2005: Progress towards sustainable forest management*. Rome: FAO.
FAO. (2007). *State of the world's forests*. Rome: FAO.
FAO. (2015). *Global forest resources assessment 2015. How are the world's forests changing?*
Forneri, C., Blaser, J., Jotzo, F., & Robledo, C. (2006). Keeping the forest for the climate's sake: avoiding deforestation in developing countries under the UNFCCC. *Climate Policy, 6*, 275–294. https://doi.org/10.1080/14693062.2006.9685602.
Hamilton, K., Ricardo, B., Guy, T., Douglas, H. (2007). *State of the voluntary carbon market 2007 picking up steam*. New Carbon Finance, a service of New Energy Finance Ltd, and Ecosystem Marketplace.
Hamrick, K., Goldstein, A. (2015). *AHEAD OF THE CURVE: State of the voluntary carbon markets 2015*. Forest Trends Ecosystem Marketplace. P. 3.
Harris, N. L., Brown, S., Hagen, S. C., et al. (2012). Baseline map of carbon emissions from deforestation in tropical regions. *Science, 336*, 1573. https://doi.org/10.1126/science.1217962.
Houghton, R. A. (2003). Revised estimates of the annual net flux of carbon to the atmosphere from changes in land use and land management 1850–2000. *Tellus Series B, 55*, 378–390. https://doi.org/10.1034/j.1600-0889.2003.01450.x.
IPCC. (2014). In Core Writing Team, R. K. Pachauri, & L. A. Meyer (Eds.), *Climate change 2014: Synthesis report. Contribution of working groups I, II and III to the fifth assessment report of the intergovernmental panel on climate change*. Geneva: IPCC 151 pp. IPCC.
Jantz, P., Goetz, S., & Laporte, N. (2014). Carbon stock corridors to mitigate climate change and promote biodiversity in the tropics. *Nature Climate Change, 4*, 138–142.
Kindermann, G. E., Obersteiner, M., Rametsteiner, E., & McCallum, I. (2006). Predicting the deforestation-trend under different carbon-prices. *Carbon Balance and Management, 1*, 15. https://doi.org/10.1186/1750-0680-1-15.
Kindermann, G., Obersteiner, M., Sohngen, B., et al. (2008). Global cost estimates of reducing carbon emissions through avoided deforestation. *Proceedings of the National Academy of Sciences, 105*, 10302–10307. https://doi.org/10.1073/pnas.0710616105.
Kissinger, G., Herold, M., de Sy, V. (2012). *Drivers of deforestation and forest degradation: A synthesis report for REDD+ policymakers*. 48p.
Koziell, I., & Swingland, I. R. (2002). Collateral biodiversity benefits associated with "free–market" approaches to sustainable land use and forestry activities. *Philosophical Transactions of the Royal Society of London Series A, Mathematical, Physical and Engineering Sciences, 360*, 1807. https://doi.org/10.1098/rsta.2002.1033.
Kurg, T. (2007). *Positive incentives for reducing emissions from deforestation*. Cairns: National Institute for Space Research – INPE, Inter-American Institute for Global Change Research – IAI.
Milesa, L., Kate, T., Matea, O., et al (undated). *REDD+ and the 2020 Aichi biodiversity targets promoting synergies in international forest conservation efforts*.
Lee D., & Pistorius T. (2015). *The impacts of international REDD+ finance*.
Madsen, B., Becca, N., Nathaniel, C., et al. (2011). *Update: State of biodiversity markets*. Washington, DC: Forest Trends.
Neeff, T., Eichler, L., et al. (2007). *Updates on markets for forestry offsets*. The FORMA project, CATIE.
Norman, M., & Nakhooda, S. (2015). *The state of REDD+ finance*.
Panfil, S. N., Harvey, C. A. (2015). *REDD+ and biodiversity conservation: A review of the biodiversity goals, monitoring methods, and impacts of 80 REDD+ projects*.
Parker, C. (2014). *Overview of REDD+ financing landscape, sources and types of funds*.

Phelps, J., Friess, D. A., & Webb, E. L. (2012a). Win–win REDD+ approaches belie carbon–biodiversity trade-offs. *REDD Conserv, 154*, 53–60. https://doi.org/10.1016/j.biocon.2011.12.031.
Phelps, J., Webb, E. L., & Adams, W. M. (2012b). Biodiversity co-benefits of policies to reduce forest-carbon emissions. *Nature Climate Change, 2*, 497–503. https://doi.org/10.1038/nclimate1462.
Robertson, J. M., & Chan, L. M. (2011). Species richness in a tropical biodiversity hotspot. *Journal of Biogeography, 38*, 2043–2044. https://doi.org/10.1111/j.1365-2699.2011.02619.x.
Santilli, M., Moutinho, P., Schwartzman, S., et al. (2003). *Tropical de-forestation and the Kyoto protocol: A new proposal*. Milan.
Stern, N. H. (2007). *The economics of climate change: The Stern review*. Cambridge, UK: Cambridge University Press.
Stocker, T. F., Qin, D., Plattner, L. V., Alexander, S. K., Allen, N. L., Bindoff, F.-M., Bréon, J. A., Church, U., Cubasch, S., Emori, P., Forster, P., Friedlingstein, N., Gillett, J. M., Gregory, D. L., Hartmann, E., Jansen, B., Kirtman, R., Knutti, K., Krishna Kumar, P., Lemke, J., Marotzke, V., Masson-Delmotte, G. A., Meehl, I. I., Mokhov, S., Piao, V., Ramaswamy, D., Randall, M., Rhein, M., Rojas, C., Sabine, D., Shindell, L. D., Talley, D. G., Vaughan, & Xie, S.-P. (2013). Technical summary. In T. F. Stocker, D. Qin, G.-K. Plattner, M. Tignor, S. K. Allen, J. Boschung, A. Nauels, Y. Xia, V. Bex, & P. M. Midgley (Eds.), *Climate change 2013: The physical science basis. Contribution of working group I to the fifth assessment report of the intergovernmental panel on climate change*. Cambridge, UK: Cambridge University Press.
Strassburg, B., Turner, R. K., Fisher, B., et al. (2009). Reducing emissions from deforestation – the "combined incentives" mechanism and empirical simulations. *Tradit Peoples Climate Change, 19*, 265–278. https://doi.org/10.1016/j.gloenvcha.2008.11.004.
Streck, C. (2016). Mobilizing finance for + after Paris. *Journal of European Environment Plan Law, 13*, 146–166. https://doi.org/10.1163/18760104-01302003.
Swingland, I. R., Bankoff, G., Frerks, G., Hilhorst, D., Royal Society Staff, et al. (2003). *Capturing carbon and conserving biodiversity: The market approach*. New York: Routledge, Florence: Taylor & Francis Group [Distributor].
UN. (2014). *New York declaration on forests (in forests: Action statements and action plans)*.
UNFCCC. (2015). *Paris agreement*.
UNFCCC. (2016). *Technical assessment process for proposed forest reference emission levels and/or forest reference levels submitted by developing country parties*.
UNFCCC, UNCCD, CBD. (2016). *Report of the fourteenth meeting of the Joint Liaison Group of the Rio Conventions*. Bonn.
Warren, R., VanDerWal, J., Price, J., et al. (2013). Quantifying the benefit of early climate change mitigation in avoiding biodiversity loss. *Nature Climate Change, 3*, 678–682.
Wolosin, M., Breitfeller, J., & Schaap, B. (2016). *The geography of REDD+ finance deforestation, emissions, and the targeting of forest conservation finance*. Washington, DC: Forest Trends.

Lokesh Chandra Dube is a Programme Officer with the National Communications Cell at the Ministry of Environment, Forest and Climate Change, Government of India. He was instrumental in preparing India's first Biennial Update Report to UNFCCC. Earlier, as a Consultant at Emergent Ventures India, he developed a Framework for CDM Program of Activities in renewable energy and authored a new CDM methodology approved by the UNFCCC. As GHG Auditor with TÜV NORD group, he validated/ verified several mitigation projects. Lokesh is also a PhD candidate at TERI School of Advanced Studies, New Delhi.

Chapter 16
Carbon Dynamics at Harvard Forest: Ecological Responses to Changes in the Growing Season

Lauren Kathleen Sanchez

16.1 Introduction

Forest ecosystems occupy over one third of the terrestrial land area, covering over 4.1 billion hectares worldwide (Dixon et al. 1994). The earth's forest ecosystems have high ecological, social, and economic values as they greatly affect the abiotic conditions and biotic communities around them (Perry 1994). Globally, forests provide water filtration and soil erosion control, cleaning our water supply in protected forests and urban communities. Forests also provide habitat for other species, promoting plant and animal biodiversity around the world (Seymour and Hunter 1999). One of the most critical ecosystem services provided by forests is gas and nutrient cycling. Nitrogen, phosphorus, and sulfur cycle through forests and their soils as they interact with organic matter and other minerals.

Forests are critical participants in complex biogeochemical cycles such as the global carbon cycle. Due to observed increases in atmospheric CO_2, the carbon cycle has recently become a focus of ecological and atmospheric research (Schimel et al. 2000). Carbon occurs in all organic life and is a crucial element to the global ecosystem. As atmospheric CO_2 levels continue to rise, climate warming is projected to increase in severity and result in ecological ramifications globally. A deeper understanding of the carbon cycle, including its pools and fluxes, is needed in order to accurately address the issue of increased carbon in the atmosphere and develop potential solutions.

L. K. Sanchez (✉)
Yale University School of Forestry and Environmental Studies, New Haven, CT, USA
e-mail: lauren.sanchez@yale.edu

M. Behnassi et al. (eds.), *Climate Change, Food Security and Natural Resource Management*, https://doi.org/10.1007/978-3-319-97091-2_16

16.1.1 Global Carbon Cycle

Carbon is stored on earth in five main pools in the following order of magnitude: the lithosphere, the oceans, the soil, the atmosphere, and the biosphere (all organisms, living or dead). Terrestrial exchanges and feedbacks allow for transfer between these pools, or carbon fluxes (Schimel 1995). Primary production and changing land use are the predominant fluxes between the atmosphere and the biosphere, while decomposition is the primary flux between the soil and the atmosphere. Carbon fluxes are dynamic, varying on seasonal, annual, and decadal scales (Schimel 1995). Thus, continuous carbon monitoring is needed for a more accurate portrayal of the cycle and its pools. Long-term data on carbon fluxes and pools may provide insight into how carbon sinks and sources have changed over time, as well as improving predictions of their future trajectories.

Difficulty arises in quantifying carbon pools due to the complex biological processes that are involved and the spatial heterogeneity of vegetation and soils (Parmesan 2006). The amount of carbon within the pools is dynamic and calculations of the five carbon pools result in a “missing sink” (Schimel et al. 2000). Carbon sources, which are known with a relatively high degree of accuracy, cannot be fully accounted for by measured sinks (Wofsy et al. 1993). The processes by which sinks accumulate carbon include uptake from oceans and the biosphere, and storage in the lithosphere and soil (Parmesan 2006). The failure to account for all of the carbon emitted highlights the importance of expanding our knowledge of forest carbon dynamics.

16.1.2 Forest Carbon Dynamics

Forest ecosystems are of particular importance in the carbon cycle of the terrestrial biosphere, containing up to 80% of all aboveground carbon in the terrestrial realm (Pregitzer and Euskirchen 2004). Forests play an integral role in the global carbon cycle; acting as both sinks and sources around the world (Dixon et al. 1994). The four main carbon pools in a forest ecosystem are the aboveground live biomass, downed woody debris, forest floor, and the soil. Carbon flows between the forest ecosystem and the atmosphere through photosynthesis and respiration (Perry 1994). As plants photosynthesize, they take in carbon that can be allocated for the production of new biomass. Live biomass also respires, releasing CO_2 and water into the atmosphere.

There are several other important carbon fluxes in forest ecosystems. Carbon in live biomass is converted into downed woody debris through disturbances, logging, and natural mortality. Downed woody debris, either in log, stump, or snag form, decomposes over time and carbon is transferred to the forest floor. Further decomposition by microbes results in carbon release into the atmosphere and carbon stored in the soils. Woody debris accounts for 10–20% of the aboveground biomass in mature forests (Brown 2002). In estimating carbon stored in dead wood, the volume of the fine woody debris and coarse woody debris are converted into biomass using allometric equations (Brown 2002). Dead wood is classified into different decay classes to determine its carbon storing capacity and density (Zheng et al. 2008). Decay classes vary across biomes and are affected by soil composition and acidity.

Fluxes between these pools are integral components of forest carbon dynamics. The net flux of carbon between the forest and the atmosphere is the net ecosystem exchange (NEE). This exchange of CO_2 is measured as the difference between total ecosystem respiration and photosynthesis, or gross ecosystem production (Wofsy et al. 1993). Analyses of NEE allow for investigations of forest carbon dynamics over several time scales. The carbon balance also varies due to daily and seasonal changes in the canopy. Further, forests in New England are carbon sources through the winter months when photosynthetic activity is limited.

Different forest management policies affect carbon dynamics and storage capacity, either mitigating or contributing to climate change issues worldwide. Forests play a crucial role in the carbon cycle and have the capacity to reduce the ramifications from climate change both regionally and globally (Brown 2002). Worldwide, forests contribute over 80% of the aboveground carbon storage in the biosphere despite acting as carbon sources to the atmosphere at times (Houghton 1995). The vast discrepancy between storage capacities is based on factors such as forest age, species composition, climate and soil composition (Dewar 1990). For example, forests in the Pacific Northwest of the United States store over 300 mega grams of carbon per hectare (MgC/ha) annually. In comparison, the New England forests store fewer than 200 MgC/ha per year (Heath et al. 2011).

Capacity for carbon storage is affected by numerous forest characteristics and abiotic factors. The observed differences in these factors in New England are impacted by the extent of land cover change, as well as differences in soil composition and acidity. Variable land use and management histories have resulted in regional differences in carbon storage. Forest age, or successional stage, significantly affects the rate of carbon sequestration and respiration. Globally, these factors vary by region as 49% of forest and soil carbon is at high latitudes (50–75°), with 37% and 14% at low-latitude and mid-latitude forests (0–25° and 25–50°), respectively (Dixon et al. 1994).

16.1.3 Disturbance Impacts on Carbon Dynamics

Forest carbon balance is heavily affected by disturbances, both natural and anthropogenic. Human activities and changes in land use have significant impacts on forest carbon cycling (Perry 1994). Changes in land cover and use, a frequent forest disturbance, affect the ecosystem functions and services alike. This factor has been determined as the dominant influence in governing forest composition and carbon accumulation in New England (Casperson et al. 2000). Historically, forest clearing for agricultural land development removed carbon from the live biomass pool and reduced carbon uptake rates. The regrowth of forests after farm abandonment has resulted in increases in carbon accumulation in New England forested regions (Casperson et al. 2000).

Other disturbances associated with human activities also significantly impact forest carbon dynamics. These activities include the introduction of invasive species and pathogens and anthropogenic climate change (Guest 2010). The hemlock woolly adelgid and ash borer, for example, are both heavily impacting the Eastern forest ecosystems (Orwig et al. 2002). Further, the observed climatic changes have resulted in CO_2 enriched forested regions that are experiencing changes in air and soil temperatures (Hopkins 2009). These changes are impacting the forest carbon cycle and will continue to do so as emissions and atmospheric CO_2 levels continue to rise.

Impacts from disturbances associated with human activities are variable between different forests (Barnes et al. 1998). Following a clear-cut or fatal pathogen introduction, a flux of carbon from live to dead biomass results in an increase of the carbon flux to the atmosphere. The subsequent changes in carbon dynamics are affected by forest recovery time and factors such as decomposition and recruitment rates. In addition to anthropogenic disturbances, the forest carbon cycle and other ecosystem services are also affected by natural disturbances. Forest disturbance regimes vary in both scale and frequency, ranging from infrequent, large-scale introduced pathogens to frequent small-scale fire events. Natural disturbances generally renew ecosystems, creating niches and initiating species recruitment, and thereby promote diversified landscapes (Barnes et al. 1998; Perry 1994). In contrast, anthropogenic disturbances typically differ in intensity and scale from natural disturbance regimes to which the forests are adapted.

New England forests are subject to several different natural disturbance types. Wind and ice storms frequently affect forest ecosystems, varying in frequency and scale from small disturbances that may kill only a few trees to regionally more severe disturbances such as hurricanes (Barnes et al. 1998). These disturbances typically result in a flux of carbon to the woody debris pool as the storms bring down branches, boles and whole trees. Forest recovery and regenerative processes are also observed periods of carbon accumulation increase (Barnes et al. 1998). The influential role that wind and ice storms play in determining carbon dynamics is variable across New England. Differences in disturbance impacts on carbon pools and fluxes arise due to the geographic differences in disturbance regimes, climate, and forest soils (Dixon et al. 1994).

16.1.4 Carbon Monitoring Techniques

In the face of global climate change, carbon dynamics are an increasingly important lens through which forest management decisions are viewed. Forest management has become an influential tool for carbon-related policies as countries and regions seek to comply with the United Nations Framework Convention on Climate Change

(Brown 2002). The ability of forests to capture and store carbon may be an integral part of our battle against climate change. Through land practices and policies, forests may be managed to increase carbon sequestration and storage worldwide (Schimel et al. 2000). One study estimates that through global forest carbon plantations and afforestation, an additional 104 gt of carbon could be sequestered. A crucial element to this strategy is our understanding of forest carbon dynamics and the ability to quantify and monitor carbon through a forest ecosystem.

Three different approaches are commonly used to measure forest carbon pools and fluxes: biometry, remote sensing, and the eddy flux technique. Biometry includes field measurements of the aboveground, belowground, and debris biomass (Zheng et al. 2008). Remote sensing entails the use of sensors and satellite imagery throughout the canopy (Gibbs et al. 2007). The approach used in this research, the eddy flux technique, relies on physics, atmospheric gas concentrations, and simulations of eddies (Wofsy et al. 1993).

Biometric estimates are based on field measurements of the live biomass, woody debris, forest floor, and soil content. For aboveground biomass, field measurements are directly converted to biomass levels using allometric biomass regression equations (Brown 2002). These equations are commonly based on tree species and exist for all forests around the world (Brown 2002). Variables in the equations include tree diameter, height, and species. Further, biological and physical characteristics such as the presence of moss, lichen, leaves, and bark are noted. Equations for woody debris calculations often include variables such as decay class and several diameter measurements (Zheng et al. 2008).

Estimations of woody biomass provide the basis for determining the amount of carbon stored in aboveground biomass. With these measurements, estimates of changes in woody biomass over time allow the rate of carbon transfer between the biomass pools to be determined. Changes in live biomass are representative of forest growth and health. Further, woody debris surveys allow tracking of forest recovery and response to both disturbances and natural life processes (Liu et al. 2006). In terms of limitations and difficulties, sampling enough aboveground biomass through live tree diameters and woody debris dimensions measurements may be time-consuming and costly (Brown 2002). Soils also play an integral role in forest carbon dynamics, though they are not heavily explored in this study. Estimating carbon amounts in woody debris, or dead wood, is integral. Discrepancies in woody debris carbon estimations arise from the diversity of the stand and the variable nature of the dead wood.

Though biomass estimates and calculations are useful for accurate measurements of carbon pools in the forest, they are difficult to obtain and monitor for larger, landscape-scale estimates. For landscapes and global estimates, numerous remote sensing techniques are used (Vincent and Saatchi 1999). This technique often involves using aerial sensor technologies to classify vegetation on earth. Though these technologies may be quite expensive, they allow for a larger scale understanding of global carbon dynamics (Brown 2002). A similar expensive and efficient carbon monitoring approach is the eddy flux technique, relying on physics and air simulations throughout the forest (Wofsy et al. 1993).

The eddy correlation method provides continuous measurements of net forest fluxes by modeling air movement throughout the canopy as eddies (Wofsy et al. 1993). Direction and speed of these eddies are measured on three dimensions to determine the net ecosystem flux. The eddy covariance method allows hourly, daily, seasonal, and annual analyses of forest nutrient and gas fluxes. Modeled eddies are used to calculate the vertical flux from the covariance of fluctuations and estimate forest fluxes (Wofsy et al. 1993). The data collected from the eddy flux method include the carbon flux and additional energy and nutrient fluxes (Wofsy et al. 1993). Net ecosystem exchange is an important measurement of the carbon entering and leaving the forest ecosystem in an attempt to quantify the forest carbon balance.

Eddy correlation relies on a sonic anemometer, a fast-response infrared gas analyzer, and the covariance equations. The method provides useful data using these technologies, most notably continuous measurements of the carbon flux, used to calculate net ecosystem exchange. Biometric measurements, remote sensing, and the eddy flux technique are three carbon monitoring approaches that are key to our understanding of the global carbon pool. When field biomass measurements, remote sensing, and the eddy covariance method are utilized in different regions worldwide, a more accurate representation of forest carbon fluxes and pools is achieved (Brown 2002). While each of these methods has its limitations, the estimates that they provide are critical for our understanding of carbon dynamics.

16.2 This Study

In this research, I focused on the carbon dynamics of Harvard Forest, a New England forest that has experienced an increase in carbon storage over the past two decades (Urbanski et al. 2007). Similar increases at nearby forests in the region were determined to be due to forest regrowth and growth enhancement (Casperson et al. 2000). In an effort to explain the carbon storage increase, several different explanations have been explored and analyzed. These hypotheses include increased nitrogen deposition, climate change impacts, historic land use and recovery, soil variability, and increases in the growing season length. The observed lengthening of the growing season is a direct result of warmer air temperatures and high levels of humidity (Richardson 2010). As a result, New England forests are experiencing an earlier leaf out date and longer annual periods of photosynthetic productivity and growth (Richardson 2010; White et al. 1999). These phenological and ecosystem function responses may explain the measured increase in carbon storage.

Given the importance of forest carbon dynamics both historically and in the face of future climatic changes, this study was undertaken to further analyze carbon dynamics at Harvard Forest. In an effort to investigate possible explanations for the observed increase in carbon storage, this study included biometric, ecological, and atmospheric analyses of the mid-latitude temperate forest. The main objective of this study was to answer the following question: what are the phenological and ecosystem function responses to changes in the growing season length at Harvard Forest? Further, do the observed changes in the growing season explain the systematic increase in carbon uptake over the past two decades?

It is hypothesized that as growing season length increases, carbon uptake, or NEE, will also increase. Further, leaf area index peaks will increase and the date at which they occur will be earlier in the year. The lengthening of the growing season will also result in an increase in aboveground woody increment over the year. Another component of this study is the drivers of the growing season length, primarily atmospheric and meteorological parameters. It is hypothesized that the increase in the growing season is primarily determined by the increase in air temperature at the forest, as well as warmer soil temperatures.

16.3 Methods

16.3.1 Study Area

Located near Petersham in north-central Massachusetts (42.530°N, 72.190°W), Harvard Forest was established in 1907 with the mission of research and education. The research described here was conducted on the Prospect Hill tract, the location of the Harvard Forest Environmental Measuring Site (HFEMS, 42.538°N, 72.171°W). The 3000-acre Harvard Forest research area became a Long Term Ecological Research (LTER) site in 1988 to further explore how physical, biological, and human systems interact to change our earth.

The HFEMS is at an elevation of 340 m and has measured ecosystem fluxes and meteorological variables since 1992. The area surrounding HFEMS is a mixed deciduous forest, dominated by red oak (*Quercus rubra*) and red maple (*Acer rubrum*). The forest composition also includes Eastern hemlock (*Tsuga canadensis*) and white pine (*Pinus strobus*), along with northern hardwood species. The forest stand at HFEMS is 80–115 years old, with one-third of the red oaks each established prior to 1895, 1930, and 1940, respectively (Urbanski et al. 2007).

For the past two decades, forest recovery and response to disturbance regimes has been monitored and evaluated (Goulden et al. 1996a; Irland 2008; Olthof et al. 2003). Harvard Forest, similar to other New England forests, experiences influential winter wind and ice storms. Other disturbances include invasive pests and pathogens, as well as direct human disturbance (Foster et al. 2010). In 1993, in an effort to further understand how these disturbances affect the forest ecosystem, the Wofsy Group of Harvard University established 40 circular plots for biometric measurements. These plots were placed in the predominant wind directions, northwest and southwest, of the HFEMS for accurate eddy simulation (Barford et al. 2001). Past beaver activity and selective logging in the study site reduced biometric measurements to 34 plots in 2001.

16.3.2 Biometric Measurements

Field measurements have been conducted in the biometric plots at the HFEMS since 1993 (Fig. 16.1). The plots were established in stratified-random positions along eight different 500-m transects from the HFEMS (Urbanski et al. 2007). Each of the circular 34 plots, with a diameter of twenty meters, has an area of 314.16 m^2. Biometric measurements are taken on monthly and annual timescales.

The biometric measurements include measurements of both live biomass and woody debris. Aboveground woody increment and aboveground woody biomass are estimated using dendrometer measurements of tree growth. Dendrometers are placed on all live trees with diameters larger than 10 cm within each plot and measurements are recorded four times per year. Tree mortality and recruitment are also recorded. Leaf litter baskets collect litter over the course of the year to study the output of litter from trees at the end of each growing season. Leaf area index (LAI) is used as an indicator of gas and nutrient exchange between the atmosphere and the forest canopy (Olthof et al. 2003). The index is measured using a plant canopy analyzer, LAI-2000. LAI has been monitored monthly during the growing season, allowing for analyses of changes in peaks, trajectories, and monthly and annual means.

Periodic measurements at the HFEMS include woody debris surveys, leaf litter chemistry, soil and debris respiration rates, and green foliage nutrient content (Urbanski et al. 2007). The woody debris surveys are conducted every 3 years in order to estimate the flux between the live and dead forest biomass. The coarse woody debris (CWD, diameter over 10 cm) carbon pool is estimated by measuring downed wood tagged in each of the plots. Using the CWD measurements and allo-

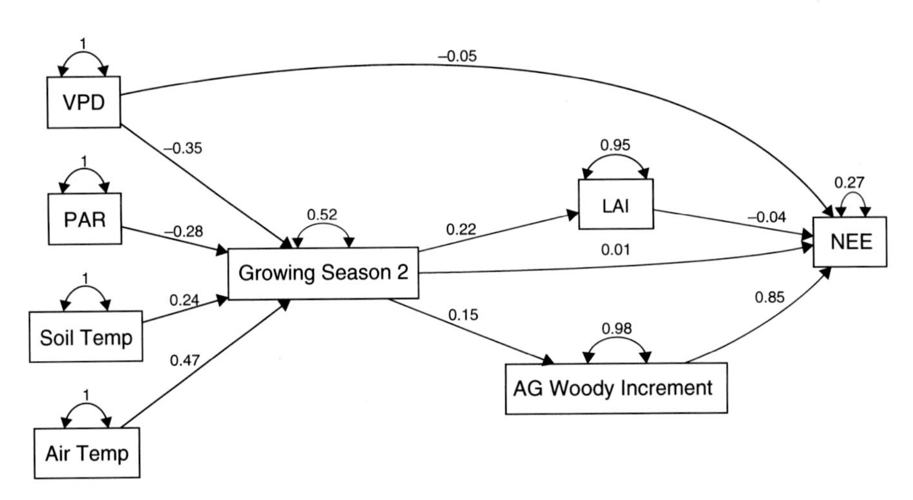

Fig. 16.1 The path diagram for the second eddy flux and biometry 1-year ecological time lag SEM analysis (GFI = 0.78, RMSEA = 0.38, χ^2 = 355.8, df = 18). For path diagram conventions, consult Fig. 16.2

metric equations, the carbon stored in the wood is estimated (Harmon and Sexton 1996). Fine woody debris (FWD, diameter 2.5–10 cm) is also a notable contributor to the woody debris carbon pool.

16.3.3 Measurements of CO_2, Water, and Energy Fluxes

The eddy covariance technique allows for continuous measurements of CO_2, water, and energy fluxes between the atmosphere and biosphere (Wofsy et al. 1993). The technique relies on measurements of carbon dioxide, latent and sensible heat, and momentum of air movement (Goulden et al. 1996a). At HFEMS, eddy flux tower measurements are used to calculate net CO_2 flux using measurements of CO_2 at eight heights (29, 24, 18, 13, 7.5, 4.5, 0.8, 0.3 m) above the forest floor (Urbanski et al. 2007). Net ecosystem exchange is the difference between respiration and gross ecosystem exchange and is used as a measure of the forest carbon balance. Other environmental variables, including air temperature and daily precipitation are gathered from the Harvard Forest meteorological station. Data acquisition is automated at HFEMS and stored on-site before processing.

16.3.4 Measurements of the Growing Season

For this study, two different definitions of the growing season were used to analyze the relationship between the changing growing season and ecosystem phenology and function. The first measure was calculated as the number of Julian days over the course of the year that the photosynthetic capacity of the forest canopy was over 12 μmol $m^{-2}s^{-1}$ (micromoles of carbon dioxide per meter squared per second; Richardson 2010). This definition is a length of the photosynthetic growing season based on a phenological parameter of the forest. The second definition of the growing season is cumulative ecological degrees, based on meteorological data at Harvard Forest. Cumulative ecological degrees is a measure of accumulated degrees Celsius over the course of the year and is thus an estimate of both growing season length and growing season warmth (Richardson 2010).

16.4 Data Analysis

In this study, structural equation modeling (SEM) was used to provide a model for interpreting whole ecosystem interactions, including direct and indirect relationships and pathways (Grace 2006). The observed variables in the model represent the ecological and atmospheric components that were directly measured. The latent variables represent unmeasured variables, the underlying causes behind ecosystem interactions and relationships (Fox 2006). These unmeasured variables are estimated from patterns of correlations and variances among the observed variables. Thus, the latent variables are not directly quantified from a data set, but are simulated by the SEM. For this study, the observed variables included the biometric and eddy flux data, as well as the two growing season definitions.

For Harvard Forest analyses, SEM models were developed separately for the biometric and eddy flux data because of the incommensurate time scales of data collection. The biometric data are on monthly and annual time scales, whereas eddy flux data are collected continuously. A third integrated model was then developed that used the annual means of the eddy flux data along with the biometric data. All models contained different combinations of the observed variables in an effort to better represent the ecological and atmospheric interactions of the forest.

The biometry SEM model included an analysis of the following observed variables: LAI mean, LAI peak (julian day) aboveground woody increment, aboveground woody biomass, and annual NEE. Observed variables in the eddy flux SEM model included air temperature, soil temperature at the surface and 20 mm below ground, photosynthetically active radiation (PAR) just below the main canopy, vapor pressure deficit (VPD), and annual NEE. The model that integrates both SEMs included all of the described observed variables.

Initial integrated SEM modeling was conducted with all variables and all possible interactions between them. The model specifications were changed in an effort to maximize the overall model fit. In further modeling, the integrated SEM was simplified by first removing the surface soil temperature component and additionally, the aboveground woody increment parameter. These variables were removed because of preliminary data investigation, a Pearson's correlation test, showing a weak relationship between these parameters and the growing season. Further, in comparison to surface soil temperature, correlation tests determined soil temperature at 20 mm below the surface to be a more influential driver of the biometric components.

The integrated SEM was also specified to simulate a 1- and 2-year ecological time lag. For these models, the biometry variables were lagged 1 or 2 years behind the eddy flux data. This ecological time lag was simulated because of the potential for climatic and meteorological changes, such as increases in air temperature, to have more direct impacts on the ecosystem in the subsequent growing season. This time lag is suggested in the literature to be a key part of forest dynamics (Menzel and Fabian 1999). To simulate the time lag, NEE, aboveground woody increment, and LAI peak from 1994 were analyzed in a model with the growing season and eddy flux components from 1993. For the 2-year lag models, these biometry variables were analyzed with eddy flux components from 1992.

Processing and analysis of the data were conducted using R (Version 2.12.2; http://www.r-project.org/). Data filling and gap partitioning was completed by HFEMS research technicians prior to data analysis. Preliminary data analysis included regression analyses and Pearson's correlations tests. For the structural equation modeling (SEM) analyses, SEM in R was used (Fox 2006).

16.4.1 Growing Season Calculations

Growing season length tended to increase over the studied period from 1992–2008. The length of the photosynthetic growing season increased significantly over the two decades (Pearson's correlation test; R = 0.72, P < 0.002) but, despite an upward trend, the rise in cumulative ecological degrees was not significant (R = 0.40, P > 0.05). Longer growing seasons were associated with higher NEE totals over the two decades (Pearson's correlation test; R = 0.69, P = 0.003 and R = 0.47, P = 0.06, respectively). Further, initial correlation tests between the growing season and the eddy flux parameters did not result in any significant relationships. A significant relationship was found between the photosynthetic growing season length and aboveground woody increment and biomass. Further, NEE was significantly correlated to both aboveground biomass measurements and LAI peak values (Pearson's correlation test; R = 0.68, P = 0.003; R = 0.82, P = 0.0009; R = 0.69, P = 0.003; respectively).

16.4.2 SEM Analyses

The model specifications and outputs from the 18 different SEM analyses are summarized in this study (Table 16.1). The four different characterizations of models resulted in different model strengths and applications. Overall, the highest goodness-of-fit index in this study was the first eddy flux model. The four eddy flux models were all better representations of ecosystem interactions than the biometry models. The models with more components and interactions, such as the first two integrated models, resulted in the lowest GFI values. Simplifying these integrated models had variable impacts on the fit indices. The eddy flux and biometry models with the best goodness-of-fit were the models that simulated the 1-year ecological time lag. Overall, cumulative ecological degrees had higher strengths of relationships with changes in carbon uptake. Further, models with the photosynthetic growing season generally had higher goodness-of-fit values.

16.4.3 Eddy Flux SEM

Four different eddy flux SEMs were specified. The first model, eddy flux 1, hypothesized relationships among seven different parameters (Fig. 16.2). The model includes 20 specified relationships and provided an overall GFI of 0.90 (χ^2 = 66.3, df = 8, RMSEA = 0.25). The two strongest relationships in this model were the relationship between air temperatures and surface soil temperatures and the relationship between growing seasons and NEE totals (Standard estimate = 0.65 and 0.60, respectively). Warmer soil temperatures at 20 cm below the surface are correlated to a longer growing season and increased NEE. Conversely, soil temperature at the surface is negatively correlated to both parameters.

Table 16.1 Outputs from the SEM analyses of the 18 specified models. The output includes the measured P value, the goodness of fit index (GFI), chi-square, degrees of freedom, and root mean square error of approximation (RMSEA)

Model	Growing season	P value	GFI	χ^2	Df	RMSEA
Biometry 1	Photosynthetic	<0.01e–14	0.68	205.8	6	0.56
Biometry 2	Ecological degrees	<0.01e–14	0.64	288.4	6	0.67
Biometry 3	Photosynthetic	<0.01e–14	0.70	188.5	5	0.59
Biometry 4	Ecological degrees	<0.01e–14	0.69	228.9	5	0.65
Eddy flux 1	Photosynthetic	0.02e–9	0.90	66.3	8	0.25
Eddy flux 2	Ecological degrees	0.07e–12	0.88	79.2	8	0.27
Eddy flux 3	Photosynthetic	0.01e–14	0.85	104.2	12	0.25
Eddy flux 4	Ecological degrees	<0.01e–14	0.83	125.9	12	0.28
EFB1	Photosynthetic	<0.01e–14	0.56	952	32	0.43
EFB2	Ecological degrees	<0.01e–14	0.53	971.5	32	0.42
EFB3	Photosynthetic	<0.01e–14	0.76	317.6	20	0.30
EFB4	Ecological degrees	< 0.01e–14	0.74	316	20	0.32
EFB5	Photosynthetic	<0.01e–14	0.71	237.8	18	0.30
EFB6	Ecological degrees	<0.01e–14	0.72	247.8	18	0.34
EFB Lag 1	Photosynthetic	<0.01e–14	0.77	325.9	18	0.33
EFB Lag 2	Ecological degrees	<0.01e–14	0.78	355.8	18	0.38
EFB Lag 3	Photosynthetic	<0.01e–14	0.68	305.9	18	0.35
EFB Lag 4	Ecological degrees	<0.01e–14	0.72	250.3	18	0.32

The SEM analysis of eddy flux 2, the second flux model, estimated growing season length using cumulative ecological degrees. The overall GFI decreased to 0.88, indicating a slightly less accurate description of the ecosystem interactions ($\chi^2 = 79.2$, df = 8, RMSEA = 0.27). In this model, the significant positive relationship between growing season and NEE decreased in comparison to the eddy flux 1 model (Standard estimate = 0.40). Another key difference between the models is the strengthened relationship between PAR and NEE in the second model (Standardized estimate = 0.6).

16.4.4 Eddy Flux and Biometry SEM

The SEM analysis was then expanded upon to include components from both the biometry and the eddy flux data. The first of these models was hypothesized with ten different parameters and 27 relationships. The low GFI for this model indicates a poor fit for these specified components and relationships (GFI = 0.56, $\chi^2 = 952$, df = 32, RMSEA = 0.43). Most notable was the decrease in the strength of the relationship between growing season and NEE (Standard estimate = −0.05). The integration of biometry and eddy flux also affected the correlations between the eddy flux data components and the growing season. The biometry variable most strongly correlated to the photosynthetic growing season was aboveground woody increment, though a strong significant relationship was also found with LAI peak (Standard estimate = 0.74 and 0.51, respectively).

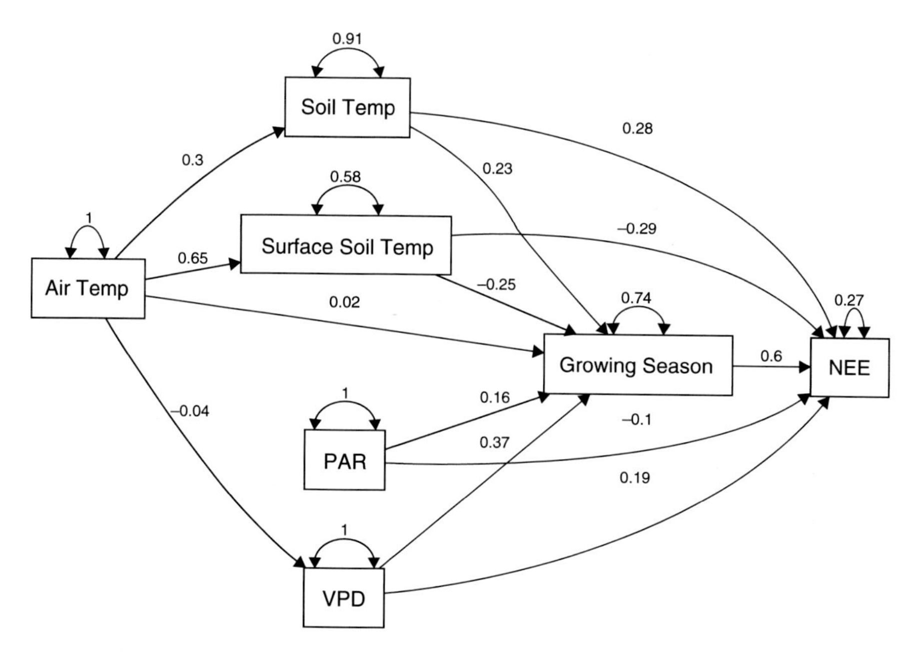

Fig. 16.2 The path diagram for the eddy flux 1 SEM analysis (GFI = 0.90, RMSEA = 0.25, $\chi^2 = 66.3$, df = 8). Arrows between components indicate specified relationships; the driving component is represented by the direction of the arrow. The numbers above the arrows indicate the standardized estimates, or the strength of the relationship. The double-headed arrows above each component indicate the variance within the parameter, the interannual variability ranging from 0 (low) to 1 (high). The growing season component indicates photosynthetic growing season, while growing season two is modeling using the cumulative ecological degrees parameter

The second eddy flux biometry model uses cumulative ecological degrees as the growing season component. Despite interesting significant relationships between the parameters, the model is a poor overall fit with a GFI of 0.53 ($\chi^2 = 971.5$, df = 32, RMSEA = 0.42). These analyses indicate that cumulative ecological degrees warmth is correlated with NEE in the eddy flux and biometry SEMs (Standard estimate = 0.16). Further, the strength of the relationships between the growing season and the biometry components decreased when analyzing cumulative ecological degrees. The positive correlation between surface soil temperature and photosynthetic growing season, as well as the negative correlation between PAR and cumulative ecological degrees, both change in direction when this growing season measure is used.

16.4.5 Ecological Time Lag SEM

Four SEM analyses were conducted to account for an ecological time lag, based on the EFB5 model of 8 parameters and 18 relationships. The first two models were analyzing a 1-year time lag affecting NEE, LAI peak, and aboveground woody increment. These integrated models provided the best fit for the biometric and eddy flux data sets and parameters compared to the other models (Fig. 16.2). The models using the photosynthetic growing season and cumulative ecological degrees both resulted in strong model fits for reality (GFI = 0.77, χ^2 = 325.9, df = 18, RMSEA = 0.33 and GFI = 0.78, χ^2 = 355.8, df = 18, RMSEA = 0.38, respectively). Further, the ecological time lag improved the correlations between the eddy flux parameters and their impacts on photosynthetic growing season length. The analyzed lag also improved the strength of the relationship between increases in cumulative ecological degrees and increases in LAI, aboveground woody biomass, and NEE.

16.5 Discussion

The results from this study suggest that changes in the growing season length and warmth have important implications for ecosystem function in the mixed deciduous forests at Harvard Forest. The investigated ecological and ecosystem function responses to these changes are crucial to understanding the future of forest carbon dynamics. These responses are even more important given observed increases in the growing season length. The longer photosynthetic growing season values support previous studies that warmer air temperatures and higher humidity levels are key factors in influencing growing season length in temperate forests (Goulden et al. 1996b; White et al. 1999). In comparison, the upward trend in cumulative ecological degrees suggests that this growing season definition was less representative of the climatic and meteorological changes at Harvard Forest. The photosynthetic growing season allowed for a better model fit than cumulative ecological degrees in the majority of the analyzed models.

16.5.1 Eddy Flux SEM

The eddy flux SEM analyses provided the best statistical model fit in this study. More specifically, the first eddy flux model resulted in the best model fit to the data. The eddy flux SEM analyses were based on data that were taken on consistent time scales from the same tower at the Environmental Measurement Site. Aside from the calculated growing season, these data were annual means of hourly measurements from the same site. Variability is therefore reduced in terms of methodology in comparison to the biometry components. Given the high statistical evidence for these models, these results provide interesting information for the key drivers of changes in the growing season.

The parameter that was most heavily correlated to the length of the growing season was vapor pressure deficit (VPD). As a function of relative humidity and air temperature, VPD is used as a measure of how dry the air is (Boisvenue and Running 2006). VPD is an integral driver of growing season length because it is a control of stomatal conductance and a major limitation to photosynthesis (Ollinger et al. 1998). Because of this, the relationship between increases in VPD, or drier air, and

NEE is a crucial aspect of carbon dynamics. VPD was found to be an indicator of fundamental physiological limits to primary production, in addition to being characterized by high regional variability (Boisvenue and Running 2006).

Leaf physiology would lead to the assumption that VPD and NEE were negatively correlated, though that result was not found in this study. The adverse effect of dryer air resulting in more carbon uptake was explored in a similar study (Boisevenue and Running 2006). Higher night temperatures potentially affected the carbon balance of dryer, warmer sites. The effects of night temperature fluctuation were not represented in taking the annual mean of air temperatures. Further, other physiological responses to increased VPD and CO_2 levels may have resulted in the closure of stomata and other parameters that affect NEE. Analyses of VPD incorporate several limiting factors that were not significantly correlated to NEE in this study.

Increases in PAR were positively correlated to longer growing seasons but lower carbon uptake years. Photosynthetically active radiation is a measure of the amount of radiation within the spectral range that is used for photosynthesis. The relationship between PAR and the growing season reaffirms the concept of the photosynthetic growing season as a measure of days of active forest productivity and growth (Perry 1994). One possible explanation for the negative relationship between PAR and NEE is the relative inaccuracy of calculating an annual mean for PAR. These models use the annual means of the parameters, such as PAR, that may need to be analyzed on a timescale that addresses its daily and seasonal fluxes. Observations of the relationship between PAR daily and seasonal means and NEE suggest a positive correlation with carbon uptake (Ollinger et al. 1998). That relationship was not found in this study, highlighting the need to analyze PAR on different time scales.

Similar patterns are seen in the second eddy flux model that used cumulative ecological degrees as the growing season parameter. However, while surface soil temperature was negatively correlated with photosynthetic growing season, it had a positive relationship with cumulative ecological degrees. This discrepancy is further complicated because surface soil temperature had a negative relationship with NEE in both models. The negative relationship may highlight the importance of meteorological factors on cumulative ecological degrees, due to the high correlation between air temperature and surface soil temperature. PAR is also negatively correlated to cumulative ecological degrees, affirming the more meteorological influence relative to phenological photosynthetic capacity as a measure of the growing season.

Simplifying the eddy flux models by removing four relationships resulted in a lower GFI, again speaking to the complexity of the forest ecosystem and its interactions. Though the model is a slightly poorer fit, the model simplification strengthened the influence that the growing season length has on annual carbon uptake. Fewer variables resulted in stronger relationships among the remaining components, though the model fit had decreased. The removal of weak relationships revealed an even stronger correlation between photosynthetic growing season and NEE, the strongest in the eddy flux model. This change suggests that model simplification can serve as an important tool to highlight or mitigate particular relationships between the eddy flux and biometry components.

16.5.2 Eddy Flux and Biometry SEM

The first two eddy flux and biometry models were the most intricate models in this study, defined by 27 different relationships between the measured 10 parameters. While the complexity of these models may better account for ecological interactions, the low GFI indicates a poor data representation of reality. The described relationships in these specified models might, in fact, be the most representative of the forest ecosystem. However, modeling the annual means from these data resulted in low fit indices. Further, the complication of these models by integrating both biometry and eddy flux variables decreased several important relationships.

Most notably, the integration of these models resulted in a decreased strength of relationship between the growing season length and carbon uptake. This lower standardized estimate could be the result of several interacting factors. First, because the relationships are standardized, the relationship between NEE and the growing season is less influential relative to the interactions between the biometry and eddy flux components. Secondly, the incorporation of the variable biometry components potentially affected the covariance values to a degree that affected the significance of the length of the growing season driving NEE totals.

The weaker relationship between growing season length and carbon uptake was also seen in a similar study (Saxe et al. 1998). This study found that the positive correlation between the growing season and forest productivity, including NEE, was reduced in nutrient limited ecosystems. Harvard Forest, like other New England forests, is nitrogen limited as low nitrogen levels prohibit the forests from reaching their full productivity capacity (Foster et al. 2010). Given this limitation, gradual increases in nitrogen input into forest soils may better explain the two-decade increase in carbon storage.

This potential explanation has prompted many studies about increased nitrogen deposition in New England (Bedison and McNeil 2009; Cha et al. 2010; Horri et al. 2006). Nitrogen deposition is the process by which atmospheric nitrogen is input into the biosphere, including forest ecosystems (Bedison and McNeil 2009). Elevated atmospheric nitrogen deposition has shown to increase plant growth and carbon sequestration in nutrient limited ecosystems (Cha et al. 2010). One study found that increased nitrogen deposition in a New England forest interacted with other co-varying factors, such as soil temperature and acidification, to result in the increase in primary production and sequestration (Bedison and McNeil 2009). However, other studies have presented different conclusions about how increased N deposition has affected New England, including a negative relationship with carbon uptake and an increase in tree mortality.

Through integrating the hypothesized SEM models, analyses were conducted about how the eddy flux variables directly influence the biometry components, not just their effects on the growing season length. The first integrated model suggests a negative relationship between soil temperature and aboveground woody increment. This low negative correlation may highlight soil water content as a more important driver of woody biomass increment than soil temperature. Soil water content is an important limiting factor in forest growth (Perry 1994). Further, the model also resulted in strengthened relationships between higher NEE totals and increases in soil temperature and VPD.

Analyses of the second integrated model result in a stronger relationship between cumulative ecological degrees and the increases in carbon uptake. The discrepancy between photosynthetic growing season and cumulative ecological degrees is contradictory to what is found in the prior separate models. The standardized estimate is, however, much lower than any of the growing season and NEE relationships in the other analyzed models. Thus, cumulative ecological degrees is not suggested to be better correlated to carbon uptake overall, given the decreases in GFI and other important relationships in the first two integrated model SEM analyses.

In an effort to increase GFI by analyzing only the strongest relationships, aboveground woody biomass was removed from the specified third and fourth SEM analyses. Reducing the models to nine parameters increased the overall fit of the integrated analyses. Though aboveground woody biomass was an influential driver of NEE, its variability reduced overall model fit of the first two integrated SEMs. Further model variability was addressed through the removal of soil temperature in the fifth and sixth integrated analyses. Removal of this parameter further improved the model fit and the fifth eddy flux and biometry SEM provided the best-fit integrated model in this study.

Aboveground woody biomass is a component that measures total woody biomass in the biometric plots. Because this study is investigating the growing season, it is more appropriate to include woody increment, which is the annual forest growth. Aboveground woody increment was more related to both the eddy flux and the biometry variables, further justifying the removal of aboveground woody biomass in the integrated models. Further, when analyzing how the eddy flux parameters drive growing season length, the removal of surface soil temperature is biologically and statistically justified. The previous 12 models all found weak and variable relationships between surface soil temperature and growing season length because of the overpowering influence of soil temperature below the surface. Of utmost importance, as previously described, are the relationships between the length of the growing season and NEE, as well as aboveground woody increment. The integration of the model also affirms the role that the eddy flux components play in influencing growing season length.

16.5.3 Ecological Time Lag SEM

The components and specified relationships from the fifth integrated model were used to simulate ecological time lags of 1 and 2 years. The results from these analyses provide interesting implications for the timescale of forest responses to climatic and environmental changes. The GFI values for models with simulated time lags were higher than any other integrated model, supporting the existence and importance of the lag in forest ecosystems. As hypothesized, the 1-year lag models were better fit than the 2-year lag.

The improved correlation between growing season and aboveground woody increment indicates that a longer growing season in 1999 results in an increase in aboveground woody increment measured in 2000. Given our knowledge of tree physiology and growth, this lag is a better approximation of reality. The overall increased GFI of the model also strengthened the relationships between the eddy flux components and growing season length. These improved relationships allow for a more accurate portrayal of the eddy flux components as drivers of the growing season.

Interestingly, the 1-year lag models changed the relationship between LAI peak and carbon uptake. The weak negative relationship is counterintuitive and contradictory to what has been seen in the other models. The result that increased LAI values are correlated to lower NEE totals may be an indirect consequence from the 1-year lag of the integrated models. That is, it highlights the difference between the accuracy of the 1-year lag for ecosystem function responses rather than phenological responses. Further, the 1-year lag models greatly improved the relationships between cumulative ecological degrees and the biometry components. This demonstrates the importance of the time lag, given the heavy influence of air temperature on cumulative ecological degrees. These results suggest that increases in air temperature and other eddy flux parameters will have the most significant effect on the biometry variables the year after the growing season is affected.

Analyses of the 2-year time lag SEMs resulted in a lower goodness of model fit. Defining the ecological time lag as 2 years was suggested to be inappropriate and relatively less accurate than the 1-year models. This observation was supported by a New England study that also found stronger relationships and trends for long-term ecological time lags (Curtis et al. 2002). Further data are needed to investigate a potential 5, 10, or 50 year ecological time lag to fully understand forest response to changing environmental conditions and factors. A better understanding of the ecological time lag between the eddy flux components and the biometry parameters is necessary to understand how disturbance regime shifts and climatic changes will influence forest growth, productivity and carbon storage.

16.5.4 Explanations for Changes in Carbon Dynamics

As previously mentioned, increased atmospheric nitrogen deposition is one possible explanation for the increase in carbon storage at Harvard Forest. It has been shown as an influential factor in driving carbon dynamics at similar New England forests (Bedison and McNeil 2009). Further, the increase in nitrogen availability would allow for forest growth and productivity to reach its capacity in nutrient-limited environments. However, one study suggests that high levels of nitrogen addition have also resulted in carbon storage decline and tree mortality. Elevated levels of nitrogen in the atmosphere and the observed increase in nitrogen deposition highlight our need for an understanding of how these changes will affect the forest ecosystems, particularly nutrient-limited systems.

Climate change is a second possible explanation for the changing rates of C uptake at the Harvard Forest. In one tropical forest, climate was proved to be a stronger driver of carbon dynamics changes than soil type or disturbances (Toledo et al. 2010). The decreased rainfall and increased air temperature in the tropical forests may therefore result in significant negative implications for tree growth and carbon sequestration in the tropical biomes. In New England forests, climatic changes may act directly, by changing rates of plant growth, or indirectly, by altering disturbance regimes. For example, future climate change may result in an increased frequency and severity of wind and ice storms (Irland 2008).

Historical land use change and recovery from clearing may also explain the observed increase in carbon storage at Harvard Forest. Further knowledge about land use history in relation to carbon dynamics is important given the recent suggestions to convert old-growth forests to younger, intensively managed forests (Brown 2002; Harmon et al. 1990). This suggestion ignores the key difference between annual carbon uptake rates, which are higher for younger forests, and overall carbon accumulation, which is typically higher in old-growth forests. The balance between a forest that is both increasing in woody increment and recruitment, as well as total carbon accumulation, is the peak of carbon storage. The 100-year-old Harvard Forest may be reaching that level, as it is dominated by red oaks that are reaching their peak carbon sequestration and storage rates.

In Southern New England, land use history was determined to be less influential than soil variability on carbon dynamics (Kulmatiski et al. 2004). It was determined that carbon storage is largely driven by soil characteristics and stand composition, which would allow for regional and global estimates of carbon dynamics. However, 15 years of chronic nutrient addition to stands at Harvard Forest resulted in increased mortality and a loss of inorganic nitrogen (Magill et al. 2004). The observed tree mortality may be indicative of a nitrogen input threshold in forest ecosystems. Changes in soil acidity, temperature, and substrate affect various components of carbon dynamics in our forests. Changes in soil and other biotic characteristics may play an important role in the increase in carbon accumulation observed at several New England forests.

These factors, including increased atmospheric nitrogen deposition, climatic changes, land use history, and soil variability, are all likely to be important drivers of forest carbon dynamics. While this study did not directly incorporate these potential explanations into the models and SEM analyses, they must be taken into consideration. This study did, however, model the indirect effects that these phenomena are having on the measured eddy flux and biometry variables.

16.6 Future Research

To better understand the phenological and ecosystem function responses to changes in the growing season at Harvard Forest, carbon-monitoring research should be continued. While two decades of biometry and eddy flux data provided for interesting SEM analyses, further investigations into carbon dynamics require more than 17 annual data points. With a longer study period, cyclical fluctuations in parameters such as air temperature could be investigated. Further, analyses of meteorological patterns such as El Nino and their impacts on forest carbon dynamics could be studied. Longer-term studies would also allow for a more complete analysis of the ecological time lag. While 1 and 2 year lags provided interesting preliminary results, ecological time lags of 10 or 50 years should be simulated in further investigations.

Further research should also be conducted to investigate the other potential explanations for the observed carbon increase at Harvard Forest. Continued extensive studies of atmospheric nitrogen deposition, land use and cover changes, and climatic conditions changes could result in significant implications for our understanding of forest carbon dynamics and suggested land management policies. To further investigate using SEM as a tool to analyze whole ecosystem relationships and interactions, more components could be integrated into the model. For example, further research could include a measure of quantification of nitrogen deposition impacts on forest ecosystems. Similarly, quantification of more forest components would generally improve the accuracy of these models and the specified interactions.

As discussed, further research should incorporate altered versions of the growing season definitions to assess accuracy of the parameters used. While photosynthetic growing season and cumulative ecological degrees provided an excellent basis for this research, further studies would redefine the parameters used for these lengths. Avoiding the cyclical nature of using the photosynthetic growing season in these models is an important issue to address for continued research. Further, more accurate measures of the growing season would allow for important implications for the forest carbon cycle and land management policies.

16.7 Conclusions

This study provides an understanding for changes in carbon dynamics at Harvard Forest. The phenological and ecosystem function responses to increases in the growing season were hypothesized and modeled in this study. The whole ecosystem analyses of direct and indirect relationships between biometry and eddy flux data provided interesting results for forest ecosystem interactions. Both photosynthetic growing season and cumulative ecological degrees were highly correlated to aboveground woody increment, which is an important driver of carbon uptake.

Changes in the growing season were most strongly correlated to soil temperatures below the surface. Incorporation of annual means of VPD and PAR provided both expected significant relationships and counterintuitive correlations in various models. Model simulations of the 1-year ecological time lag resulted in stronger relationships between the eddy flux parameters and the lagged biometry components. This time lag may be crucial to our understanding of how carbon dynamics will change in New England in response to global climate change. This study provides a baseline for Harvard Forest carbon investigations using SEM and highlights the importance of our understanding of forest carbon dynamics.

References

Barford, C. C., Wofsy, S. C., Goulden, M. L., Munger, J. W., Pyle, E. H., Urbanski, S. P., Hutyra, L., Saleska, S. R., Fitzjarrald, D., & Moore, K. (2001). Factors controlling long- and short-term sequestration of atmospheric CO_2 in a mid-latitude forest. *Science, 294*, 1688–1692.

Barnes, B. V., Zak, D. R., Denton, S. R., & Spurr, S. H. (1998). *Forest ecology*. New York: Wiley.

Bedison, J. E., & McNeil, B. E. (2009). Is the growth of temperate forest trees enhanced along an ambient nitrogen deposition gradient? *Ecology, 90*, 1736–1742.

Boisvenue, C., & Running, S. W. (2006). Impacts of climate change on natural forest productivity-evidence since the middle of the 20th century. *Global Change Biology, 12*, 862–882.

Brown, S. (2002). Measuring carbon in forests: Current status and future challenges. *Environmental Pollution, 116*, 363–372.

Casperson, J. P., Pacala, S. W., Jenkins, J. C., Hurtt, G. C., Moorcroft, P. R., & Birdsey, R. A. (2000). Contributions of land-use history to carbon accumulation in U.S. forests. *Science, 290*, 1148–1152.

Cha, D. H., Appel, H. M., Frost, C. J., Schultz, J. C., & Steiner, K. C. (2010). Red oak responses to nitrogen addition depend on herbivory type, tree family, and site. *Forest Ecology and Management, 259*, 1930–1937.

Curtis, P. S., Hanson, P. J., Bolstad, P., Barford, C., Randolph, J. C., Schmid, H. P., & Wilson, K. B. (2002). Biometric and eddy-covariance based estimates of annual carbon storage in five eastern North American deciduous forests. *Agricultural and Forest Meteorology, 113*, 3–19.

Dewar, R. C. (1990). A model of carbon storage in forests and forest products. *Tree Physiology, 6*, 417–428.

Dixon, R. K., Brown, S., Houghton, R. A., Solomon, A. M., Trexler, M. C., & Wisniewski, J. (1994). Carbon pools and flux of global forest ecosystems. *Science, 263*, 185–193.

Foster, D. R., Donahue, B. M., Kittredge, D. B., Lambert, K. F., Hunter, M. L., Hall, B. R., Irland, L. C., Lilieholm, R. J., Orwig, D. A., D'Amato, A. W., Colburn, E. A., Thompson, J. R., Levitt, J. N., Ellison, A. M., Keeton, W. S., Aber, J. D., Cogbill, C. V., Driscoll, C. T., Fahey, T. J., & Hart, C. M. (2010). *Wildlands and Woodlands: A vision for the New England landscape*. Cambridge, MA: Harvard Forest, Harvard University Press.

Fox, J. (2006). Structural equation modeling with the SEM package in R. *Structural Equation Modeling, 13*, 465–486.

Gibbs, H. K., Brown, S., Niles, J. O., & Foley, J. A. (2007). Monitoring and estimating tropical forest carbon stocks; making REDD a reality. *Environmental Research Letters, 2*, 1–13.

Goulden, M. L., Munger, J. W., Fan, S. M., Daube, B. C., & Wofsy, S. C. (1996a). Exchange of carbon dioxide by a deciduous forest: Response to interannual climate variability. *Science, 271*, 1576–1578 In text: Goulden, 1996a.

Goulden, M. L., Munger, J. W., Fan, S. M., Daube, B. C., & Wofsy, S. C. (1996b). Measurements of carbon sequestration by long-term eddy covariance: Methods and a critical evaluation of accuracy. *Global Change Biology, 2*, 169–182 In text: Goulden, 1996b.

Grace, J. B. (2006). *Structural equation modeling and natural systems*. New York: Cambridge University Press.

Guest, R. (2010). The economics of sustainability in the context of climate change: An overview. *Journal of World Business, 45*, 326–335.

Harmon, M. E., & Sexton, J. (1996). Guidelines for measurements of woody detritus in forest ecosystems. U.S. *LTER Network Office, 20*, 1–73.

Harmon, M. E., Ferrell, W. K., & Franklin, J. F. (1990). Effects on carbon storage of conversion of old-growth forests to young forests. *Science, 247*, 699–702.

Heath, L. S., Smith, J. E., Woodall, C. W., Azuma, D. L., & Waddell, K. L. (2011). Carbon stocks on forestland of the United States, with emphasis on USDA Forest Service ownership. *Ecosphere, 2*, 1–21.

Hopkins, R. (2009). *Peak oil and climate change. In: The transition handbook: From oil dependency to local resilience*. White River Junction: Chelsea Green Publishing.

Horii, C. V., Munger, J. W., Wofsy, S. C., Zahniser, M., Nelson, D., & McManus, J. B. (2006). Atmospheric reactive nitrogen concentration and flux budgets at a Northeastern U.S. forest site. *Agricultural and Forest Meteorology, 136*, 159–174.

Houghton, R. A. (1995). Land-use change and the carbon cycle. *Global Change Biology, 1*, 275–287.

Irland, L. C. (2008). Ice storm 1998 and the forests of the Northeast. *Pathology, 1*, 32–39.

Kulmatiski, A., Vogt, D. J., Siccama, T. G., Tilley, J. P., Kolesinskas, K., Wickwire, T. W., & Larson, B. C. (2004). Landscape determinants of soil carbon and nitrogen storage in Southern New England. *Soil Science Society of America Journal, 68*, 2014–2023.

Liu, W. H., Bryant, D. M., Hutyra, L. R., Saleska, S. R., Hammond-Pyle, E., Curran, D., & Wofsy, S. C. (2006). Woody debris contribution to the carbon budget of selectively logged and maturing mid-latitude forests. *Ecosystem Ecology, 148*, 108–117.

Magill, A. H., Aber, J. D., Currie, W. S., Nadelhoffer, K. J., Martin, M. E., McDowell, W. H., Melillo, J. M., & Steudler, P. (2004). Ecosystem response to 15 years of chronic nitrogen additions at the Harvard Forest LTER, Massachusetts, USA. *Forest Ecology and Management, 196*, 7–28.

Menzel, A., & Fabian, P. (1999). Growing season extended in Europe. *Nature, 397*, 659.

Ollinger, S. V., Aber, J. D., & Federer, C. A. (1998). Estimating regional forest productivity and water yield using an ecosystem model linked to a GIS. *Landscape Ecology, 13*, 323–334.

Olthof, I., King, D. J., & Lautenschlager, R. A. (2003). Overstory and understory leaf area index as indicators of forest response to ice storm damage. *Ecological Indicators, 3*, 49–64.

Orwig, D. A., Foster, D. R., & Mausel, D. L. (2002). Landscape patterns of hemlock decline in New England due to the introduced hemlock woolly adelgid. *Journal of Biogeography, 29*, 1475–1487.

Parmesan, C. (2006). Ecological and evolutionary responses to recent climate change. *The Annual Review of Ecology, Evolution, and Systematics, 37*, 637–639.

Perry, D. A. (1994). *Forest ecosystems*. Baltimore: John Hopkins University press.

Pregitzer, K. S., & Euskirchen, E. S. (2004). Carbon cycling and storage in world forests: Biome patterns related to forest age. *Global Change Biology, 10*, 2052–2077.

Richardson, A. D. (2010). Influence of spring and autumn phenological transitions on forest ecosystem productivity. *Philosophical Transactions of the Royal Society, 365*, 3227–3246.

Saxe, H., Ellsworth, D. S., & Heath, J. (1998). Tansley Review No. 98: Tree and forest functioning in an enriched CO_2 atmosphere. *New Phytologist, 139*, 395–426.

Schimel, D. S. (1995). Terrestrial ecosystems and the carbon cycle. *Global Change Biology, 1*, 77–91.

Schimel, D., Melillo, J., Tian, H., McGuire, A. D., Kicklighter, D., Kittel, T., Ojima, D., Parton, W., Kelly, R., Skyes, M., Neilson, R., & Rizzo, B. (2000). Contribution of increasing CO_2 and climate to carbon storage by ecosystems in the United States. *Science, 287*, 2004.

Seymour, R., & Hunter, M. (1999). Principles of ecological forestry. In: *Managing biodiversity in forested ecosystems*. New York: Cambridge University Press.

Toledo, M., Poorter, L., Pena-Claros, M., Alarcon, A., Balcazar, J., Leano, C., Licona, J. C., Llanque, O., Vroomans, V., Zuidema, P., & Bongers, F. (2010). Climate is a stronger driver of tree and forest growth rates than soil. *Journal of Ecology, 10*, 1–11.

Urbanski, S., Barford, C., Wofsy, S., Kucharik, C., Pyle, E., Budney, J., McKain, K., Fitzjarrald, D., Czikowsky, M., & Munger, J. W. (2007). Factors controlling CO_2 exchange on timescales from hourly to decadal at Harvard Forest. *Journal of Geophysical Research, 112*, 1–25.

Vincent, M. A., & Saatchi, S. S. (1999). Comparison of remote sensing techniques for measuring carbon sequestration. *Jet Propulsion Laboratory, California Institute of Technology, 1*, 36.

White, M. A., Running, S. W., & Thornton, P. E. (1999). The impact of growing-season length variability on carbon assimilation and evapotranspiration over 88 years in the eastern US deciduous forest. *International Journal of Biometeorology, 42*, 139–145.

Wofsy, S. C., Goulden, M. L., Munger, J. W., Fan, S. M., Bakwin, P. S., Daube, B. C., Bassow, S. L., & Bazzaz, F. A. (1993). Net exchange of CO_2 in a mid-latitude forest. *Science, 260*, 1314–1318.

Zheng, D., Heath, L. S., & Ducey, M. J. (2008). Spatial distribution of forest aboveground biomass estimated from remote sensing and forest inventory data in New England, USA. *Journal of Applied Remote Sensing, 2*, 1–18.

Lauren Kathleen Sanchez was a junior researcher at the Yale University School of Forestry and Environmental Studies in New Haven, CT. Her graduate research in Environmental Science investigated global climate change impacts in the Middle East and North Africa with Dr. Mark Bradford. More specifically, an analysis of how climate change will affect the ecological and geopolitical dynamics of the region. Her previous research with Harvard University was conducted with the Wofsy Group at Harvard Forest and in Cambridge, MA. This research was formulated into an independent honors thesis analyzing how changes in the growing season length are impacting the ecological components of forests in New England. She is a native of Seattle, WA and is very involved in climate campaigns both at Yale University and her undergrad institution, Middlebury College. In January 2015 – July 2015, Lauren was a Fulbright Scholat in Abu Dhabi, United Arab Emirates. She conducted an independent research on the geopolitics of energy policies in the UAE at Masdar Institute of Science and Technology, a leading global institution in renewable energy research. From September 2015 so far, she works as a Climate Negotiator at the US Department of State.

Chapter 17
Ensuring Sustainability in Forests Through the Participation of Locals: Implications for Extension Education

Mirza Barjees Baig, Juhn Pulhin, Loutfy El-Juhany, and Gary S. Straquadine

17.1 Introduction

With very thin stands, the forests and other woody vegetation lands spread over a limited area of 9770 km^2, covering more or less 1.0% of the landmass of the Kingdom of Saudi Arabia (KSA), making it really a forest deficient country (World Bank 2014). The kingdom is blessed with a variety of landscapes like deserts, mountains and wadis representing a rich diversity of vegetation. Some stands of forests of varying degrees of density, volume, and age and species composition also thrive in the valleys of the desert areas, and some scattered scrubs can be noticed on the rangelands. According to the report submitted to UNCCD (2006), the Kingdom sustains a huge diversity of 2300 plant species on an area of about 2.7 million hectares representing roughly 1.35% of the total area of the Kingdom.

In Saudi Arabia, major forests with prime tree species are widely distributed in the southwestern regions of the Kingdom (El-Juhany 2009a, b) as depicted in the

M. B. Baig (✉)
Department of Agricultural Extension and Rural Society, College of Food and Agriculture Sciences, King Saud University, Riyadh, Kingdom of Saudi Arabia
e-mail: mbbaig@ksu.edu.sa

J. Pulhin
Department of Social Forestry and Forest Governancem College of Forestry and Natural Resources, University of the Philippines, Los Baños College, Laguna, Philippines
e-mail: jmpulhin@up.edu.ph

L. El-Juhany
Prince Sultan Institute for Environmental, Water and Desert Research, King Saud University, Riyadh, Saudi Arabia
e-mail: ljuhany@ksu.edu.sa

G. S. Straquadine
Utah State University – Eastern, Price, UT, USA
e-mail: gary.straquadine@usu.edu

M. Behnassi et al. (eds.), *Climate Change, Food Security and Natural Resource Management*, https://doi.org/10.1007/978-3-319-97091-2_17

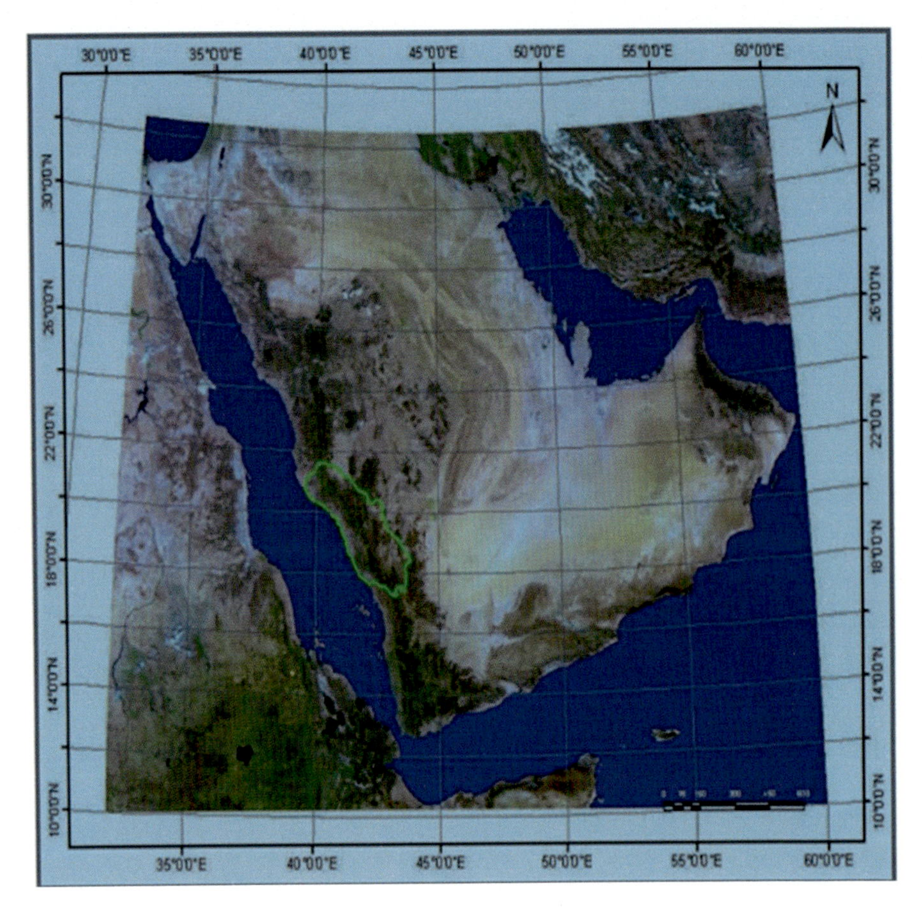

Fig. 17.1 A satellite image for the map of Saudi Arabia shows the location of the natural forest in the southwestern region. (Source: Department of Natural Resources 2007)

Fig. 17.1. The southwest region of the country consists of high mountains called Al-Sarawat Mountains, and these run parallel to the Red Sea, covering an area of about 1800 km (NCWCD and JICA 2006). In general, the Kingdom's forests are thin, weak, and mostly located at places damaged due to anthropogenic activities and natural climatic conditions. Some of the forestlands are severely degraded and cannot sustain woody vegetation. These forests are further threatened by harsh environmental conditions, which are characterized by low fluctuating rainfall, severe and prolonged droughts, high temperatures, and anthropogenic factors including urban and agricultural expansion as well as fuelwood extraction (Al-Subaiee et al. 2014). However, at some places, both indigenous and exotic tree species have proven to grow well under the existing environmental conditions. Plantations of those species can potentially increase green areas, which are suitable for recreational activities, improve the climate and environment (i.e. provide a suitable environment for wildlife and birds), and help control soil erosion.

In this perspective, the Ministry of Agriculture has taken numerous initiatives including afforestation programs to rehabilitate degraded natural forest areas and to stabilize sand dunes. In addition, due to the scarcity of fresh water resources, treated sewage water has also been used to irrigate the planted trees in order to increase the green area and combat desertification in the Riyadh and Taif regions. Also, water from dams has been employed to irrigate trees in the reforestation of degraded forest areas. Despite all these efforts, Saudi Arabia, with its low roundwood production, still remains an extremely wood deficit country, relying on significant imports of broad range of wood and paper products (FAO 2011).

The state forests are under severe stress due to several natural and anthropogenic factors. In addition, local people and communities living in the neighborhood of the forests cause severe and heavy damage to them. Some local people meet their domestic fuelwood needs through illegal harvesting of valuable trees whereas others damage them by unauthorized entering into the forests for picnics, littering and lighting fires at the undesignated areas. The interaction of natural and anthropogenic factors is further compounded for not having a forest management plan. In Saudi Arabia, forests were managed by local people under a traditional management system, but the system was abolished. Recently participatory forestry has come up with an appropriate forest management system to ensure the long-term sustainability. Participation, support and cooperation of local residents are an indispensable factor for sustainable woodland management (Agrawal and Gibson 1999; Ostrom 1999; Ferraro 2002; Wiggins et al. 2004; Robertson and Lawes 2005; Tesfaye et al. 2012; Matiku et al. 2013; Ameha et al. 2014). Similar views have been expressed in a series of studies conducted by Al-Subaiee (2014a, b, 2015a, b) indicating that local people have very valuable knowledge on forests, however the forest department is not utilizing this beneficial resource. The studies further revealed that local people are very eager to participate in the conservation, maintenance and management of local forests. In return, all they wanted is to have the regulated access to the services and benefits offered by forests; a demand which sounds quite reasonable.

In this scenario, it seems appropriate to examine the issues and challenges faced by the Kingdom's forest resources so that, through the active participation of the local communities, suitable solutions be tailored by combining indigenous experience with the scientific knowledge. It has been anticipated that meager and fragile forest resources of the Kingdom can be promoted, protected and conserved with the involvement of local people and communities. In this chapter, an effort has been made to review the possible role of local people in managing forests to ensure their protection and sustainability on a long-term basis. Vibrant forestry extension programs can enhance the local participation of all stakeholders. Implications of such initiatives on extension education programs have also been examined in the chapter.

17.2 Forest Ecosystems in the Kingdom of Saudi Arabia

Forests in Saudi Arabia are one of the scarcest natural resources, spreading over an area of 977,000 ha (FAO 2013) and cover only 1.0% of the total area of the country (FAO 2011). The growing stock of forests and woodlands is less than 10 m^3 per hectare as depicted in Fig. 17.1 (FAO 2005). According to the data of the Ministry of Agriculture, there are 27,000 km of woodland or 1.2% of the country's national land area. Natural forests occupy a significant area in the southwestern region. Approximately 80% (21,000 km) woodlands exist throughout the Sarawat mountain range in the southwestern region (NCWCD and JICA 2006). ***Juniperus procera*** as the climax species represents approximately 95% of the tree species grown in these forests (Abo-Hassan et al. 1984). With the potential of producing over 18 million tons of biomass annually (FAO 2005), the forests are concentrated in the Sarawat Mountains (2.7 million ha) in the south west, however, they also exist in the Hejaz Mountains in the north and the Asir Mountains in the south in Saudi Arabia (El Atta and Aref 2010).

Since centuries, forests are known by its innumerable associated environmental, economic, and social benefits (FRA 2010; FAO 2016). Forests and trees stabilize soils and climate, regulate water flows, give shade and shelter, and provide a habitat for pollinators and the natural predators of agricultural pests (FAO 2016). The productive and protective roles have been discussed at length at different fora to establish their importance and recognized them as fundamental for sustainable development. Forests play pivotal role in: protecting soils from wind and water erosion and flood hazards and damages; facilitating downward movement of water; increasing soil moisture; and maintaining reasonable underground water reserves. Forests also have an important role in minimizing greenhouse gas emission effect, through carbon sequestration and Oxygen release, and in minimizing dust and air pollution which are a major environmental problem in many countries, including Saudi Arabia. In addition, forests have an important role in conserving biodiversity and wildlife and enhancing environmental tourism (Al-Subaiee et al. 2014; FAO 2016). In Saudi Arabia, forests are considered one of the renewable natural resources that play an important role in the country's ecosystem given its extensive area and diverse environment. This renewable natural resource provides protection to such areas, namely by: preserving the soil from water and wind erosion; enabling the distribution of water and the control of its flow; and consequently increase the moisture in the soil. This is in addition to the economic, recreational, scenic, tourist, and climatic moderating values of the forests (FAO 1997). However, the gathering of commercial firewood, the uprooting of plants and the consequent soil disturbance has caused a significant degradation of forests and rangeland in Saudi Arabia (Al-Rowaily 1999). The important forests of the kingdom, their status and the major types of degradation are discussed in the proceeding paragraphs (Table 17.1):

Table 17.1 Land use indicators in the Kingdom of Saudi Arabia

Item	Value
Country area	214969.00
Land area	214969.00
Agricultural area	173295.00
Area under crops	
Arable land	3068.00
Arable land and permanent crops	3295.00
Permanent crops	227.00
Temporary crops	468.00
Area under Irrigation	
Total area equipped for irrigation	1620.00
Area under organic agriculture	
Agricultural area organic, total	36.60
Agricultural area certified organic	33.70
Agricultural area in conversion to organic	2.90
Area under forests	
Forest area	977
Primary forest	360.00
Planted forest	0.00
Other naturally regenerated forest	617.00
Primary forest	360.00
Planted forest	0.00
Other naturally regenerated forest	617.00

Data presented for the year 2013; unit for area = 1000 ha
Source: http://faostat3.fao.org/download/R/RL/E

17.2.1 Forests in Taif

With an estimated area of 24,428.60 km^2, forests of Taif are located between latitudes 20°00′ south and 21° 36′ north, and between longitudes 39° 31′ west and 41°56′ east (Fig. 17.1) According to another estimates made by Abo-Hassan et al. (1984), productive forests of Taif almost spread over an area of 43.02 ha. The area receives seasonal rains in summer and continental rains in winter, not exceeding 500 mm per year. Taif is the part of the southwestern region of the country, characterized by mountainous terrain and some arable lands. With climax species ***Juniperus procera***, forests are patchy and sparse, concentrated in the mountainous region however, other species like acacia, wild olive, local Neem, can be noticed as well (Natural Resource Department 2007; El-Juhany and Aref 2012a). A recent study conducted by (El-Juhany 2014) revealed that juniper trees in Taif were 56–132 years old with an average of 83 years. El-Juhany et al. (2013) mentioned that the present status of Taif forests suggests that they have been exploited over a long period of time, causing the obvious visible effects on the existing trees. They

Table 17.2 Growing stock of the ten most common species

FRA 2010 category/species name		
Rank	Scientific name	Common name
1st	*Juniperus procera*	Arar
2nd	*Acacia* spp.	Akasiat
3rd	*Ziziphus spina-christi*	Al sidir
4th	*Olea europaea*	Zaitoon bari
5th	*Prosopis juliflora*	Miskeet
6th	*Tamarix* spp.	Al tarfa
7th	*Avicennia marina*	Mangroves
8th	*Hyphaene thebaica*	Al dom
9th	*Capparis decidua*	Tontob
10th	*Ficus* spp.	Al labakh

Source: FRA (2010)

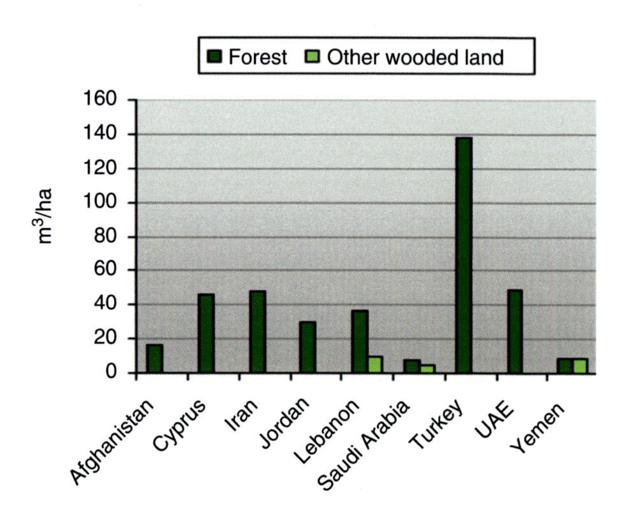

Fig. 17.2 Growing Stock of forests and woodlands in the Kingdom of Saudi Arabia. (Source: FRA 2005)

believed that disappearing of most of the larger trees is an indication of the degradation of forests in Taif (Table 17.2 and Figs. 17.2 and 17.3).

Taif forests exhibit degradation in the form of low capacity of natural regeneration of the dominant tree species, wide distance among the trees, high proportion of irregular trees, disappearing of the bigger trees, noticeable eroded soils, die-back and die-off of trees, insects break out, decreasing volume of tree species, loss of wild animals species with an increase of monkeys, expansion of urbanization and accompanying activities, stretching of recreational areas, etc. These forests have been subjected to misuse, intensive grazing and deforestation. Consequently, forest soils have lost their properties such as organic matter, profile depth, and water holding capacity that helps seeds to germinate and grow. These issues warrant immediate interventions to halt the forest degradation. Forest mapping could be an essential and effective tool to launch further rehabilitation programs (El-Juhany and Aref 2012a) (Tables 17.3 and 17.4).

Fig. 17.3 Map of the Kingdom of Saudi Arabia. (Source: http://www.worldatlas.com/webimage/countrys/asia/lgcolor/sacolor.htm)

Table 17.3 Extent of forest and other wooded lands (area in thousand hectares)

FRA 2005 categories			
	1990	2000	2005
Forestlands	2728	2728	2728
Other wooded land	34,155	34,155	34,155
Forest and other wooded land	36,883	36,883	36,883
Other lands	178,086	178,086	178,086
Total area of country	214,969	214,969	214,969

Data source: FAO, Global Forest Resources Assessment (2005) (http://www.fao.org/forestry/country/32185/en/sau/)

17.2.2 Forests in Al-Baha Region

Al-Baha region is positioned in the southwest part of the country with its coordinates between the longitudes 41° and 42′ east and between 19° and 20° latitudes south (Doha 2009). It is bounded on the north and west by Makkah. With an area of 12,000 km², representing 0.6 of the total land area of the Country, Al-Baha is

Table 17.4 Production of roundwood and its imports in the Kingdom of Saudi Arabia

	Volume (1000 CUM)				Annual growth rate (%)		
	1998	2003	2008	2013	1998–2003	2003–2008	2008–2013
Production	159.21	198.33	232.90	267.77	4.49	3.27	2.83
Imports	8.20	25.20	28.77	21.32	25.17	2.69	−5.82
Exports	0.00	0.00	0.00	1.74	n.a.	n.a.	n.a.

Source: FAOSTAT, FAO of the UN source: http://faostat.fao.org/CountryProfiles/Country_Profile/Direct.aspx?lang=en&area=194
http://faostat.fao.org/site/626/default.aspx#ancor (Accessed on December 18, 2014)

Table 17.5 Areas covered by different land cover and land use in Al-Baha

Land cover/land use	Area (km^{2})	% of total area
Dry land cropland and pasture	2870.7	23.5
Sparsely vegetated	2769.3	22.7
Forest and shrub land	3703.2	30.3
Bare soil	2863.9	23.5
Total	12,207.1	100

Source: Mahmoud et al. (2014)

Table 17.6 Tree and shrub species grown in Al-Baha forests

Scientific name of the tree species	Family
Juniperus procera Hochst. ex Endlicher	Cupressaceae
Acacia sp.	Memosaceae
Barbeya oleoides Schweinf	Barbeyaceae
Olea europea ssp. Africana (Mill.) P. Green	Oleaceae
Rhus retinorrhea Steud. Ex. A. Rich	Anacardiaceae
Teclea nobilis Del	Rutaceae
Euclea shimperi (A. DC.) Dandy	Ebanaceae
Pistacia falcata Becc. ex Martelli	Anacardiaceae
Dodonaea viscosa Jacq	Sapindaceae
Azedarachta indica A. Juss	Meliaceae
Zizyphus spina-christi (L.) Willd	Rhamnaceae
Salvadora persica (L.)	Savadoraceae
Tamarix sp.	Tamaricaceae

Source: El-Juhany and Aref (2012b)

situated between the Holy Makkah and Asir Regions (Doha 2009) and is the smallest region among all the 13 administrative regions of the country (Saudi Geological Survey 2012) (Tables 17.5 and 17.6).

Al-Baha region occupies some parts of the mountains of Al-Sarawat (plural of Sarah). The region has a very pleasant climate due to its location above the sea level; moist winds coming from the narrow plains of Tihama in the western Saudi Arabia make the climate moderate in summer and winters cool and wet. The region is divided by huge and steep rocky mountains into two main sectors: a) 'Tihama', the lowland coastal plain lying in the west; and b) 'Al-Sarat or Al-Sarah', a mountainous area with an elevation of 1500–2450 m above sea level, occupying some parts of Al-Sarawat Mountains range (Ibrahim and Abdoon 2005; Alahmed et al. 2010). The scenic beauty of the region attracts tourists from within the Kingdom and the neighboring states due to its green forests (about 53 forests), wildlife areas, agricultural plateaus, valleys, and high mountains. Pleasant climatic features like: rainfall throughout the year, low temperatures in summer and winter and lot of fog in the mountains have helped the region making the major tourist resort (El-Juhany and Aref 2012b).

17.2.3 Forests in Asir Area

Asir region occupies the significant part of Al-Sarawat Mountain range in southwestern Saudi Arabia, where its northern border begins from the southern border of Baha region and its southern borders end at the northern border of Jazan region. Asir region includes large tracts of land stretching across the eastern slopes of these mountains to the borders of the Empty Quarter and across its western slopes until the Red Sea coast (El-Juhany et al. 2013). The trees in Asir forests have witnessed exploitation for the long time and the existing trees clearly show such effects. For example, the trees in the whole Asir and East Jazan are irregular, degraded, stricken by dieback and completely dead. The main threats to these forests include: low capacity of natural regeneration of the main forest species, die-back and die-off diseases of trees, spread of forest fires, wood cutting, grazing, insects break out and spread of recreation areas (El-Juhany et al. 2013). The forests had been heavily exploited or severely subjected to pests. A large number of trees are either irregular or cull trees in Asir region (El-Osta 1983). El-Juhany et al. (2008) attributed the existence of irregular and deteriorated juniper trees in Ridah Reserve at Asir Mountains to various factors such as the mechanical effects of wind, soil erosion, competition between trees, animal grazing and others. However, Abo Hasan et al. (1984) reported that the juniper trees in Asir region were healthy and of good quality due to the relatively high rainfall and the deep soil. In the most recent study conducted by El-Juhany et al. (2013), they noticed that trees from Asir forests were healthier than those from both Taif and Baha forests.

17.3 Problems, Issues and Challenges Faced by the Forests in the Kingdom

Forest and rangelands are facing serious challenges in Saudi Arabia as a result of severe natural and climatic conditions (dry climate, high temperatures, poor soil) and anthropogenic activities (such as excessive utilization of woodfuel, overgrazing, expansion of uncontrolled urbanization, sprawling of agricultural and residential areas in the forests and intensive recreation) (El-Juhany 2009a, b; Darfaoui and Al Assiri 2010; El-Juhany and Aref 2012a). These challenges and barriers prevent the realization of forest sustainable development in the country (Ministry of Economy and Planning 2005). Some of the prominent problems impeding the development and threatening the protection of meager forest resources of the Kingdom are discussed below.

17.4 Forest Degradation and Deforestation in KSA

Natural forests in Saudi Arabia cover an area of about 762,474 ha and extend from Yemen territory in the south to Taif in the north. However, El-Juhany and Aref (2012a, b) are of the opinion that a large part of these areas are deteriorated and need development and maintenance. Forestlands in the country have a long history of overexploitation and degradation. Most of the forests are now classified as other wooded land, but there are still a small percentage of closed forests in the hills and mountains (ACSAD 2003). At present, there are indications of forestland degradation in various parts of the Kingdom; however, the available scientific data on its extent and severity is insufficient. Al-Shomrany (1980) reported that excessive deforestation in the Kingdom was clearly evident and vegetation loss could be due to various reasons. The following are considered the most important indicators of forestland degradation in Saudi Arabia (El-Juhany 2009a, b):

Vegetation in forests and forestlands seems under stress with the following symptoms and features

- Woody species are decreasing due to low natural regeneration among the main tree species in the forests of the Kingdom. Forest vegetation prevails in patches and small pockets with high number of dead and irregular trees (curved, twisted, cleft, multi-stemmed, dwarf, leaning trees etc.). A higher number of deteriorated and damaged trees (partly cut, fully cut, burned, etc.) with dried tree branches and limbs are noticed. Forest trees are with dull color and the tree leaves do not show shiny appearance. Larger trees are disappearing from the forests and invasion of alien plant species are replacing the native species;
- Forest trees suffer from die-back and die-off diseases and the vegetative cove is infested insects attacks;
- Forests experience more frequent outbreak of forest fires;
- Forestlands are exposed to soil erosion due to loss of vegetation cover;
- Ruthless removal of forest vegetation results in loss of biodiversity as some wild animal species are losing their habitats and native tree species are low in number and the disappearance of flora and fauna is causing an imbalance in the ecosystem.

Forestlands in the country have been exposed to overexploitation and degradation for the last many decades (El-Juhany 2009b). Many interacting factors also result degradation of forestlands and forests. The following are main causes responsible for forest degradation in Saudi Arabia.

17.4.1 Overgrazing

The acacia woodlands in the southwestern region of Saudi Arabia are a unique example of forest degradation caused by overgrazing. Overgrazing has been recognized as one of the primary reasons for forest and rangeland degradation, and the chief impediment to the regeneration of vegetation (Darfaoui and Al-Assiri 2010; Ministry of Agriculture 2002; Ministry of Economy and Planning 2010). In most of arid and semiarid zones, livestock farmers keep more number of animal heads and use their rangelands above their carrying capacities for most of the year (Ma 2008) causing severe damage to the vegetation and the grazing lands. The trampling of the grazing livestock prevents regeneration and the survival of the emerging seedlings. The magnitude of the negative impact caused by overgrazing on forests depends on the intensity of grazing, which is determined by the number of animals and the time spent grazing (El-Juhany 2009b). At places where the palatable shrubs and herbs witness severe overgrazing often sustain vegetation in thin small and individual clusters (Chaudhary and Le Houerou 2006). Similarly, due the absence of restricted grazing laws on using the woodlands in the Asir Mountains results vegetation degradation and, in principle; anyone is allowed grazing his livestock in the area.

17.4.2 Low Natural Regeneration

The Juniper forests inhibit low capacity for the natural regeneration. Low natural regeneration has been observed in the natural woodlands and rangelands due to harsh climate, human interferences and pressure of overgrazing (El-Juhany et al. 2008). Natural regeneration is known to improve vegetation cover that, in turn, protects the degraded land and accelerates forest succession by re-establishing and recovering a healthy forest in a short time (Al-Subiee 2015). Natural regeneration is more efficient than artificial, as it avoids transplant shock. Moreover, natural regeneration can be assisted by protecting the area from grazing, fire, and removing weeds around new seedlings to accelerate the process. Assisting natural regeneration has been a successful technique in all the ecological zones of sub-Saharan Africa.

17.4.3 Illegal and Unlawful Cutting and Removal of Trees and Shrubs

Due to illegal and unlawful cutting and removal of trees, vegetative cover in the Kingdom is disappearing from more than 120,000 ha of the country every year (Al-Abdulkader et al. 2009; El-Juhany 2009b; El-Juhany and Aref 2012a, b). People in the Kingdom traditionally use wood and charcoal for heating in winter and cooking food. This is still customary despite the availability of electricity, (butane) gas, and other petroleum derivatives at generous and highly subsidized prices. In addition, selective species cutting may cause erosion of genetic diversity, particularly when regeneration is slow or completely absent (El-Juhany 2009b). To regulate or restrict this process, the Ministry of Agriculture (MAW) introduced a licensing system for utilizing dry (dead) plants for firewood, charcoal production, and for transporting either of these. The impact of these new regulations is evident, as a result of public awareness campaigns on the importance of maintaining trees and shrubs, and applying penalties for violators. However, there are still unlicensed operations that are felling and transporting trees and shrubs, and efforts should be intensified and coordinated to enforce the rules that aim at protecting natural vegetation resources from extinction (El-Juhany and Aref 2012a, b).

17.4.4 Harsh Environmental Conditions

Among the main impediments to forest development in Saudi Arabia, the country location in the desert belt region is of utmost importance. Geographically speaking, the Kingdom is dominated by an arid desert climate, with a semi-arid climate along the Red Sea (FAO 2009). This Kingdom is characterized by a very low and variable rainfall, high temperature, low humidity, lack of rivers, and diminishing underground water. These climatic and geological factors make the natural growth and regeneration of forests very slow, and the afforestation and tree planting programs very expensive (El-Juhany 2009b). This is especially visible in the interior parts of the country such as Riyadh area where the woodlands cover an area of about 290,000 ha with the most of the prevalent trees species like: *Acacia spp., Tamarix spp. and Haloxylon persicum* (Ministry of Agriculture 2002; Badai and Aldawoud 2004).

Forestry development initiatives like afforestation programs are adversely affected due to high temperatures and harsh environmental conditions due to the location of the Kingdom in the dry desert belt (Darfaoui and Al-Assiri 2010). Al-Subaiee et al. (2014) identified several factors preventing natural forests from flourishing, such as climate (characterized by scarcity of rain, and dominated by drought throughout the year), high temperatures (especially in summer), lack of adequate water supply, and the absence of permanent water bodies such as rivers. Similarly, Abido (2000) of the opinion that harsh environmental conditions may result in low regeneration and similarly the southwestern region of the Kingdom of Saudi Arabia dominating forests may experience biological stress.

17.4.5 Climate Change

The current harsh and dry climate in the KSA already posed serious challenges to the forests and rangelands management. With the anticipated increase in global average temperature brought about by climate change, natural ecosystems, such as mountain forests and woodlands, wadis (valleys), rawdas (gardens), wetlands, and coastal areas, are at risk. Therefore, in the absence of appropriate adaptation, more detrimental and damaging effects of climate change on rangelands and forestlands could be expected in the future. For example, an increase in the frequency and changes in the patterns of natural disturbances (drought, sandstorms, fires, and floods) could lead to increased die-back and die-off in forests and woodlands, spread of diseases, change in species composition and richness, drop in productivity, and loss of biodiversity (Darfaoui and Al-Assiri 2010). Gardner and Fisher (1994) maintain that the poor natural regeneration in Juniper forests and prevalent biological stresses could be attributable to climate change.

17.4.6 Pests and Diseases

Decline and dieback occur extensively in the natural forests of the Kingdom. Trampling due to overgrazing causes forestlands degraded and it is one of the causes of die-off and die-back phenomena (JICA 2002). The overall health of *Juniperus procera* woodlands in the Sarawat Mountains in the Kingdom is generally not good, exhibiting an extensive decline and dieback. In the old tree stands, the pods are often infested with insects, with seeds being partially or completely damaged. In general, various forms of dieback result to poor stands with loss of vigor (FAO 2006, 2008). Similarly in Saudi Arabia, seeds of native acacia species are heavily infested by the bruchid beetle (Bruchidius) that reduces natural regeneration. Abdullah and Abulfatih (1995) found that the rate of seed predation by bruchid beetles in south-western Saudi Arabia varied among the Acacia spp., but the highest rate of perdition were recorded in *A. tortilis* and *A. ehrenbergiana*. Ernst et al. (1989) found that bruchid beetles damaged cotyledons and radicular in seeds of *A. tortilis* where infected seeds had no germination. Hajar et al. (1991) reported that larvae *Strepsicrates cryptosema* belonging to the family Tortricidae adversely affected the seed production of *Juniperus excels* with an infestation level of about 64%.

Table 17.7 presents an account of the damages caused by various forms of disturbances affecting forests and other wooded land vegetation in the Saudi Arabia for the years 1990 and 2000 where pests and diseases are significant factors.

Also, Rathore (1995) reviewed some literature and listed many insects associated with *Acacia tortilis*. These insects belong to the orders Coleoptera, Homoptera and Lepidoptera and feed on bark and wood, destroy seeds, bore under bark, destroy softwood, suck sap, and cause defoliation.

Table 17.7 Disturbances affecting forest and other wooded land in Saudi Arabia for the years 1990 and 2000

	Average annual area affected (1000 ha)			
	Forest		Other wooded land	
FRA 2005 categories	1990	2000	1990	2000
Disturbance by fire	n.s.	n.s.	n.s.	n.s.
Disturbance by insects	4	4	30	30
Disturbance by diseases	2	2	15	15
Other disturbances	3	3	20	20

Data source: FAO, Global Forest Resources Assessment (2005) http://www.fao.org/forestry/country/32267/en/sau/

For native species, lack of information about seed harvesting, storage, treatment and germination requirements may magnify the dilemma, but appropriate solutions of these issues can be made available through research. However, pest management strategies must be devised to maintain pest populations, if any, at the normal levels and that can be achieved with compatibility of integration of chemical, mechanical and biological control.

Parasitic plants are known to have an impact on the structure and dynamics of plant community through their negative effects on growth, reproduction and photosynthesis of the host plant (Marvier 1996; Mutikainen et al. 2000). Similarly El-Juhany (2009b) noted that different parasitic plants grow on the aerial parts of different tree species including *Juniperus procera* and *Acacia spp* in the natural forests of Saudi Arabia in southwestern region. He also found that parasitic plants seemed healthy and their host trees were weak in appearance. In Saudi Arabia, the parasitic plant *Phragmanthera austro* mainly infects Acacia spp. that grows in southwest part of the country.

17.4.7 High Cost of Reforestation

Indiscriminate removal of forests are limiting and depleting the natural tree cover in Saudi Arabia making reforestation an essential, albeit very expensive, strategy (Al-Subaiee 2015a, b). Numerous other factors, such as low and variable rainfall, low soil fertility, high temperatures, low humidity, lack of rivers, and scarcity of underground water have a great influence on forest stands. All of these factors hamper the natural growth, making regeneration of the forests very slow and afforestation and tree planting programs very expensive (Aref and El-Juhany 2000; Badai and Aldawoud 2004; Ministry of Agriculture 2002). These factors are also responsible for limiting the afforestation initiatives. Importing of suitable exotic species capable of growing in local conditions is predicted to compensate for the lost trees to some extent. However, such a venture could be expensive and considerable financial resources would be required to launch this type of afforestation project (FAO

1977). In addition, the exotic species can aggressively compete with and displace native plant communities and invasive plants impact the ecosystems. Replacement of native species with the exotic could result destruction of forage for wildlife and their habitat, loss of available grazing lands, forest productivity, reduced groundwater levels, soil degradation, increased risk of devastating wildfires, and diminished recreational resources, and decreased land values (US Forest Services 2009).

17.4.8 Lack of Water Harvesting Techniques

Moisture provision to vegetation sites enhances the ability of plants to grow and regenerate. In the central region of Saudi Arabia, water-harvesting techniques are not applied and consequently rainwater is being lost as runoff. However, there are many basic and low cost methods that can be used to achieve regeneration and return vegetation cover to the degraded lands. Reducing water runoff with physical structures, such as stone bunds, ridges, half-moons, and the zaï (a basin to keep water available), is used in different countries to enhance land productivity (Al-Subaiee et al. 2014). The most widespread agricultural system, practiced since ancient times in the southwestern region of Saudi Arabia involves the man-made terraces, meant for production as well as water harvesting. The water collected due to these terraces move into the soil. Walls at the edge of the terraces prevent runoff from flowing down to the next terrace except during intense rainfall. Unfortunately, use of terracing has been declining due to lack of maintenance, migration of labor and emphasis on large-scale agricultural development (El-Juhany 2009b).

17.4.9 Expansion of Agriculture

Forests and trees also contribute to the food security and are important sources of food, energy and income for millions of people in the developing countries. Yet, agriculture remains the major driver of deforestation globally, and agricultural, forestry and land policies often do not support each other (FAO 2016). Similarly, rapid development in the Arabian Peninsula has caused severe environmental damage (Nasroun 1993). For example, in Saudi Arabia, agricultural development has been largely achieved at the expense of forest and pasture areas as trees were cut to expand agricultural areas (FAO 2008). However, the absence of regulations for environmental protection and poor environmental awareness has resulted in the deterioration of renewable natural resources. The huge agricultural expansion has been achieved at the expense of rangelands, forests, and marginal areas, the intensive use of groundwater resources, and with the excessive use of pesticides and fertilizer (FAO 2006, 2008). Similarly, Al-Subaiee (2013) conducted a study in Al-Baha (Saudi Arabia) and reported that forests are shrinking and being cleared to expand the agricultural fields. The Johannesburg Summit (2002) asserted that expanding farms play a major role in forestland degradation in Saudi Arabia.

17.4.10 Urban Expansion into Forest Areas

The discovery of oil wealth brought real boom in the construction of new buildings and an expansion of residential areas, particularly in the southwestern region, resulting in the shrinking of forest areas. Extensive urban planning of towns, construction of residential centers, sprawling of towns into the villages in the forestlands reduced the area under forests and the associated vegetation (FAO 2008; El-Juhany and Aref 2012a; Al-Subaiee 2013). Similarly, NCWCD and JICA (2007) reported that juniper woodlands have declined mainly due to human activities, such as tree felling, overgrazing, expanding farmlands, building of recreational facilities, and construction of roads and housing etc. Similarly the Johannesburg Summit (2002) asserted that expanding farms, establishing infrastructure, constructing houses, building roads, and developing recreational areas were the major indicators of urbanization, causing forestland degradation in the KSA. Consequently, degradation of forests, deforestation and deterioration of natural pastures, have been recognized among the main challenges and impediments to sustainable development in the country in the eighth National development plan (Ministry of Economy and Planning 2005).

17.4.11 Deforestation Causes the Loss of Biodiversity

Deforestation has been designated a prominent menace destructing the habitats for animals, plants, and microorganisms and resultantly reducing biodiversity. Deforestation leads to desertification and it is known as a major land degradation process responsible for eroding the genetic diversity of plant varieties and species existing in these fragile ecosystems.

17.4.12 Deforestation Causes Desertification

Deforestation (loss of trees along with their multiple functions) is known to cause desertification (Darfaoui and Al-Assiri 2010; El-Juhany 2009b; El-Juhany and Aref 2012a, b; Ministry of Economy and Planning 2010). Both desertification and deforestation are considered among the most detrimental, severe, and critical environmental problems faced by many parts of the world, with serious long-term economic and social consequences. The Kingdom also suffers from an extreme degree of

Table 17.8 Breakup of land areas in the KSA (1000 ha)

Irrigated farmlands (IF)	Rainfed farmlands (RF)	Rangelands (R)	Extremely arid lands (D)	Total
415	760	112,345	126,480	240,000

Source: Dregne and Chou (1992)
Available at: http://www.ciesin.columbia.edu/docs/002-186/tab1b.gif

Table 17.9 Degree of desertification in KSA in (1000 ha)

Degree of desertification						
Irrigated farmlands	Sleight	Medium	Severe	Very severe	M = medium + severe + very severe	%
415	155	200	40	20	260	63
Rainfed farmlands						
760	300	420	38	2	460	61
Rangelands						
112,345	22,345	60,000	29,800	200	90,000	80

s slight, *m* medium, *sv* severe, *vsv* very severe, *m+* medium + severe + very severe
Source: Dregne and Chou (1992)
Available at: http://www.ciesin.columbia.edu/docs/002-186/tab2c.gif

desertification due to its location in the continental west dry desert belt, and exposure to frequent drought cycles, increased population activity, overgrazing, logging, expansion of agriculture, and urban expansion etc. (Darfaoui and Al-Assiri 2010; El-Juhany 2009b; El-Juhany and Aref 2012a, b; Ministry of Economy and Planning 2010). Extent of desertification in the Kingdom of Saudi Arabia is depicted in Tables 17.8 and 17.9.

17.4.13 Deforestation Causes Soil Erosion

The disastrous impacts of deforestation appear in the accelerated soil erosion. The menace of soil erosion results in the permanent loss of agricultural productivity, and further leads to desertification accompanied by drought and food shortage (FAO 2007). Deforestation is noticed in the form of excessive tree felling for wood fuel and the expansion of agriculture (crop and livestock production) into the forests by encroaching forestlands (Assaeed and Al-Doss 2002). About 75% of deforested areas are used for agriculture, as witnessed in most developing countries (El-Lakany 2004). According to Clarke and Thaman (2006) deforestation and clearing of forestlands result the loss of soil due to erosion process.

17.4.14 *Loss of Soil Fertility*

It is well-documented that deforestation adversely affects soil fertility, declines crop yields, and brings about a change in the quality, quantity and diversity of vegetation. Negative impacts of deforestation could be observed in the form of less number of perennial trees and grasses; and palatable and nutritious plants are replaced by less number of edible species, or even toxic shrubs (Echholm 1976). Consequently, the natural potential of the land gets reduced along with a decrease in surface and ground water. Thus, the reduced volume of biomass and low carrying capacity of land realized due to deforestation negatively impact the on living conditions of livestock and the economic development of affected areas.

17.5 KSA Initiatives to Bring Sustainability in the Forests Taken So Far

In the light of the aforementioned scenario, it becomes imperative to manage the low, thin and patchy vegetative cover of the Kingdom in a sustainable manner to ensure the continued production of forest goods and environmental services through time. The Kingdom of Saudi Arabia attaches great importance to its natural forests and makes efforts to address the issues causing forests degradation. Accordingly, the degraded or over-grazed forests are now in the rehabilitation process through an extensive management plan in the entire Kingdom. The main purpose of planting forest trees is to protect the environment (Abo-Hassan 1983). In order to rehabilitate degraded natural forest areas several afforestation programs have also been undertaken by the Ministry of Agriculture (Al-Mosa 1999). Several indigenous and exotic tree species tried so far to increase the green area, showed good growth under the existing environmental conditions in these afforestation programs. They improved the climate and environment, established picnic areas, helped in controlling soil erosion and created a suitable habitat and environment for wildlife and birds. The Initiatives to bring sustainability in the forests in the Kingdom taken so far are presented in the Table 17.10.

17.6 6 Participatory Approach: A Viable Option to Realize Sustainable Forest Management

Participatory forest management is defined as "an arrangement where key stakeholders enter into mutual agreements that identify their respective roles, assign responsibilities, share benefits and the designate authority in the managing the nearby woodland resources" (Matiku et al. 2013). Participatory forestry has emerged and widely accepted as a viable sustainable forest management system (Gilmour

Table 17.10 Forestry initiatives and programs undertaken so far in the kingdom

Project, the implementing agency and the year of execution	Activities	Resultant benefits and achievements
Afforestation and sand fixation Ministry of Agriculture (1999)	Several indigenous and exotic tree species are grown to increase the green area;	Improved the climate and environment, established picnic areas; helped controlling soil erosion;
	Afforestation programs and sand dune stabilization projects were in place on 54 sites, covering an area of 3582 hectares (ha) of degraded forests. The rehabilitation sites consisted of about 265 ha in the Al- Baha region, 98 ha in the Bisha region, almost 143 ha in Asir region, 1131 ha in Riyadh region, 91 ha in the Taif region and 1800 ha in the sand dune stabilization project at Al-Elsa region;	Created a suitable habitat and environment for wildlife and birds;
		Established trees meant to increase the green area and for combating desertification in the Riyadh and Taif regions and were irrigated with treated sewage waters
Inventory of the Natural Forests in the Southwestern Region of Saudi Arabia (2001)	The project funded by The Ministry of Agriculture was supervised by The Space Research Institute at King Abdulaziz City for Science and Technology. The researchers from different Saudi Universalities, Research centers and the Natural Resources Department at the Ministry of Agriculture implemented the project and carried out the activities	The results have described the present status of the natural forests in the southwestern region of Saudi Arabia in terms of their areas, trees and shrubs, soil, disease and insect outbreak, social and economic properties etc.
Juniper Ecosystem Rehabilitation Project FAO and Ministry of Agriculture (2006)	Through the implementation of project, more than 10,000 juniper trees were planted in Asir area to rehabilitate the degraded Juniper Ecosystem	The project helped controlling runoff, improving water harvesting; To combat the juniper die back in the Sarawat Mountains, branches of dead trees were removed; local juniper saplings were planted; Measures to rehabilitate the rangelands were undertaken

(continued)

Table 17.10 (continued)

Project, the implementing agency and the year of execution	Activities	Resultant benefits and achievements
Development of drought resistant fodder seeds Ministry of Agriculture and FAO (2006)	Developed applicable technologies to produce good quality seeds for the development of rangelands and seedlings to enhance green cover	The seed developed is being used to produce highly drought-resistant fodder on private farms and rangelands
Ministry of Agriculture (2010–2014)	Identified 40 sites in the various regions to protect pastoral plants from overgrazing;	Efforts towards the rehabilitation of the vegetation cover in rangelands and forests were successful;
	Setup 3 stations for the propagation of pastoral-plant seeds;	Establishment and supervision of many national parks promoted eco-tourism in the Kingdom, that in turn helped preserving the environment;
	Established 27 pastoral and forestry-plant nurseries, and a plant gene bank;	
	Established many national parks that are important for tourism and entertainment have been offered for investment, in an effort to enhance private-sector participation in their development and highlight the role of national parks in preserving the environment	
	Established a center to receive satellite images of movement of sand dunes, to monitor and study such movements that threaten areas suitable for human settlement and agricultural production;	
	Established a special centre to study deserts and combat of desertification; Launched project for protection and rehabilitation of the mangrove areas; Sand dunes fixation in the areas of Al- Gonfoda, Wadi Dawassir, Al-Hassa and Najran was made	

National Forest Strategy and Action Plan, National Action Program to Combat Desertification 2010	The projects carried out the activities related to combating desertification and to mitigate the effect of drought in the framework of the UNCCD (United Nation Convention of Combat Desertification) were initiated	Lands were made productive by fitting the plant life in the degraded lands
Indigenization of forests development technologies to enhance forests environmental role in Riyadh Area; Al-Subaiee (2014b)	In Saudi Arabia forests are scarce and are faced with an acute issue of indiscriminate feeling and removal of trees at a rate that exceeds its natural regeneration. The situation warranted immediate attention and required actions to explore and identify alternatives for forest products, substitutes of wood fuel needed by the different social groups. Further viable and workable strategies were required to create awareness on these wood substitutes and their dissemination through vibrant extension education programs. Such initiatives are of paramount importance to help people adoption of the suitable and available substitutes in order to reduce their dependence on forests as a source for such products	The project identified the reasons of forest degradation in the Riyadh area; locals were illegally felling neighborhood forests for woodfuel, therefore alternate clean and easily available alternate energy sources were identified and concrete suggestions were made to protect the meager forest resources. The project established the need for the formulation of new forestry extension system that could involve all the stakeholders to ensure sustainability and save the local forests

1995; FAO 1998). Participatory forestry concepts are drawn on the principles of the forest management – a complex, valuable, natural resource system that has several crucial functions and multiple benefits. Participatory forest management is based on the experience and knowledge of professional foresters and the indigenous knowledge of locals in a partnership arrangement whereas other stakeholders may be engaged, in an essentially new management paradigm (FAO 2016; Carter 1999). The concepts of participatory forest management complement the forest development strategies with new strategies focusing on basic needs, equity, and active participation of locals, accommodating national interests (Carter 1999; Wiersum 1995; Wily 2002; FAO 2016).

17.6.1 Participatory Approach in Forest Management Is Beneficial

Participatory forestry is a valuable management system that enjoys the support and cooperation of locals, aimed at improving woodland conservation and ensuring sustainability (Ameha et al. 2014; FAO 2016; Ferraro 2002; Matiku et al. 2013; Robertson and Lawes 2005; Tesfaye et al. 2012; Wiggins et al. 2004). Voluminous literature exists that suggests the efficiency of government as a forest resource manager improves when authorities and responsibilities are shared with different user groups (Jodha 1992; Kumar et al. 2000; World Bank 2001). Several studies have indicated that sustainable forest management also requires meaningful collaboration among local communities, government agencies, concession holders, NGOs, and other institutions involved in forest management to assess, plan, monitor, and manage operations according to locally defined concerns, needs, and goals (Elsiddig et al. 2001; FAO 1998; Kobbail 1996).

17.6.2 The Role of Participatory Forestry in Realizing Sustainability

Participatory forestry enables both the locals and the nation to realize goods and services that improves the livelihoods of the locals, without compromising the long-term sustainability of the jointly-managed forests (FAO 2003). Although inhabitants of local communities form a distinct group in relation to their neighboring forests (Western and Wright 1994), other stakeholders include remotely based collectors of specific forest products, forest products merchants, sellers, and logging companies (Carter 1999; Elhassan 2000). In the past, forests were managed solely by governmental forest departments and this practice has now been reappraised.

17.6.3 Participation of Locals in Forest Management

Locals are knowledgeable enough to make positive and beneficial contributions to the neighborhood forest management if involved, and that's why they have a legitimate right to participate (Thomson and Coulibaly 1995; FAO 2016). Participation of the locals constitutes the prime component of participatory sustainable forest management (FAO 1998, 2016). Locals have close interaction with the forests, and with their first-hand experience and perspectives, have viable and valid ideas on the management of forests. Ozturk et al. (2010) reported that to ensure sustainable management of forests, it is important for the forest managers to know the local people and their basic socioeconomic characteristics. Similarly, Jewitt (1996) believed during the past many decades, the government staff and forest fringe villagers are managing forest resources under various cost-benefit sharing arrangements and the sort of joint forest management approach has gained wide acceptance to practice at the global level. The views expressed in the most recent document of FAO (2016) are similar to the opinion of Jewitt (1996). Furthermore, FAO (2016) maintains that such types of management are more efficient and successful since they involve the different forest user groups that bring great wealth of information about local forests, indigenous experienced-based knowledge on sustainable use and agreed procedures for access to neighboring woodlands.

17.6.4 Willingness of Locals to Participate Brings Success

Forest management in Saudi Arabia used to be the single important responsibility of the local people till its abolishment in 1953. This historically prevailed forest management system known as "Tribal Hima" proved to be an efficient indigenous system for providing detailed information and thoughtful considerations to the sustainable use of forest and rangelands in the Kingdom of Saudi Arabia. In the system, forest management, protection, and conservation remained the responsibility of the tribes in their respective areas of residence (Assaeed and Al-Doss 2002; Aref and El- Juhany 2000). Even today locals living in the neiborhood of the forests express their interest and willingness to participate in managing the wood and rangelands in the KSA. Studies by Schindler et al. (2011) and Kobbail (2012) also indicated that locals demonstrated an eagerness to participate in the protection and management of the woodlands. Ozturk et al. (2010) also confirmed that people living around the forests (locals) in the neighboring villages were more willing to protect the forest resources, together with the state forest department, through the participative and cooperative approach, rather than protecting and managing the forest alone and reaping the benefits. Many researchers have reported that local people residing in nearby woodlands are critical to conservation and management efforts (Agrawal and Gibson 1999; Ferraro 2002; Ostrom 1999; Robertson and Lawes 2005; Wiggins et al. 2004). Many studies (see for instance, Burley 2004;

Gilmour et al. 2000; United Nations Commission on Sustainable Development 1998; El-Juhany 2009b) stress that it is important that forest managers need to identify the needs of all key stakeholders and they must make sure their needs through the research and development process are adequately addressed and accommodated. The technical constraints that may limit rehabilitation efforts should also be addressed as part of a comprehensive program launched through decentralized structures with an active participation of key stakeholders and interest groups (Gilmour et al. 2000).

17.6.5 Participation of the National Scientific Institutions

Involvement of the national scientific institutions, in addition to locals, has been viewed as equally important (El-Juhany 2009b). The researcher considers that participation of the national scientific institutions all the stages of the projects would improve the overall quality of the technical procedures undertaken in rehabilitation of the degraded forests. Furthermore, that collaboration and involvement of the local community combined with scientific input improves the resource management. Similarly, Teketay (2004) suggested that research organizations and their researchers could help improving and promoting indigenous knowledge, and generating new knowledge and appropriate technologies having practical application in the overall rehabilitation process towards sustainable production, protection, conservation and utilization of forest resources. Many researchers consider participatory forestry an appropriate option and sound strategy for the local communities for the conservation and sustainable use of forests (Arnold 1998; Wellstead et al. 2003; Wily 2002).

17.7 Role of Extension Education

Extension education programs have been able to change the negative attitudes and behaviors of local people (Ban and Hawkins 1990). Primarily, extension is associated with the dissemination of necessary, required, and pertinent information and advice to the farmers, through an appropriate disseminating mode towards modern farming. Thus, the primary function of extension is to help farmers by providing sufficient and suitable information to enable them to make their own decisions so that they are able to help themselves. Onumadu et al. (2001) maintain that extension education designs and provides educational inputs to the farmers to help themselves. Researchers, such as Onumadu et al. (2001) and Williams et al. (1984) consider extension education is a voluntary, out-of-school an informal educational programs for primarily meant for adults, whereas the contents of these programs are derived from research conducted in the physical, biological, and social sciences. In turn, they generate applicable knowledge in the form of concepts, principles, and procedures for the farmers. Similar views were expressed by Agbogidi and Ofuoku (2009), indicating that forestry extension education can assist in modifying the behaviors of individuals, groups, and the community as a whole towards the forests and the associated benefits, services, environment. Their participation would help improving the knowledge of other stakeholders, creating awareness, and making the attitudes of the participants favorable towards local communities and nearby forests. The productive role extension to enhance knowledge of locals on the importance of forests and their conservation at relatively low costs have been reported (FAO 2003, 2016).

17.7.1 Economic Profitability Enhances Participation

Acceptance of an innovation by the users depends upon its relative advantage and they measure its relative advantage in terms of its profitability. Increased participation of locals in forest management and afforestation programs could be enhanced if they perceive that their participation is beneficial to them and has a relative advantage. Therefore, it seems important to encourage locals to participate effectively in afforestation programs; they have to be persuaded that trees and forests are beneficial to the society and the environment. Forestry extension can make the public aware of the benefits of forests and foster an understanding that the shared benefits of forests will be better enjoyed if sustainably managed with basic principles of natural and social sciences.

17.7.2 Forestry Extension in Relation to Sustainable Forest Management

Forestry extension has emerged as an important tool to achieve sustainable forest management (SFM), and that remains the prime priority of the forestry professionals (Agbogidi et al. 2005). These researchers also further maintained that forestry as a profession is capable of addressing many more areas of concern for ecosystems, such as biodiversity conservation; community participation, and the need for forestry extension education and capacity building programs at all levels within the vicinity of the forests.

Forestry extension programs are meant and designed to meet the demands and satisfy the needs of small-scale farmers through educational and capacity building programs by providing information on appropriate technologies, biodiversity conservation, and new concepts on using, managing and conserving all the forest resources. This can only be achieved through focused and effective forestry extension. While addressing the issue of forest degradation and its rehabilitation in China, Wenhua (2004) stressed the need for training of forestry professionals and the development of human resources, as future education and training activities will further open new horizons of education. However, keeping in view the constructive and positive role of forestry extension and education, the following practical measures were suggested:

- Introduce reforms in forestry education that would in turn enhance the quality and help realize its potential benefits;
- Improve the conditions for forestry education;
- Support technical and vocational forestry education;
- Develop on-the-job training and continuing education programs for adults.
- Develop and launch the public forestry education system to enhance public understanding of the multiple functions of forests.

All these measures would create awareness on importance of the forests and help locals participate in the sustainable forest development.

17.8 Implications for Forestry Extension

According to Agbogidi and Ofuoku (2009), forestry extension has many implications but the most prominent is to create awareness on economic, social, and ecological benefits of the forests resources and the how these benefits are impact the urban and rural areas and their dwellers.

Forestry extension furnishes an opportunity to the locals and the other stakeholders to gain knowledge, know the values, modify attitudes, make commitments, and develop skills needed to protect and improve both the environment and forest resources. In addition, extension education modifies the behavior of individuals groups, and the whole community towards the forests and the associated environment. Forestry extension could have many implications to promote and enhance the participation of all the stake-holders including locals however, the most important are discussed as under:

17.8.1 Awareness and Understanding

Creation of awareness amongst the locals regarding the multiple values and services of the forestlands, the gravity of shrinking forests, and degradation of forestlands from multiple factors is extremely important. Therefore, El-Juhany (2009b) stresses the need for the launching of awareness campaigns amongst the locals who could support the rehabilitation activities. However, awareness activities should take into account the needs and concerns of locals, and conflicts of interests among the other stakeholders. Because once the awareness is achieved, the participation of the local people in the sustainable development of forests can be expected (El-Lakany 2004). Forestry extension can assist and help locals to enhance their understanding and increase their awareness of improved forest technologies, and awareness in-turn improves their decision-making skills towards environmental issues. Forestry extension creates awareness among the stakeholders on the importance of forests and promotes their understanding. However, the success of forestry extension will depend upon strong public understanding and support from the stakeholders. In order to achieve these, it is necessary to initiate the communication, provide information about values, status, and conservation of biological diversity, and arrange professional training at various public fora. Therefore, it is imperative that forestry extension be strengthened to obtain maximum cooperation from rural communities toward SFM.

17.8.2 Forestry Extension Helps Changing Behavior

Forestry extension upgrades the knowledge levels and skills and brings the positive change in attitudes of the locals. The attitude of locals toward forests has been regarded as a key determinant of their adoption of any forest conservation practice or innovation. For the reason, Ahnstrom et al. (2009) advised the forestry extension workers to pay special attention and make greater efforts to improve the locals' attitude toward forests and their sustained conservation.

This form of improvement in the skills and change in attitudes can help to achieve sustainable natural resource management and rural development (Rogers 2003). Behaviors are shaped with the phase and stage of a particular development initiative. Forestry extension can also contribute to social development, youth development, and subsequently, reduce poverty (Eke 2001; Ekpere and Durant 1999; Reid et al. 2006).

17.8.3 Protecting, Conserving and Managing Forests

Extension services could play a vital role towards the protection and conservation of the forests by linking researchers, local communities, and the end users/consumers. Without extension, even the most valuable research undertaken by eminent researchers would be useless, unless it is put into practice by the local communities. However, the protection of forests would primarily depend upon the effectiveness of the extension services, guidance, and advice. Forestry extension service must take up the faced challenges seriously to be an effective limb of extension in the future. While highlighting the importance of locals in forest management, Agbogidi et al. (2005) argued that in order to protect and conserve forests extensionists are expected to demonstrate the locals on making a reasonable, sustainable livelihood from the forests of the vicinity Thus, the best way to protect forests and their vast diversity is by creating awareness of their value and importance amongst the locals, and involving them in taking protective measures through people-friendly extension initiatives. Deforestation has also been reported as a primary contributing factor to the high rate of loss and erosion of genetic diversity (Agbogidi et al. 2005). Issues such as erosion of genetic diversity and deforestation can also be addressed appropriately by devising suitable extension activities (Agbogidi and Dolor 2002; Kola-Olusanya 2000). Today forestry extension programs have received recognition and are becoming increasingly popular worldwide, due to their strong inherent potential to overcome the challenges and impediments faced to conserve natural resources in general, and natural forests in particular. Consequently, Al-Zabee (1997) established the requirement for extension education programs to convince locals to adopt environmentally-friendly technologies and practices, and be recognized by all those who are involved and concerned with sustainable development. For example, May (1991) argued that the extension professionals need to focus on environmental issues and extension education programs should be developed on a strong research foundation.

17.8.4 Training of the Stakeholders Through Forestry Extension Programs

For further strengthening the relationships of foresters with the locals, suitable forestry extension programs should also place emphasis to train the locals by conducting short–term courses, field visits, and practical demonstrations on various aspects of forestry like sustainable harvesting practices and interrelationships of the forest resources with the other components (Eke 2001; FAO 1997). Also all the groups of forestry personnel require trainings to equip themselves to face the challenges of the sustainability of the resources they manage and make them familiar on the approaches to adopt to better respond to the recent developments (Agbogidi and Ofuoku 2009). Specifically, they require training on, among others, extension practices, data collection and storage techniques, and sustainable forest management practices (Ogunwale et al. 2006).

17.8.5 Training in Environmental Education

After a comprehensive review on forest management and extension, Agbogidi and Ofuoku (2009) stressed the need for equipping the public and rural inhabitants on environmental protection, which is vitally important for their survival. Their review also pointed out that forestry extension had great implications to protect and conserve the forests as environment and forest ecosystems are extremely important for human survival. The authors also maintained that scientific information on biodiversity, and the concepts and technologies in conservation, need to be disseminated for the managing and development the forest on sustainable basis and to halt the present rapid erosion of genetic diversity.

17.8.6 Training in Forests Conservation and Sustainable Management

Forestry extension aims at helping rural communities to satisfy their needs, address environmental problems, and provides appropriate solutions to the problems they face. Foresters need to strengthen the extension system to facilitate the dissemination of information and improve knowledge of locals on conservation, the sustainable development, management of forests, harvesting of tree crops and processing of forest products (Agbogidi and Ofuoku 2009) to achieve sustainable management of forests with the help and cooperation of the locals.

17.9 Conclusions and Recommendations

This chapter elucidates that Saudi Arabia has meager forest resources and the forest and rangelands vegetation is under stress due to climatic and physical features of the land, and operative anthropogenic forces, resulting in thin, weak, stressed, and poor-condition vegetation.

Primary challenges for the natural resources and vegetation cover in the Kingdom arose from unregulated overgrazing and logging that, in turn, caused increased pressure on the forests and rangelands, destruction of natural vegetation cover, and increased rates of desertification. Participatory forest management emerged and proved to be a very useful conceptual tool for conserving forest resources and ensuring their long-term sustainability. Foresters and researchers are in agreement that cooperation and support from the communities living in the neighborhood of woodlands is vital, and an indispensable factor for their sustainable management. A voluminous literature indicates that participation of locals residing in the neighborhood are extremely important to the initiatives related conservation and management of the nearby woodlands. Locals are an un-explored and under-utilized source of knowledge and experience; hence, there is high expectation that their participation in managing the nearby forests will be highly beneficial. Therefore, it is imperative that the Kingdom of Saudi Arabia develops a woodlands management system that could harness the participation of local residents to ensure the sustainability.

- Locals are closely and culturally attached to the use of wood, meeting their domestic requirements either from the neighboring forests or buying it from local sellers. Both of these sources exert undue pressure on the meager, local, natural woodlands. A significant proportion of the local population extracts huge volumes of wood to meet their domestic needs for cooking and heating. In this situation, it would be important to identify reasons preventing them using the alternative cleaner and better energy sources. Efforts are needed to make suitable energy sources available to the consumers.
- A huge volume of literature indicates that communities living in the neighborhood of forests appreciate their presence and are willing to participate in the

management of local forests. Better forest management plans can be evolved by blending the local knowledge with the modern research-based principles and practices of forest sciences.

- The development of vibrant extension education programs is a valid, practicable, and probably the best strategy to improve the participation of local residents and other stakeholders in developing the management plan. To move in this direction, forestry extensionists must identify the factors that encourage the participation of locals in managing the national forests.
- Extension services also need to facilitate the dissemination of information and knowledge related to conservation initiatives; the sustainable development and the management of forests; and harvesting and processing of forest goods amongst the locals and other stakeholders to formulate an acceptable forest management plan.
- Locals primarily harvest the trees to meet their domestic energy needs. Forestry extensionists need to create awareness on alternatives of wood and educate the communities living in the neighborhood of forests on use of clean energy sources.
- Extension education programs are needed to enable consumers to make informed choices in selecting the right type of energy sources. Launching such educational, capacity building and awareness raising programs would help conserving natural resources of the Kingdom.
- Extension education is an important channel for informal education of adults; therefore it seems logical to explore the significant role forestry extension programs in educating the locals towards forest management, development, and protection in the KSA.

The discussions made in the chapter lead to conclude that Kingdom of Saudi Arabia needs forest management plan that primarily focuses on conserving the woodlands and emphasizing on the elements like: sound technical plans, friendly policies, and vibrant extension education/capacity building programs. The planners and the policy-makers must ensure the participation of locals at all stages of formulation, implementation, monitoring, and evaluation. Participation of local residents would ensure the long-term sustainability of forests, while sharing the benefits and services of the forests with the locals.

Acknowledgements The authors are extremely grateful and express their gratitude to the Saudi Society of Agricultural Sciences for extending us all the possible help and sincere cooperation toward the completion of this piece of work and research.

References

Abdullah, M. A. R., & Abulfatih, H. A. (1995). Predation of Acacia seeds by bruchid beetles and its relation to altitudinal gradient in south-western Saudi Arabia. *Journal of Arid Environments, 29*(1), 99–105.

Abdulaziz, M., & Al-Doss, A. A. (2002). Soil Seed Bank of a Desert Range Site Infested with Rhazya stricta in Raudhat al-Khafs, Saudi Arabia. *Arid Land Research and Management, 16*(1), 83–95. https://doi.org/10.1080/153249802753365340.

Abido, M. S. (2000). *State of the environment and policy restorative: 1972–2002* (Forests: West Asia, pp 113–115). Available at: http://www.unep.org/dewa/WestAsia/assessments/national_SOEs/Other%20SOE%20Reports/GEO3/GEO3_fores t.pdf.

Abo-Hassan, A. A. (1983). Forest resources in Saudi Arabia. *Journal of Forestry, 81*(4), 239–241.

Abo-Hassan, A. A., El-Osta, M. L. M., & Sabry, M. M. (1984). *Natural forests in the Kingdom of Saudi Arabia and the possibility of exploiting them economically*. Riyadh: Scientific Research Directory, National Center for Science and Technology (Now: King Abdulaziz City for Science and Technology) Book No. 1 (in Arabic) (p. 182).

ACSAD. (2003). Inventory study and regional database on sustainable vegetation cover management in West Asia (TN2), Prepared for the sub-regional action program (SRAP) to combat desertification and drought in West Asia under the memorandum of understanding signed with UNEP/ROWA 21st August 2001. Damascus: The Arab Center for the Studies of Arid Zones and Dry Lands.

Agbogidi, O. M., & Dolor, D. E. (2002). Deforestation and the Nigeria rural environment. *African Journal of Environmental Studies, 3*(1&2), 26–29.

Agbogidi, O. M., & Ofuoku, A. U. (2009). Forestry Extension: Implications for Forest Protection. *International Journal of Biodiversity and Conservation, 1*(5), 98–104 Available at: http://www.academicjournals.org.

Agbogidi, O. M., Okonta, B. C., & Dolor, D. E. (2005). Participation of rural women in sustainable forest management and development. In E. Okoko, V. Adekline, & S. Adeduntan (Eds.), *Environ. sustainability and conserve. in Nigeria* (pp. 264–270). Akure: Jubee–Niyi Publisher.

Agrawal, A., & Gibson, C. (1999). Enchantment and disenchantment: The role of community in natural resource conservation. *World Development, 27*(4), 629–649.

Ahnstrom, J., Hockert, J., Bergea, H. L., Francis, C. A., Skelton, P., & Hallgren, L. (2009). Farmers and nature conservation: What is known about attitudes, context factors and actions affecting conservation? *Renewable Agriculture and Food Systems, 24*(01), 38–47. https://doi.org/10.1017/S1742170508002391.

Al-Abdulkader, A. M., Shanavaskhan, A. E., Al-Khalifah, N. S., & Nasroun, T. H. (2009). The economic feasibility of firewood plantation enterprises in Saudi Arabia. *Arab Gulf Journal of Scientific Research, 27*, 1–6.

Alahmed, A. M., Kheir, S. M., & Al Khereiji, M. A. (2010). Distribution of Culicoides latreille (Diptera: Ceratopogonidae) in Saudi Arabia. *Journal of Entomology, 7*, 227–234. https://doi.org/10.3923/je.2010.227.234.

Al-Mosa, K. N. (1999). *Report on forests in Saudi Arabia Kingdom*. Riyadh: Department of Forests, Ministry of Agriculture and Water Available at: D:\fao Saudi Arabia forests.mht. Accessed on 3 Apr 2011.

Al-Rowaily, S. L. (1999). Rangeland of Saudi Arabia and the tragedy of commons. *Rangelands, 21*(3), 27–29.

Al-Shomrany, S. A. A. (1980). *Types, distribution and significance of agricultural terraces in Assarah, Southwestern Saudi Arabia*. M.Sc. thesis, Michigan State University. USA.

Al-Shomrany, S. A. A. (1994). Methods of protecting soil from drifting in Al-Sarah region in the Southwestern Saudi Arabia. In *The proceedings of desert studies in Saudi Arabia, Symposium held at The Desert Studies Centre, King Saud University, Riyadh, Saudi Arabia, 248–312, from 2 to 4 October 1994*. Riyadh, Saudi Arabia: King Saud University.

Al-Subaiee, F. S. (2014a). Attitudes of locals and their dependence on the natural forests: A Case study of Al-Baha area- Saudi Arabia. *The Journal of Animal & Plant Sciences, 24*(2), 643–650.

Al-Subaiee, F.S. (2014b). Indigenization of forests development technologies to enhance forests environmental role in Riyadh area. Final Report Submitted for National Plan for Science and Technology. King Saud University, Riyadh, Saudi Arabia.

Al-Subaiee, F. S. (2015a). Socio-Economic Factors Affecting the Conservation of Natural Woodlands in Central Riyadh Area–Saudi Arabia. *Saudi Journal of Biological Sciences, 23*(3), 319–326. https://doi.org/10.1016/j.sjbs.2015.02.017.

Al-Subaiee, F. S. (2015b). Local participation in woodland management in the Southern Riyadh area: Implications for agricultural extension. *Geographical Review, 105*(4), 408–428.

Al-Subaiee, F. S. (2016). Socio-economic factors affecting the conservation of natural woodlands in Central Riyadh area, Saudi Arabia. *Saudi Journal of Biological Sciences, 23*(3), 316–329. https://doi.org/10.1016/j.sjbs.2015.02.017.

Al-Subaiee, F. S., Al-Shuhrani, T. S., Muneer, S. E., & Shalaby, M. Y. (2014). *Final technical report of the project "indigenization of forests development technologies to enhance forests environmental role in Riyadh area" (ENV 514–02– 08).* Submitted to the National Plan for Science and Technology. Kingdom of Saud Arabia, Riyadh.

Al-Zabee, A. (1997). Future roles of agricultural extension within policies and programs of economical improvement. The National Conference about Enhancing the Roles of Agricultural Extension in Sustainable Agriculture Development, Held in Algiers. Arab Organization for Agricultural Development. Sudan, Khartoum. Accessed.

Ameha, A., Nielsen, O. J., & Larsen, H. O. (2014). Impacts of access and benefit sharing on livelihoods and forest: Case of participatory forest management in Ethiopia. *Ecological Economics, 97*, 162–171. https://doi.org/10.1016/j.ecolecon.2013.11.011.

Aref, I. M., & El-Juhany, L. I. (2000). Natural and planted forests in Saudi Arabia: Their past, present and future (in Arabic). *Arab Gulf Journal of Scientific Research, 18*(10), 64–72.

Arnold, M. (1998). *Managing forests as a common property* (FAO Paper No. 136). Food and Agricultural Organization of the United Nations: Rome.

Assaeed, A. M., & Al-Doss, A. A. (2002). Soil seed bank of a desert range site infested with Rhazya stricta in Raudhat al-Khafs, Saudi Arabia. *Arid Land Research and Management, 16*, 83–95.

Badai, K. H., & Aldawoud, A. N. (2004). *The natural forests in Saudi Arabia*. Riyadh: Samha Press.

Ban, V. D., & Hawkins, H. S. (1990). *Agricultural extension*. New York: Wiley.

Burley, J. (2004). The restoration of research. *Forest Ecology and Management, 20*(1), 83–88.

Carter, J. (1999). *Recent experience in collaborative forest management approaches: A review of key issues*. A paper prepared as part of the forestry policy review process of the World Bank, Washington, DC.

Chaudhary, S. A., & Le Houérou, H. N. (2006). The rangelands of the Arabian peninsula. *Science etchangements planétaires/Sécheresse, 17*(1), 179–194.

Clarke, W. C., & Thaman, R. R. (2006). *Agroforestry in the Pacific islands: Systems for sustainability*. Suva: The University of the South Pacific.

Darfaoui, E., & Al Assiri, A. (2010). *Response to climate change in the Kingdom of Saudi Arabia*. Cairo: FAO-RNE Available at: www.fao.org/forestry/291570d03d7abbb7f341972e8c6ebd2b25a181.pdf.

Department of Natural Resources. (2007). *Forest inventory project in the southwestern region of Saudi Arabia*. Department of Natural Resources, Ministry of Agriculture. Implemented by The Space Research Institute at King Abdul Aziz City for Science and Technology in collaboration with King Saud University, Riyadh, Saudi Arabia.

Doha, S. A. (2009). Phlebotomine sand flies (Diptera, Psychodidae) in different localities of Al-Baha province, Saudi Arabia. *Egyptian Academic Journal of Biological Sciences, 1*(1), 31–37.

Dregne, H. E., & Chou, N. -T. (1992). *Global desertification dimensions and costs*. Center for International Earth Science Information Network 1992. http://www.ciesin.columbia.edu/docs/002-186/tab2c.gif.

Eckholm, E. P. (1976). *Losing ground: Environmental stress and world food prospects*. New York: W.W. Norton & Co Inc ISBN 10: 0393091678 ISBN 13: 9780393091670.

Eke AC. (2001). Communication: A tool for forest management and development in environmental conservation. Proceedings of the 27th Annual Conference of the Forestry Association of Nigeria, Abuja In: Popoola L, Abu JE ,Oni PI (eds). Abuja, Nigeria, Federal Capital Territory (FCT) held between 17th and 21 st of Sept., 2001. pp. 214–216.

Ekpere, J. A., & Durant, T. J. (1999). Agricultural extension and rural sociology. In A. Youdeowei, O. C. Ezedinma, & O. C. Onazi (Eds.), *Introduction to tropical agriculture* (pp. 282–300). China: Longman Group Ltd.

El Atta, H., & Aref, I. M. (2010). Effect of terracing on rainwater harvesting and growth of *Juniperus procera* Hochst. *International Journal of Environmental Science and Technology, 7*(1), 59–66.

Elhassan, N. G. (2000). *Stakeholders approach for sustainable management of forest resources: Case study Jebel Marra forest circle*. M.Sc. Thesis. University of Khartoum, Sudan.

El-Juhany L. I. (2009a). Forest land degradation in the Southwestern Region of Saudi Arabia. In *The proceedings of "land degradation in dry environments", an International conference held at Kuwait University on March, 8–14, 2009*, Kuwait.

El-Juhany, L. I. (2009b). Forest land degradation and potential rehabilitation in Southwest Saudi Arabia. *Australian Journal of Basic and Applied Sciences, 3*(3), 2677–2696.

El-Juhany, L. I. (2014, November). Cross section characteristics and age estimation of *Juniperus procera* trees in the natural forests of Saudi Arabia. *Journal of Pure and Applied Microbiology, 8*(Spl. Edn. 2), 657–665.

El-Juhany, L. I., & Aref, I. M. (2012a). The present status of the natural forests in the Southwestern Saudi Arabia: 1-taif forests. *World Applied Sciences Journal, 19*(10), 1462–1474. https://doi.org/10.5829/idosi.wasj.2012.19.10.1877.

El-Juhany, L. I., & Aref, I. M. (2012b). The Present Status of the natural forests in the Southwestern Saudi Arabia 2-baha forests. *World Applied Sciences Journal, 20*(2), 271–281. https://doi.org/10.5829/idosi.wasj.2012.20.02.1899.

El-Juhany, L. I., Aref, I. M., & Al-Ghamdi, M. A. (2008). The possibility of ameliorating the regeneration of juniper trees in the natural forests of Saudi Arabia. *Research Journal of Agriculture and BiologicalSciences, 4*(2), 126–133.

El-Juhany, L. I., Aref, I. M. (2013). The present status of the natural forests in the Southwestern Saudi Arabia: 3- Asir and East jazan forests. *World Applied Sciences Journal 21*(5):710–726. https://doi.org/10.5829/idosi.wasj.2013.21.5.2841.

El-Lakany, H. (2004). *Improvement of rural livelihoods: The role of agroforestry*. Rome: FAO.

El-Osta, M. L. M. (1983). Estimation of Volume of Standing Timber in the South-Western Region of Saudi Arabia. *Journal of the College of Agriculture King Saud University, 5*, 115–123.

Elsiddig, E., Goutbi, A. N., & Elasha, B. (2001). *Community based natural resource management in Sudan* (p. 121). Sudan: Intergovernmental Authority on Development. Final Draft.

Ernst, W. H., Tolsma, D. J., & Decelle, J. E. (1989). Predation of seeds of *Acacia tortilis* by insects. *Oikos, 54*, 294–300.

FAO. (1977). *Savanna afforestation in Africa*. Food and Agriculture Organization of the United Nations (FAO) Forestry Paper 11. Lecture Notes for the FAO/DANIDA training course for the nursery and establishment techniques for African Savannas and papers from the symposium on Savanna afforestation with the support of the Danish International Development Agency. Kaduna, Nigeria. 1976.

FAO. (1997). Improving agricultural extension: A reference manual Rome: Italy.

FAO. (1998). *Guidelines for the management of tropical forests: The production of wood* (FAO forestry Paper No. 135). Rome: FAO.

FAO. (2003). *State of the world's forests*. Rome: Food and Agriculture Organization of the United Nations http://www.fao.org/docrep/005/y7581e/y7581e00.htm.

FAO. (2005) *Global forest resources assessment (2005)*. Available at: http://www.fao.org/forestry/country/32185/en/sau/.

FAO. (2006). *Current and future challenges for forest extension*. Rome: FAO.

FAO. (2007). *State of the world forests 2007*. Rome: Food and Agriculture Organization of the United Nations (FAO).

FAO. (2008). *The status and trends of forests and forestry in West Asia. Sub regional report of the forestry outlook study for West and Central Asia, by Q. Ma. Forestry Policy and Institutions* (Working Paper No. 20). Rome.

FAO (2009), FAO. (2010). Forests and forestry sector—Saudi Arabia. http:/www.fao.org/forestry/country/57478/en/sau.

FAO. (2011). *Forests – For improved nutrition and food security*. Rome: Food and Agriculture Organization ofthe United Nations (FAO).

FAO. (2013). *Managing forest for the future*. http://www.fao.org/docrep/014/am859e/am859e08.pdf.

FAO. (2016). *State of the world's forests 2016. Forests and agriculture: Land-use challenges and opportunities*. Rome: Food and Agriculture Organization of the United Nations Available at: http://www.fao.org/publications/sofo/2016/en/.

Ferraro, P. J. (2002). The local costs of establishing protected areas in low-income nations: Ranomafana National Park, Madagascar. *Ecological Economics, 43*(2–3), 261–275.

FRA. (2005). Global Forest resources assessment country reports Saudi Arabia. Food and Agriculture Organization of the United Nations, 2005 FRA 2005/088 Rome. Available at: http://www.fao.org/tempref/docrep/fao/010/aj002E/aj002E00.pdf

FRA. (2010). Global Forest resources assessment country reports Saudi Arabia. Food and Agriculture Organization of the United Nations, 2010 FRA 2010/185 Rome, 2010 Available at: http://www.fao.org/docrep/013/al620E/al620E.pdf

Gardner, A. S., & Fisher, M. (1994). How the forest lost its trees: Just so storytelling about ***Juniperus excelsa*** in Arabia. *Journal of Arid Environments, 26*(3), 299–301. https://doi.org/10.1006/jare.1994.1031.

Gilmour, D. (1995). Conservation and development: Seeking the linkages. In Ø. Sandbukt (Ed.), *Management of tropical forests: Towards an integrated perspective* (pp. 255–267). Oslo: Center for Development and the Environment, University of Oslo.

Gilmour, D. A., Van San, N., & Tsechalicha, X. (2000). *Rehabilitation of degraded forest ecosystems in Cambodia, Lao PDR, Thailand and Vietnam: An overview. conservation issues in Asia*. A joint publication of IUCN/WWF/GTZ Working Group on Community Involvement in Forest Management (IUCN), In Andrew Ingles (Ed.), Head, Regional Forest Programme. Pathumthani, Thailand: IUCN, The World Conservation Union Asia Regional Office, Klong Luang. Available at: assets.panda.org/downloads/lowermekongregionaloverview.pdf.

Hajar, A. S., Faragella, A. A., & Al Ghamdi, K. M. (1991). Impact of biological stress on *Juniperus excelsa* M. Bieb in south-western Saudi Arabia: Insect stress. *Journal of Arid Environments, 21*, 327–330.

Ibrahim, A. A., & Abdoon, M. A. (2005). Distribution and population dynamics of Phlebotomus Sandflies (Diptera, Psychodidae) in an endemic area of cutaneous leishmaniasis in Asir Region, Southwestern Saudi Arabia. *Journal of Entomology, 2*, 102–108. https://doi.org/10.3923/je.2005.102.108.

Jewitt, S. (1996). Agro-ecological knowledges and forest management in the Jharkhand, India: Tribal development or populist impasse? Doctoral dissertation. University of Cambridge, Cambridge.

JICA-Japan International Cooperation Agency. (2002). *The Joint study project on the conservation of Juniper woodlands in Saudi Arabia*. Draft Final Report of Japan International Cooperation Agency, submitted to NCWCD, Riyadh, Kingdom of Saudi Arabia.

Jodha, N. S. (1992). *Common property resources: A missing dimension of development strategies* (Discussion Paper no. 169). Washington, D.C: The World Bank.

Johannesburg Summit. (2002). *Saudi Arabia, country profile*. Summit on Sustainable Development, held in Johannesburg, South Africa, 26 August to 4 September 2002, The United Nations.

Kobbail, A. A. (1996). Managerial and social aspects of community forestry in kosti area (Central Sudan). M.Sc. Thesis, University of Khartoum.

Kobbail, A. A. R. (2012). Local people attitudes towards community forestry practices: A case study of kosti province-central Sudan. *International Journal of Forestry Research, 2012*, 1–7. https://doi.org/10.1155/2012/652693.

Kola, O. A. (2000). Towards sustainable development: Reinventing education for enriching environmental knowledge and consciousness. In M. A. G. Akale (Ed.), *Proceeding of the 41st*

annual conference of the Science Teachers' Association of Nigeria (STAN) held on Awka, Anambra State between 21–26 August, 2000 (pp. 191–195).
Kumar, N., Saxena, N., Alagh, Y., & Mitra, K. (2000). *India: Alleviating poverty through forest development (country case study, operations evaluation department)*. Washington, DC: The World Bank Available at: www.worldbank.org, accessed on 15 June 2001.
Ma, Q. (2008). *The status and trends of forests and forestry in West Asia. Sub-regional report of the forestry outlook study for West and Central Asia*. Rome: Food and Agriculture Organization of the United Nations.
Mahmoud, S. H., Mohammad, F. S., & Alazba, A. A. (2014). Determination of potential runoff coefficient for Al-Baha region, Saudi Arabia using GIS. *Arab Journal of Geosciences, 7*, 2041. https://doi.org/10.1007/s12517-014-1303-4.
Marvier, M. A. (1996). Parasitic plant-host interactions: Plant performance and indirect effects on parasite-feeding herbivores. *Ecology, 77*, 1398–1409.
Matiku, P., Caleb, M., & Callistus, O. (2013). The impact of participatory forest management on local community livelihoods in the in the Arabuko-Sokoke forest, Kenya. *Conservation and Society, 11*(2), 112–129.
May, T. (1991). Extension and the environment. *Journal of Extension, 29*(1), 1 Available online: http://www.joe.org/joe/1991spring/let2.html.
Ministry of Agriculture. (2002). *Forestry Strategy and Work Plan in Saudi Arabia*. Riyadh: Ministry of Agriculture.
Ministry of Agriculture. (2010). A glance on agricultural development in the kingdom of Saudi Arabia. Available at: http://www.moa.gov.sa/files/Lm_eng.pdf.
Ministry of Economy and Planning, Saudi Arabia. (2005). *The eighth development plan*. Riyadh: Ministry of Economy and Planning.
Ministry of Economy and Planning, Saudi Arabia. (2010). *Kingdom of Saudi Arabia*. The ninth development plan 1431–1435 H. Available at: http://www.mep.gov.sa/. Accessed 1 Feb 2012).
Mutikainen, P., Salonen, V., Puustinen, S., & Koskela, T. (2000). Local adaptation, resistance, and virulence in a hemi-parasitic plant-host plant interaction. *Evolution, 54*, 433–440.
Nasroun, T. H. (1993). An outline for a sustainable agricultural development plan in the Arabian Gulf Region. (In arabic). In *The Proceedings of a symposium on Desertification and Land Reclamation in CCG Countries*. Arabian Gulf University Bahrain.
NCWCD and JICA. (2006). *The management plan for conservation of Juniper woodlands*. The final report of the jointed study between National Commission for Wildlife, Conservation and Development (NCWCD) and Japan International Co-operation Agency (JICA).
NCWCD and JICA. (2007). NCWCD [National Commission for Wildlife, Conservation and Development] and JICA [Japan International Co-operation Agency]. 2007. The Management Plan for Conservation of Juniper Woodlands The final report of the joint study prepared by NCWCD and JICA. Riyadh: Ministry of Agriculture.
Ogunwale, et al. (2006). Impact of extension Service on farmers production activities in Ogbomoso agricultural zone of Oyo State, Nigeria. *Journal of Agricultural Extension, 9*, 143–149.
Onumadu F. N., Popoola L., & Akinsorotan A. O. (2001). Environmental forestry extension: The missing links. In L. Popoola, J. E. Abu, & P. I.Oni, (Eds.), *Proceeding of the 27th annual conference of FAN held in Abuja, FCT between 17th and 21th of Sept., 2001* (pp. 290–298).
Ostrom, E. (1999). *Self-governance and forest resources*. (Center for international Forestry Research (CIFOR) Discussion Paper No. 20).
Ozturk, A., Saglam, B., & Bali, O. (2010). Attitudes and perceptions of rural people towards forest protection within the scope of participatory forest management: A case study from Artvin, Turkey. *African Journal of Agricultural Research., 2*(12), 1399–1411.
Rathore, M. P. S. (1995). *Insect pests in agroforestry* (Working Paper No. 70). International Centre for Research in Agroforestry. Nairobi.
Reid, W. V., Money, H. A., & Cropper, A. (2006). Ecosystem conservation for economic development. *Id 21 Natural Resources Highlights Conservation, 2*, 2.

Robertson, J., & Lawes, M. J. (2005). User perceptions of conservation and participatory management of Gxalingenwa forest, South Africa. *Environmental Conservation, 32*, 64–75.

Rogers, E. M. (2003). *Diffusion of innovations* (5th ed.). New York: The Free Press. Simon & Schuster, Inc.

Saudi Geological Survey. (2012) *Facts and numbers* (In Arabic).

Schindler, S., Cimadom, A., & Wrbka, T. (2011). The attitude towards nature and nature conservation on the urban fringes. *Innovation–The European Journal of Social Science Research, 24*(3), 379–390.

Teketay, D. (2004). Whose responsibility dryland management? In: Rehabilitation of Degraded Lands in Sub-Saharan Africa: Les sons Learned from Selected Case Studies. P. Wood and A. M. Yapi. (Editors). Forestry Research Network for Sub-Saharan Africa (FORNESSA) and International Union of Forest Research Organizations. A Special Programme for Developing Countries (IUFRO-SPDC).

Tesfaye, Y., Roos, A., & Bohlin, F. (2012). Attitudes of local people towards collective action for forest management: The case of participatory forest management in Dodola Area in the Bale Mountains, Southern Ethiopia. *Biodiversity and Conservation, 21*, 245–265.

Thomson, J. T., & Coulibaly, C. (1995). Common property forest management systems in Mali: Resistance and vitality under pressure. *Unasylva, 46*(180), 16–22.

UNCCD. (2006). *Executive summary*. The third national report for the Kingdom of Saudi Arabia. Submitted to the Implementation of the United Nations Convention to Combat Desertification (UNCCD). Paris. France. Available at: www.unccd-prais.com/.../GetReportPdf/798b2cc4-1772-4726-a3f7-a0fa.

UNCSD. (1998). *United Nations Commission on sustainable development information note on assessing, monitoring and rehabilitation of forest cover in environmentally critical areas*. Background Paper. The United Nations Department of Economic and Social Affairs (DESA). Available at: http://www.un.org/documents/ecosoc/cn17/iff/1998/background/ecn17iff1998-bgamrfc.htm.

US Forest Services. (2009). *Invasive plants*. Northeastern Area. Available at: http://na.fs.fed.us/fhp/invasive_plants/index.shtm.

Wellstead, A. M., Stedman, R. C., & Parkins, J. R. (2003). Understanding the concept of representation within the context of local forest management decision making. *Journal of Forest Policy and Economics, 5*, 1–11.

Wenhua, L. (2004). Degradation and restoration of forest ecosystems in China. *Forest Ecology and Management, 201*(1), 33–41.

Western, D., & Wright, M. (1994). *Natural connections; perspectives in community based conservation*. Washington, DC: Island Press.

Wiersum, K. F. (1995). *Forestry and rural development lecture*. Wageningen: Department of Forestry. Wageningen Agricultural University.

Wiggins, S., Marfo, K., & Anchirinah, V. (2004). Protecting the forest or the people? Environmental policies and livelihoods in the forest margins of Southern Ghana. *World Development, 32*(11), 1939–1955.

Williams, S. K. T., Fenley, J. M., & Williams, C. E. (1984). *A manual for agricultural extension workers in Nigeria* (pp. 119–130). Ibadan: Shyraden.

Wily, L. (2002). Participatory forest management in Africa. *An overview of Progress and Issues* (Draft Paper 2002).

World Bank. (2001). *A revised forest strategy for the World Bank group (Draft, 30 July 2001)*. Washington, DC: The World Bank (Available from: www.worldbank.org, accessed on 20 Aug 2001).

World Bank. (2014). Saudi Arabia. Little data book. Available at: http://data.worldbank.org/sites/default/files/the-little-data-book-2014.pdf. Accessed on 19 Dec 2014.

Dr. Mirza Barjees Baig is working as a Professor of Agricultural Extension and Rural Development at the King Saud University, Riyadh, Saudi Arabia. He has received his education in both social and natural sciences from USA. He completed his Ph.D. in Extension Education for Natural Resource Management from the University of Idaho, Moscow, Idaho, USA. He earned his MS degree in International Agricultural Extension in 1992 from the Utah State University, Logan, Utah, USA and was placed on the "Roll of Honour". During his doctoral program, he was honored with "1995 Outstanding Graduate Student Award". Dr. Baig has published extensively in the national and international journals. He has also presented extension education and natural resource management extensively at various international conferences. Particularly issues like degradation of natural resources, deteriorating environment and their relationship with society/community are his areas of interest. He has attempted to develop strategies for conserving natural resources, promoting environment and developing sustainable communities through rural development programs. Dr. Baig started his scientific career in 1983 as a researcher at the Pakistan Agricultural Research Council, Islamabad, Pakistan. He has been associated with the University of Guelph, Ontario, Canada as the Special Graduate Faculty in the School of Environmental Design and Rural Planning from 2000-2005. He served as a Foreign Professor at the Allama Iqbal Open University (AIOU) through Higher Education Commission of Pakistan, from 2005-2009. Dr. Baig the member of IUCN – Commission on Environmental, Economic and Social Policy (CEESP). He is also the member of the Assessment Committee of the Intergovernmental Education Organization, United Nations, EDU Administrative Office, Brussels, Belgium. He serves on the editorial Boards of many International Journals Dr. Baig is the member of many national and international professional organizations.

Prof. Dr. Juhn Pulhin is the Professor and UP Scientist II of the Department of Social Forestry and Forest Governance (DSFFG), College of Forestry and Natural Resources (CFNR), University of the Philippines Los Banos (UPLB). He was a Visiting Associate Professor at the University of Tokyo from January to April 2007. He was Associate Dean of CFNR and a former Department Chair of DSFFG.

He served as a Lead Author in the Adaptation Chapter in the Working Group II of the Fourth Assessment Report of the Intergovernmental Panel on Climate Change (IPCC) and a Lead Author in the Forest and Woodland Systems Chapter of the 2006 Millennium Ecosystem Assessment. He was also involved in several climate change related research and development projects foremost of which is the Assessment of Impacts, Vulnerability, and Adaptation to Climate Change in Selected Watersheds and Communities in Southeast Asia. He is a co-editor and co-author in three chapters of the book Climate Change and Vulnerability and co-author in a chapter of the book Climate Change and Adaptation, both published in 2008 by Earthscan. Presently, he is the Philippine Project Leader of Improving Equity and Livelihood in Community Forestry which is a global collaborative research project of the Center for International Forestry Research (CIFOR) and the Resource Rights Initiatives (RRI).

Dr. Pulhin has completed 25 researches/studies on various topics involving socio-economic, political/institutional and environmental dimensions of forestry including forest tenure, decentralization, forest rehabilitation, and environmental justice as well as on impacts, vulnerability and adaptation of local communities and institutions to climate variability and extremes. He has also developed and applied various participatory and multi-stakeholder techniques in conducting community forestry and climate-change related researches and assessments. As outputs of his numerous researches, he has authored/co-authored 12 peer-reviewed scientific articles; co-edited 2 books; contributed 15 chapters in 9 different books; authored/co-authored 11 articles in published conference proceedings and 10 policy papers; and presented more than 40 other papers in international conferences, trainings, seminars, and professional meetings in more than 20 different countries.

He has also served as social forestry/community forestry consultant to international and national organizations such as the Asian Development Bank (ADB), Food and Agriculture Organization of the United Nations (FAO-UN), Ford Foundation, Regional Community Training Center for the Asia and the Pacific Region (RECOFTC), Japan Bank for International Cooperation

(JBIC) Department of Environment and Natural Resources (DENR), National Power Corporation, and a number of international and national NGOs in more than 10 major projects including feasibility study preparation, land use development planning, project appraisal, national assessment project, and monitoring and evaluation.

In recognition of his achievements, Dr. Pulhin is a recipient of the following awards: (1) Co-recipient of 2007 Nobel Peace Prize Award shared with Al Gore and the members of the United Nation-Intergovernmental Panel on Climate Change (IPCC); (2) Co-recipient of the 2006 ZAYED International Prize for the Environment for Scientific and Technological Achievements in Environment given to the Millennium Ecosystem Assessment (of which he is one of the Lead Authors) by the Zayed International Prize on February 6, 2006 in Dubai, United Emirates; (3) UP Scientist II, given by the University of the Philippines System covering the period of January 2008-December 2010; (4) Outstanding Scientist Award in Socioeconomics and Policy for 2006, given by the Forests and Natural Resources Research Society of the Philippines, Inc. (FORESPI); (5) UPLB Alumni Association 2006 Distinguished Alumni Award for Institutional Service for the College of Forestry and Natural Resources; (6) CFNR Outstanding Alumnus for Institutional Service for 2006; (7) UPLB 2000 Outstanding Teacher Award in Social Science and Humanities (2000); (8) UPLB-CFNR Best Teacher Award in Social Science (1999); and (9) Vicente Lu Professorial Chair from July 2006-June 2007; and 10) Dean Gregorio T. Zamuco Professorial Chair in Social Forestry from July 2000-June 2001 and July 2001-June 2002.

Dr. Pulhin obtained his Bachelor of Science Degree in Forestry and his Master's Degree at the University of the Philippines at Los Banos. He holds a Doctoral Degree in Geographical Sciences (Human Geography) in 1997 from the Research School of Pacific and Asian Studies, The Australian National University, Canberra, Australia.

Prof. Dr. Loutfy El-Juhany is working at the Prince Sultan Institute for Environmental, Water and Desert Research - King Saud University, Riyadh, Saudi Arabia since 2013. Before joining the PSIEWDR, he has served as the Professor at the College of Agriculture and Food Science, KSU Saudi Arabia. He produced dozens of MS and PhD students. He has more than 50 research publications to his credit. He is the pioneer and founding member of Research Group on forests and Forestry in the Kingdom of Saudi Arabia. He has conducted many surveys of forests of the Kingdom of Saudi Arabia to evaluate them. He has identified the problems and a challenge faced by the forests and has attempted to develop the improvement strategies through his research. His research has helped greening the desert.

Prof. Dr. Gary S. Straquadine serves as the Interim Chanceelor, Vice Chancellor (Academic Programs) and Vice Provost at the Utah State University – Eastern, Price UT, USA. He did Ph.D. from the Ohio State University, USA. Presently he also leads the applied sciences division of the USU-Eastern campus. He is responsible for faculty development and evaluation, program enhancement, and accreditation. In addition to his heavy administrative assignments, he manages to find time to teach some undergraduate and graduate courses and supervise graduate student research. Being an extension educator, he has passion for the economic development of the community through education and has also successfully developed significant relations with agricultural leadership in the private and public sector. He has also served as the Chair, Agricultural Comm, Educ, and Leadership, at The Ohio State University, USA. Before accepting the present position as the Vice Provost, he served on many positions as the Department Head; Associate Dean; Dean and Executive Director, USU-Tooele Regional Camp and the Vice Provost (Academic). His professional interests include extension education, sustainable agriculture, food security, statistics in education, community development, motivation of youth and outreach educational programs. He has also helped several under developing countries improving their agriculture and educational programs. He has been honored with numerous awards and honors to recognise his countless scientific and educational contributions.

Postface

The world is facing new challenges with changing political realities and there are persistent dilemmas in ways to achieve a climate secure future. In the light of numerous challenges linked to climate change, including food insecurity, ecosystem instability and water shortages, this volume has shed light upon the most fundamental concepts and requirements for achieving mitigation and particularly adaptation. The livelihoods of millions of people in the developing world are already threatened. The impact of climate change spans across sectors and is impacted by these sectors not hitherto brought to the collective consciousness. Ensuring food security and maintaining ecosystem resilience are salient priorities under Article 2 of the United Nations Framework Convention on Climate Change (UNFCCC).

The volume brought together different perspectives usually missed out in larger scientific and policy assessments. In describing impacts and adaptation stories, it covers Asia, Africa and Europe on water, food and forestry sector. The scope of the volume ranges from impacts to mitigation and from in-field experiments to policy implementation. The *first* part of the volume discussed issues surrounding food security and socio-economic dynamics. The section covered the role of relevant stakeholders, as both the protectors and destructors of the environment, while also discussing the challenges due changing cropping patterns and decreasing water resources and how it directly impacts the food security in areas struck with food scarcity. This has been done with the help of relevant case studies. The *second* part nicely covered the interlinkages between the climate, water, soil and agriculture. The *third* part gave a turn to the complete discussion and focussed more on the participatory governance and the positive impact of engaging local people in climate action besides the problems with finances available and loose handling of important mitigation mechanisms.

Achieving food security is a challenge in itself and climate change could make it more challenging, in turn affecting the Sustainable Development Goals (SDGs). A wide gap still exists in the understanding of climate change impacts related to food, water and forest resources, especially at the regional level. The strength of this volume lies in the fact that it covered the most neglected topics in UNFCCC negotiations

M. Behnassi et al. (eds.), *Climate Change, Food Security and Natural Resource Management*, https://doi.org/10.1007/978-3-319-97091-2

in spite of the fact that these decide the fate of millions of people around the world, especially the developing countries. These are the forests, soil health, agriculture, and water. These aspects do not directly lead to mitigation of climate change in most cases but are significant for improving the adaptive capacity of the local communities. Indeed, most cases presented in this volume capture the everyday realities of the local communities and how they react and adapt to similar situations in different geographical settings.

Although this volume is a collection of several case studies, each giving a particular message, there are four most important insights and future concerns that we would like to conclude with. *Firstly*, food security may become a difficult challenge to deal with in future, especially in poorer countries, in spite of the finances that are being provided. This is because of the declining land, water and forest resources. *Secondly*, the world is fast changing even at the local level and policy makers will have to play an active role in understanding the ever-changing ground level situation that is weakening the overall environmental resilience present today. They will have to take appropriate measures to help their people to adapt to difficult situations quickly. *Third*, developed countries need to continue supporting development programmes in the poor countries taking into account the actual effect it has brought in the situation. It would become increasingly important to assess and sense that the finance is going in the right hands. *Fourth*, forest resilience is the key to improving adaptation, especially through participatory skills. Forests provide for improved food diversity that could help in times of crop failures of a single variety along with providing pharmaceutical resources for which, the developing world acts as a storehouse and may become the strength of forest rich countries in the future.

Printed in the United States
By Bookmasters